DATE			

Applied Hydraulics for Technology

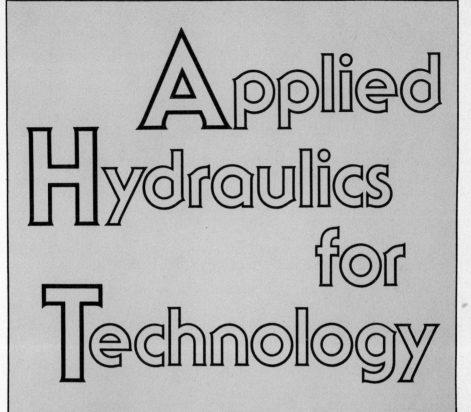

Applied Hydraulics for Technology

JOHN D. KANEN, P. Eng.

Seneca College of Applied Arts and Technology

HOLT, RINEHART AND WINSTON
New York   Chicago   San Francisco   Philadelphia
Montreal   Toronto   London   Sydney
Tokyo   Mexico City   Rio de Janeiro   Madrid

Library of Congress Cataloging in Publication Data

Kanen, John D.
    Applied hydraulics for technology.

    Includes index.
    1. Hydraulics.   2. Fluid mechanics.   3. Civil
engineering.   I. Title.
TC160.K25   1985   627   85-17544
ISBN 0-03-061707-3

Printed in the United States of America
Published simultaneously in Canada

5 6 7 8   038   9 8 7 6 5 4 3 2 1

CBS COLLEGE PUBLISHING
Holt, Rinehart and Winston
The Dryden Press
Saunders College Publishing

Contents

Preface

This text has been written for use in community college civil engineering technology curriculums. It is based on the author's more than 25 years of practical experience in hydraulics, as well as more than a decade of teaching hydraulics to technologists. Because of its practical approach to the subject, the book will also prove valuable to undergraduate students at the university level and to the practicing civil engineer.

The book is not intended for a course in a series of fluid mechanics, but it will stand alone as a text for a complete treatment of the study of fluid mechanics as it applies to civil engineering technology. The material is presented so that a student familiar with basic mathematics and physics from the secondary school level will require no other prerequisites and will be able to follow and comprehend all the topics discussed. To this end, little or no calculus has been used in the derivation of the formulas and principles. Instead, in order to illustrate these principles and the use of the formulas, numerous worked-out problems and examples are presented throughout the text.

The main body of each chapter has been developed in the SI system of units. However, where applicable, many chapters are followed by a summary of the theoretical discussions and examples of their use in the British system of units.

Most of the fourteen chapters of the text are separate entities, each dealing with a specific aspect of hydraulics. The exceptions are those dealing with pipe and open-channel flow. For these topics the basic principles of fluid flow are discussed in one chapter and then are followed by other chapters devoted specifically to civil engineering applications.

Throughout the text, and especially in the areas dealing with the cumbersome empirical relationships common in hydraulics, aids and tables have been introduced to provide easy solutions. These aids are presented with the knowledge that the electronic calculator readily allows the raising of numbers to any power without the use of logarithms. In this way, for instance, the solution of the Hazen-Williams and the Manning equations is accomplished easily by means of a one-page table, instead of with cumbersome and inaccurate traditional nomographs.

The text lends itself well for use in a one- or two-semester course in hydraulics. Students will also find the book a useful reference and aid in subsequent subjects such as municipal technology and other specialties related to the flow of water.

The author acknowledges the encouragement and help received in the preparation of this text from his colleagues, students, and many other interested persons. In conclusion, this book is dedicated to those engineers and technologists in the public works and municipal fields who, always behind the scenes, use the principles in this text to ensure that there is a constant flow of water from our taps and that waste and storm water are removed efficiently and diligently.

John D. Kanen

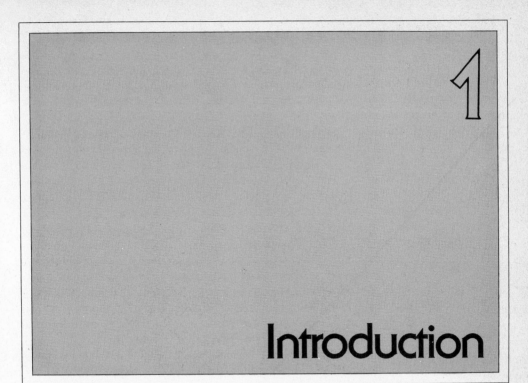

Introduction

1.1 HYDRAULICS

The science of hydraulics is not new, as is evidenced by the extensive hydraulic works constructed by ancient civilizations. Long before our era, the Egyptians, Persians, Chinese, and many other peoples used man-made structures to transport water. Not only did these civilizations devise elementary works, related to transportation of and on water, but they also developed fairly sophisticated principles concerning the behavior of water and invented complicated machinery using water power.

The word *hydraulics* comes from the Greek word for water and around 200 A.D., the Greek, Ctesibius, invented water clocks, air guns and a hydraulic organ, using waterpower to produce airflow in its pipes. The latter musical instrument was named *hydraulis*.

Since then, numerous mathematicians and scientists have devoted many lifetimes developing principles and laws related to the flow of fluids. Many of these laws—among them, Bernoulli's theorem of the conservation of energy—are still valid today and, in some form or other, represent an important part of the base on which the modern science of hydraulics is built.

A large amount of this earlier work, unfortunately, is only of historical interest. Most of the effort and experimentation by these early pioneers was carried out with water in its natural state and under very primitive laboratory conditions. As a result, much of this work related only to water or theoretical fluids without taking into account, or even being aware of the fact, that fluid properties such as density, temperature, and viscosity, play an important role in the behavior of fluids.

It is only relatively recently that hydraulics, particularly applied hydraulics, has made fast and important progress, through a judicious combination of experiment, theory, and practical experience. Hence, many laws of modern applied hydraulics are empirical in nature, and mathematical or analytical derivations of many aspects of the behavior of fluids are still lacking. As future study of fluids expands the theoretical knowledge of these complex hydraulic phenomena, some of the empirical laws will require modification or abandonment.

The present state of the art of applied hydraulics, even with its reliance on empirical relationships and formulas, nevertheless provides solutions to hydraulic problems within the usually acceptable margins of accuracy. But for complex projects, such as dams, river and harbor works, and many others it is still often necessary to find solutions by means of physical or mathematical model studies.

The science of hydraulics deals with the study of fluids in their two states: at rest or in motion. The former is generally referred to as the study of hydrostatics and the latter as fluid flow.

Hydrostatics

Hydrostatics is the study of fluids at rest and the forces they exert on surfaces or objects with which they are in contact. It solves problems related to the design of dams, the walls of swimming pools, air tanks for the scuba diver, aerosol containers, and numerous other modern daily necessities.

Fluid Flow

The study of *fluid flow* deals with the behavior of fluids in motion and their interaction with the surfaces and objects with which they come in contact. Fluid-flow studies aid the engineer in determining the size of pipes for a water distribution system, assessing the potential damage of river floods, harnessing rivers for hydroelectric power developments, and the design of lubrication, cooling, and fuel supply systems for industry, the home, and the automobile.

1.2   UNITS OF MEASUREMENT

Various systems of measurement, using a great variety of units, have developed through time and some have become standard systems in certain areas of science or geographical regions.

The two most commonly used systems of measurement today are the

Metric System and the *British System*, developed in Continental Europe and the English-speaking countries, respectively. The Metric System is based on decimal units and uses the meter and the kilogram as its base units for length and mass.

The British System is not based on decimal units and uses the foot and the pound as its base units for length and mass. The unit of time, the second, is common to both systems.

Each of the above systems of measurement has further developed into separate systems of engineering units, depending on different methods of defining units of mass and force.

Since the international acceptance of the SI (Système International) System of Units of Measurement, worldwide agreement to the gradual conversion to this system has been reached. For some considerable time, however, it will be necessary to work in, or to refer to work done in, other systems. Therefore an understanding of the most commonly used systems and their relationship to SI, will be of great value.

In order to facilitate ready interchange between the four frequently encountered systems, they are briefly reviewed here. The four systems are: The *Metric Gravitational* System, the *Metric Absolute* System, (SI), the *British Gravitational* System, and the *British Absolute* System.

The difference between the Metric and British Systems has already been mentioned and is related to the use of decimal units, such as the meter and the kilogram in the former and the nondecimal units of the foot and the pound in the latter.

The fundamental relationship between the gravitational systems is based on Newton's second law of motion. This law states that force is proportional to the rate of change of momentum, or

$$F = kMa \tag{1.1}$$

where F is force, M is mass, a is acceleration, and k is a proportionality factor depending on the definition of force or mass.

The Gravitational Systems

In the Gravitational Systems of measurement, the unit of mass is defined as *the mass that will be accelerated at the rate of standard gravitational acceleration, when acted upon by a unit of force.*

If gravitational acceleration is denoted by g and substituted in Eq. (1.1), it will yield

$$F = kMg \tag{1.2}$$

or

$$M = \frac{F}{kg} \tag{1.3}$$

On earth, standard gravitational acceleration at sea level and at $45°$ latitude is 32.17398 ft/s^2 in British units and 9.80665 m/s^2 in Metric units. For most civil engineering applications, the value of gravitational acceleration is, generally, taken as follows:

In British units,

$$g = 32.2 \text{ ft/s}^2 \tag{1.4}$$

In Metric units,

$$g = 9.81 \text{ m/s}^2 \tag{1.5}$$

When these values of g are substituted in Eq. (1.3), the basic expressions for the definition of unit of mass become as follows:
In the British Gravitational System,

$$1 \text{ lb mass} = \frac{1 \text{ lb force}}{k \times 32.2 \text{ ft/s}^2} \tag{1.6}$$

Similarly, in the Metric Gravitational System,

$$1 \text{ kg mass} = \frac{1 \text{ kg force}}{k \times 9.81 \text{ m/s}^2} \tag{1.7}$$

and $k = 1/(32.2 \text{ ft/s}^2)$ or $k = 1/(9.81 \text{ m/s}^2)$ in the British and Metric Gravitational Systems, respectively.

The Absolute Systems

In the Absolute Systems, the unit of force is defined as *the force that imparts to a unit of mass a unit of acceleration*. Therefore, in the British Absolute System, Eq. (1.1) becomes

$$1 \text{ poundal} = k \times 1 \text{ lb mass} \times 1 \text{ ft/s}^2 \tag{1.8}$$

and in the Metric Absolute System, or SI,

$$1 \text{ newton} = k \times 1 \text{ kg mass} \times 1 \text{ m/s}^2 \tag{1.9}$$

In both equations, $k = 1$.

Other Systems

Besides the four systems of measurements mentioned above, three other systems have been in use, more or less frequently. Two of these systems are, the *Metric CGS Absolute* and the *Metric CGS Gravitational Systems*. They differ from the above-mentioned Metric Systems only in the base units used, which are the centimeter and the gram, as opposed to the meter and the kilogram in the Metric Gravitational System and SI.

The third system, the *British Engineering System*, defines the unit of mass as the *mass that will be accelerated at the rate of 1 ft/s²*, when acted upon by a *1 lb force*. In this system, $k = 1$ and the unit of mass is called the *slug*.

Table 1.1 represents a comparison of all seven systems of measurement, discussed above, together with the corresponding equations for Newton's second law of motion, given by Eq. (1.1).

TABLE 1.1 Systems of Measurement and the Corresponding Equation for Newton's Second Law of Motion

| Name of System | Units | | | | F | $=$ | k | $\times$ | m | $\times$ | a |
	Length	Time	Mass	Force							
Metric Absolute SI	m	s	kg	newton	1 N		1		1 kg		1 m/s²
Metric Gravitational	m	s	kg	kg force	1 kgf		$\frac{1}{9.81}$		1 kg		1 m/s²
Metric Absolute (cgs)	cm	s	gm	dyne	1 dyne		1		1 gm		1 cm/s²
Metric Gravitational (cgs)	cm	s	gm	gm force	1 gmf		$\frac{1}{981}$		1 gm		1 cm/s²
British Absolute	ft	s	lb	poundal	1 lb		1		1 lb		1 ft/s²
British Gravitational	ft	s	lb	lb force	1 lbf		$\frac{1}{32.2}$		1 lb		1 ft/s²
British Engineering	ft	s	slug	lb force	1 lbf		1		1 slug		1 ft/s²

Mass, Weight, and Force

The terms *mass* and *weight* often cause confusion. In common use the term "weight" is frequently used to denote mass. Mass is the property of a body that is a measure of its inertia. Weight, on the other hand, is the force required to resist the movement of a mass caused by gravitational acceleration. Weight is the force required to support or lift a mass. Hence its units are those of force.

In some systems of measurement, units of mass and force are designated by the same names. In order to ensure clarity and to avoid ambiguity, it is good practice to distinguish clearly between these two units. This can be done by adding the letter "f" to the unit's symbol, so that kilogram force, gram force, and pound force are represented by kgf, gf, and lbf, respectively.

1.3 THE SI SYSTEM OF MEASUREMENTS

The system of measurements used in this text is the *Système International d'Unités* (International System of Units), known by its abbreviated form as the SI System. Also, throughout this text, the recommended SI practice of replacing the comma, separating groups of three decimals by a small space has been adopted. The latter does not apply to numbers up to 99999, which are written without a space. However, one million will be written as 1 000 000.

Also, the SI practice related to dimensioning of drawings has been followed in this text. This practice consists of dimensioning all distances in millimeters (mm). The only exceptions are long distances, and those are dimensioned in meters (m). All dimensioning is done without the use of symbols such as m or mm. By convention, dimensions shown without a decimal point are in millimeters, whereas those with a decimal point are in meters. For instance, the distance 1.35 m would preferably be shown as 1350 but could also be indicated as 1.35.

Where applicable, each chapter is followed by a summary of that chapter, together with some examples in the British Gravitational System of Measurements, which will be, throughout this text, referred to as the British System.

Numerous publications exist, dealing with SI, some of which are listed in the references and bibliography sections of this text. Consequently, only a short summary of the SI system and its units will be included here.

SI Groups of Units

There are three distinct groups of units in SI: the base units, the supplementary units, and the derived units.

The *base units* are a group of seven units, consisting of arbitrarily assigned unit values of physical quantities, which are independent of each other. Two other units defined in a manner as set out for the base units are recognized in SI. These are the *supplementary units*. All other units in SI are *derived units*, which are made up by defining relationships between base and supplementary units. The base units and supplementary units, together with some commonly used derived units, recognized in SI, are listed in Table 1.2.

Since SI units are multiplied or divided by 10 or powers of 10, prefixes are added to their names to distinguish the power of 10 used. For instance, distances on highways are not expressed in meters, but in *kilometers*, where the prefix *kilo* denotes 1000. The prefixes used in SI are shown in Table 1.3.

Conversions Between SI and British Units

Often it is necessary to convert values from one system of units to corresponding values of another system. Table 1.4 is a listing of conversion factors, of some of the more commonly used units in the SI and the British Systems of units.

It is virtually impossible to remember all the conversion factors. This is not necessary, however, since in many cases it is relatively easy to carry out conversions knowing only one or two selected factors and the use of an electronic calculator. Essentially, if the conversion from feet to meters and from

TABLE 1.2   The SI System of Measurement Units

Physical Quantity	Unit	Symbol
1. *Base Units*		1
Length	meter	m
Mass	kilogram	kg
Time	second	s
Electric current	ampere	A
Absolute temperature	kelvin	K
Amount of substance	mole	mol
Luminous intensity	candela	cd
2. *Supplementary Units*		
Plane angle	radian	rad
Solid angle	steradian	sr
3. *Derived Units with Special Names*		
Force	newton	N
Pressure	pascal	P
Energy, work	joule	J
Power	watt	W
4. *Derived Units with Compound Names*		
Area	square meter	m^2
Volume	cubic meter	m^3
Velocity	meters per second	m/s
Acceleration	meters per second per second	m/s^2
Rate of flow	cubic meters per second	m^3/s
Density	kilograms per cubic meter	kg/m^3
Unit weight	newtons per cubic meter	N/m^3
5. *Non-SI Units Commonly Used with SI*		
Time	day	d
	hour	h
	minute	min
Plane angle	degree	°
	minute (1°/60)	'
	second (1'/60)	"
Power	kilowatthour	kWh

pounds to kilograms is known, most other factors related to civil engineering applications can readily be found.

For instance, the conversion factor for square feet to square meters is found as follows:

$$1.0 \text{ ft} = 0.3048 \text{ m}$$

$$1.0 \text{ ft}^2 = (0.3048)^2 \text{ m}^2$$

TABLE 1.3   Names of Prefixes

Prefix	Symbol	Number by which Unit is Multiplied		
mega	M	10^6	=	1 000 000
kilo	k	10^3	=	1000
hecto	h	10^2	=	100
deca	da	10^1	=	10
deci	d	10^{-1}	=	0.1
centi	c	10^{-2}	=	0.01
milli	m	10^{-3}	=	0.001
micro	μ	10^{-6}	=	0.000 001
nano	n	10^{-9}	=	0.000 000 001

TABLE 1.4   Conversion Factors for Commonly Used Units in Hydraulics

British Unit	Operation	Factor	SI Unit	Operation	Factor	British Unit
in	×	25.4	mm	×	0.0394	in
		25.4			*0.039*	
ft	×	0.3048	m	×	3.2808	ft
		0.305			*3.28*	
gal (U.S.)	×	3.7860	L	×	0.2641	gal (U.S.)
		3.8			*0.26*	
gal (imp.)	×	4.5461	L	×	0.2193	gal (imp.)
		4.55			*0.22*	
lb (mass)	×	0.4536	kg	×	2.2046	lb (mass)
		0.45			*2.2*	
lbf (force, weight)	×	4.4482	N	×	0.2248	lbf (force, weight)
		4.45			*0.22*	
psi	×	6.8948	kPa	×	0.1450	psi
		6.9			*0.14*	
ft·lbf	×	1.3558	J	×	0.7376	ft·lbf
		1.36			*0.74*	
hp (550 ft·lbf/s)	×	0.7457	kW	×	1.341	hp (550 ft·lbf/s)
		0.75			*1.34*	

The conversion from pounds per square inch to pascals can be found by

$$1.0 \text{ psi} = 144 \text{ lbf/ft}^2$$

$$= 144 \div (0.3048)^2 \text{ lb/m}^2$$

$$= 144 \div (0.3048)^2 \times 0.4536 \times 9.81 \text{ N/m}^2$$

$$= 6897 \text{ N/m}^2$$

$$= 6.9 \text{ kPa}$$

Conversions should be used only when absolutely necessary. When performing work in an unfamiliar system of units there is temptation to translate the given data to a more familiar system, perform the necessary calculations, and convert the final answer to the original system of units. This method is not only time-consuming, but the many additional calculations and checking of tables leads to a substantial increase in the possibility of errors creeping into the work. Additionally, it does not increase familiarity with the system of units in question.

It is good practice to consider systems of units like languages and adopt the corresponding principles of practice through use and early ability to think in a new language, as opposed to translating already formed sentences.

Accuracy in Conversions

The conversion factors in Table 1.4 are shown in two versions. The first value shown is the regular or official value of the conversion factor. The numbers in the table shown in italics are recommended values of the conversion factor applicable to most civil engineering calculations. The official values shown have applications in the laboratory and scientific research.

In order to visualize the significance of the rounding off of these conversion factors, consider the SI unit of the pascal. It is a very small unit. One pascal roughly represents the pressure exerted by a layer of instant coffee, 3 to 4 mm ($\frac{1}{8}$ inch) in thickness. Consequently, the kilopascal or kPa is the more commonly used form of expressing pressure. From Table 1.4 it is found that

$$1.0 \text{ kPa} = 0.14 \text{ psi}$$

Hence,

$$0.1 \text{ kPa} = 0.014 \text{ psi}$$

When dealing, for instance, with pressures in a municipal water distribution system, pressures are seldom expressed in the British System in terms of tenths of a pound per square inch. Therefore expressing such pressure to the nearest tenth of a kPa in SI would be more accurate than required or justified.

Symbols

The symbols used in this text are those suggested in *Manual of SI Units of Measurement and Symbols and Abbreviations for Hydraulics and Fluid Mechanics*, compiled by C. K. Jonys for the Canadian Centre for Inland Waters, 1973. A list of these symbols and their meanings is shown in Table 1.5.

TABLE 1.5   List of SI Symbols and Their Meanings

Mechanical Quality	Symbol	SI Unit	Comments
acceleration, linear	a	m/s²	
acceleration, angular	α	rad/s²	
angle, solid		sr	
area	A	m²	hectare (ha) 1 ha=10⁴ m²
energy	E	J	
force	F	N	
gravitational acceleration	g	m/s²	
length	L	m	
mass	m	kg	ton (t), 1 t=1000 kg
moment of force		N · m	
moment of inertia	I	kg · m²	
momentum	M	kg · m/s	
power	P	W	
pressure	p	Pa	
section modulus	S	m³	
time	t	s	
velocity, linear	v	m/s	
velocity, angular	ω	rad/s	
volume	V	m³	liter (L) l m³=1000 L
work, energy	W	J	
bulk modulus of elasticity	K	N	
density	ρ	kg/m³	
specific gravity, (relative density)	s		
specific volume		m³/kg	
specific weight (unit weight)	γ	N/m³	
surface tension	σ	N/m	
viscosity, dynamic	μ	Pa · s	
viscosity, kinematic	ν	m²/s	
discharge	Q	m³/s	liters per second (L/s)

1.4 EXAMPLES IN THE USE OF SI UNITS

Area

(a) A rectangle 500 mm long and 350 mm wide has an area A of

$$A = 500 \text{ mm} \times 350 \text{ mm}$$
$$= 175\,000 \text{ mm}^2$$

or

$$A = 0.50 \text{ m} \times 0.35 \text{ m}$$
$$= 0.175 \text{ m}^2$$

(b) A circular pipe with a diameter of 750 mm has a cross-sectional area of

$$A = (750 \text{ mm})^2 \times \frac{\pi}{4}$$
$$= 441\ 786 \text{ mm}^2$$
$$= 442\ 000 \text{ mm}^2$$

or

$$A = (0.75 \text{ m})^2 \times \frac{\pi}{4}$$
$$= 0.442 \text{ m}^2$$

Volume

A tank that is 3.0 m wide, 4.5 m long, and 1.2 m deep has a volume V of

$$V = 3.0 \text{ m} \times 4.5 \text{ m} \times 1.2 \text{ m} = 16.2 \text{ m}^3$$

Since $1 \text{ m}^3 = 1000$ liters (L) of liquid volume, the capacity Q of this tank could be expressed as

$$Q = 16.2 \text{ m}^3 \times 1000 \text{ L/m}^3 = 16200 \text{ L}$$

Discharge or Capacity of Flow

A circular pipe with a diameter $d = 500$ mm and flowing full with water at an average velocity $v = 1.5$ m/s has a capacity of flow Q of

$$Q = (0.500 \text{ m})^2 \times \frac{\pi}{4} \times 1.5 \text{ m/s} = 0.295 \text{ m}^3/\text{s}$$

or

$$Q = 0.295 \text{ m}^3/\text{s} \times 1000 \text{ L/m}^3 = 295 \text{ L/s}$$

2

Properties of Fluids

2.1 FLUIDS

Definition of a Fluid

All matter can be divided into two major classes: *solids* and *fluids*. However, to define a fluid as a substance that is not a solid would be an oversimplification. Many types of matter can be encountered in the solid as well as the fluid state. Steel, for instance, when heated to a sufficiently high temperature, becomes fluid, and water, when cooled to freezing, turns solid.

Certain other types of matter appear solid, and yet they will demonstrate characteristics of flowing. As an example, tar and plasticene will, on their own, deform in time from an apparently solid ball to relatively flat "pancake."

The difference between a solid and a fluid can be more specifically defined, by considering two fundamental properties of matter, the *stress-strain relationship* and the *elasticity*.

The Stress-Strain Relationship

A solid requires considerable external forces (stress) to cause it to deform (strain). For instance, relatively large forces have to be applied to a steel bar in order to elongate it.

13

A fluid will deform, in time, without the application of external forces. Water and even the heaviest oils will eventually, of their own, take on the shape of the container in which they are held.

The Elasticity Relationship

When external forces stress and deform solids, within certain limits, the solids will regain their original shape when these external forces are removed. A steel bar elongated by external forces will regain its original length when the forces causing the elongation are removed.

Fluids, on the other hand, will continue to change shape in time even after the removal of the external forces causing the deformation. They will not, of their own, return to their original shape.

Ideal Fluid

The properties of solids that enable them to resist deformation and give them elastic properties are due to their ability to resist internal tensile and shear forces.

Resistance to internal tensile forces and shear stresses is essentially non-existent in fluids. As will be seen later, very minimal capabilities to resist tensile as well as shear stresses exist in fluids. However, they are negligible in most practical applications of applied hydraulics.

It is the apparent absence of these two properties in fluids that leads to the definition of an *ideal fluid* as a *substance that is unable to resist internal shear and tensile forces.*

2.2   LIQUIDS AND GASES

Fluids can be classified into two classes or forms of matter: *liquids* and *gases*. As is the case for solids and fluids, two fundamental properties of matter, *compressibility* and *continuity*, can be used to distinguish between and define a liquid and a gas.

Compressibility

For practical purposes, liquids are considered to be incompressible. Oil, water, and other liquids undergo only minute changes in volume, even when subjected to very high pressures.

Gases, on the other hand, are very compressible. Their volume can increase or decrease appreciably when subjected to only slight variations in pressure.

Continuity

When a liquid is held in a container, its entire mass will arrange itself so as to be in contact with the bottom and the sides of that container, and a well-defined surface of the liquid will form. The entire space bounded by the sides and the bottom of the container and the surface of the liquid will be filled by the mass of the liquid.

A gas held in a closed container will not form a well-defined surface and will tend to fill the entire container. Even when some of the gas is removed, the remaining mass of the gas will expand in an attempt to fill the container anew.

2.3 PROPERTIES OF FLUIDS

Density (ρ)

The *density* of a fluid, ρ, is the mass of the fluid per unit volume. Consequently, ρ is expressed in units of

$$kg/m^3$$

The density of water at 4°C is

$$\rho = 1000 \ kg/m^3$$

Specific Weight (γ)

The specific weight or unit weight of a fluid, γ, is the weight of the fluid per unit volume. The specific weight is equal to the force of gravity exerted on a unit volume of the fluid. Consequently, γ is expressed in units of

$$kg/m^3 \times m/s^2 = N/m^3$$

The specific weight of water at 4°C is

$$\gamma = 9810 \ N/m^3 = 9.81 \ kN/m^3$$

Relative Density or Specific Gravity (s)

The relative density of a fluid is the ratio of the density or specific weight of the fluid to the density or specific weight of water, at a temperature of 4°C. Since it is a ratio of two values with identical units, it follows that relative density s is dimensionless.

The density of mercury is 13600 kg/m^3; hence its relative density or specific gravity is

$$s = \frac{13600 \text{ kg/m}^3}{1000 \text{ kg/m}^3} = 13.6$$

Similarly, if the specific weight of a certain oil is 8.34 kN/m^3, then its relative density will be

$$s = \frac{8.34 \text{ kN/m}^3}{9.81 \text{ kN/m}^3} = 0.85$$

Dynamic Viscosity (μ)

Viscosity is the measure of a fluid's resistance to internal shear stresses. The absolute or dynamic viscosity can be visualized by considering Fig. 2.1, which represents two very thin layers of a fluid. Let A be the horizontal area of each layer and h the vertical distance between their centerlines. If the top layer is acted upon by a force F, this layer will move with a velocity v relative to the bottom layer.

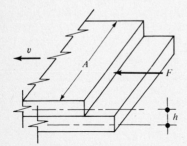

FIGURE 2.1

For a perfectly nonviscous or ideal fluid, unable to resist internal shear stresses, this relative velocity would remain constant, even after the force F is removed.

All real fluids, however, present some resistance to this movement due to internal molecular activity, closely resembling friction or shear stresses between the layers. Experiments have shown that the force F required to maintain v is proportional to v, and the area A, and inversely proportional to the distance h; that is,

$$F = \mu \frac{vA}{h} \tag{2.1}$$

where μ is a proportionality constant, the value of which depends on the type and temperature of the fluid. This constant, μ, is the *absolute* or *dynamic viscosity* of the fluid.

If the shear stress per unit area, $F/A = \tau$, is substituted in Eq. (2.1), it reads

$$\tau = \mu \frac{v}{h} \qquad (2.2)$$

For infinitely small values of h, or when $h \rightarrow 0$, Eq. (2.2) can be expressed in the form

$$\tau = \mu \frac{\Delta v}{\Delta h} \qquad (2.3)$$

Equation (2.3) is a measure of the shear stress and $\Delta v/\Delta h$ represents the instantaneous rate of deformation of the fluid. This principle was first expressed by Newton and is called *Newton's law of viscosity*.

A Newtonian fluid has a constant value for μ and there exists a linear relationship between τ and $\Delta v/\Delta h$. In a non-Newtonian fluid μ is not constant and the rate of deformation is nonlinear. The above relationships are represented graphically in Fig. 2.2.

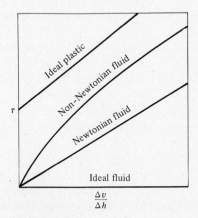

FIGURE 2.2

If μ is equal to zero, τ will also be zero. In that case the fluid is an ideal fluid, as represented by the abscissa in Fig. 2.2.

If Eq. (2.2) is solved for μ, it yields

$$\mu = \frac{\tau h}{v} \qquad (2.4)$$

Therefore the dynamic viscosity μ is expressed in units of

$$\frac{(N/m^2) \times m}{m/s} = N \cdot s/m^2 \text{ or } Pa \cdot s$$

Kinematic Viscosity (υ)

The *kinematic viscosity* of a fluid is the ratio of its dynamic viscosity to its density, or

$$v = \frac{\mu}{\rho} \tag{2.5}$$

The units of v are

$$\frac{N \cdot s/m^2}{kg/m^3} = m^2/s$$

Surface Tension (σ)

Molecular attraction in liquids causes a film to form at the interface between a liquid and a gas, or between two immiscible liquids. This film is apparently capable of resisting tensile forces. The capacity of liquids to resist tensile stresses at their surface is called *surface tension*.

Surface tension can be observed at the formation of a drop of water at a tap or by the formation of a meniscus in glass tubes. Surface tension is the force responsible for capillary action, which occurs when liquids rise in soils, for example. This phenomenon is attributable to the fact that the surface tension in the very small interstices between the soil particles exceeds the force of gravity exerted on the minute volumes of liquid contained in the individual interstices.

Even though surface tension is a demonstrable property of liquids, for most engineering applications it is ignored in practice, and liquids are considered to be unable to resist internal tensile forces. Surface tension is generally only an important consideration in the study of capillary action and for small scale-model studies. The units of surface tension, σ, are

$$N/m$$

Vapor Pressure (p_v)

The evaporation of liquids is caused by the fact that molecules in the liquid, which are in constant motion, escape into the space above the surface of the liquid. If the space above the liquid's surface is confined, as in a closed container, the evaporation will continue until saturation by the molecules of this space is reached. At this point, the pressure exerted by the escaped molecules is large enough to create an equilibrium between the molecules escaping and those reentering the liquid.

The pressure exerted by the escaped molecules above the surface of a liquid is called the *vapor pressure*. The units of vapor pressure, p_v, are

$$N/m^2 = Pa$$

Bulk Modulus of Elasticity (K)

Although, as previously mentioned, for all practical purposes liquids can be considered to be incompressible, they are actually like all other matter, compressible to some degree.

If a liquid with volume V is subjected to an increase in pressure of Δp, the volume of that liquid will decrease by a small volume ΔV. The bulk modulus of elasticity, K, is the ratio

$$K = \frac{\Delta P}{\Delta V/V} \tag{2.6}$$

Since the ratio $\Delta V/V$ is the ratio of two volumes, it is dimensionless. The bulk modulus of elasticity therefore has the dimensions of its numerator, ΔP. Consequently, K is expressed in units of pressure, or Pa.

2.4   PROPERTIES OF COMMON FLUIDS

Properties of the more common fluids encountered in civil engineering applications are listed in Table 2.1, in both SI and British units.

EXAMPLE 2.1

Determine the total mass and weight of the fluid in the container of Fig. 2.3 if the fluid is (a) water, (b) mercury, with $s=13.6$.

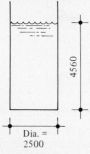

FIGURE 2.3

TABLE 2.1 Physical Properties of Certain Liquids

Fluid	Temperature °C	°F	Relative Density s	Mass Density ρ (kg/m³)	(slugs/ft³)	Specific Weight γ (N/m³)	(lb/ft³)	Viscosity Dynamic μ×10⁻⁵ (Pa·s)	(lb·s/ft²)	Viscosity Kinematic ν×10⁻⁵ (m²/s)	(ft²/s)
Water	0	32	0.999	999.8	1.940	9805	62.42	179.4	3.746	0.179	1.931
	4	39.2	1.000	1000.0	1.941	9807	62.43	156.8	3.274	0.157	1.687
	10	50	0.999	999.8	1.940	9805	62.42	131.0	2.735	0.131	1.410
	21.1	70	0.998	997.8	1.936	9785	62.29	98.15	2.050	0.098	1.059
	37.8	100	0.993	993.1	1.927	9739	62.00	68.18	1.424	0.069	0.739
	93.3	200	0.963	962.7	1.868	9440	60.10	30.50	0.637	0.032	0.341
Seawater	25	77	1.03	1030	2.0	10101	64.30				
Alcohol	25	77	0.787	787	1.53	7718	49.01	109.6	2.29	0.139	1.497
Carbon tetrachloride		77	1.59	1590	3.08	15593	98.91	90.97	1.90	0.057	0.617
Gasoline	25	77	0.721	721	1.40	7071	45.0	44.77	0.935	0.062	0.668
Glycerine	25	77	1.263	1263	2.45	12386	78.62	94803	1980	75.06	808.2
Linseed oil	25	77	0.93	930	1.80	9123	58.0	3309	69.1	3.566	38.39
Mercury	25	77	13.63	13630	26.45	133665	849.0	153.2	3.2	0.011	0.121
Fuel oil Medium	25	77	0.852	852	1.65	8355	53.16	299.3	6.25	0.352	3.788
Heavy	25	77	0.906	906	1.76	8885	56.53	10725	224	11.84	127.3

SOLUTION

The volume of the fluid is

$$V = \frac{(2.5 \text{ m})^2 \times \pi}{4} \times 4.56 \text{ m}$$

$$= 22.38 \text{ m}^3$$

(a) The mass of the water in the container is

$$M = 22.38 \text{ m}^3 \times 1000 \text{ kg/m}^3$$

$$= 22400 \text{ kg}$$

and the weight of the water is

$$W = 22400 \text{ kg} \times 9.81 \text{ m/s}^2$$

$$= 219\,600 \text{ N}$$

$$= 220 \text{ kN}$$

(b) If the fluid is mercury, the mass and the weight will be 13.6 times that of water, since $s = 13.6$, or

$$M_{\text{merc}} = 22400 \times 13.6$$

$$= 304\,600 \text{ kg}$$

and

$$W = 220 \text{ kN} \times 13.6$$

$$= 2990 \text{ kN}$$

EXAMPLE 2.2

If the total weight of the liquid in Fig. 2.3 is 505 kN, determine the height of liquid if it is (a) water, (b) oil, with $s = 0.85$.

SOLUTION

(a) The volume of water in the container is

$$V = \frac{505 \text{ kN}}{9.81 \text{ kN/m}^3}$$

$$= 51.48 \text{ m}^3$$

Consequently, the liquid height h is

$$h = \frac{51.48 \text{ m}^3}{(2.5 \text{ m})^2 \times \pi/4}$$

$$= 10.5 \text{ m}$$

(b) The volume of oil in the container is

$$V = \frac{505 \text{ kN}}{9.81 \text{ kN/m}^3 \times 0.85}$$

$$= 60.6 \text{ m}^3$$

and

$$h = \frac{60.6 \text{ m}^3}{(2.5 \text{ m})^2 \times \pi/4}$$

$$= 12.3 \text{ m}$$

EXAMPLE 2.3

Determine the dynamic viscosity of an oil with a kinematic viscosity of 0.352×10^{-5} m²/s and a relative density of 0.88.

SOLUTION

From Eq. (2.5),

$$v = \frac{\mu}{\rho}$$

It follows that the dynamic viscosity μ is

$$\mu = v \times \rho$$

$$= 0.352 \times 10^{-5} \text{ m}^2/\text{s} \times 1000 \text{ kg/m}^3 \times 0.88$$

$$= 0.00310 \text{ Pa} \cdot \text{s}$$

$$= 310 \times 10^{-5} \text{ Pa} \cdot \text{s}$$

EXAMPLE 2.4

A fluid has a specific weight of 9.345 kN/m³ and a dynamic viscosity of 3.31×10^{-2} Pa·s. Determine its relative and mass density and its kinematic viscosity.

SOLUTION

The relative density is

$$s = \frac{9.345 \text{ kN/m}^3}{9.81 \text{ kN/m}^3}$$

$$= 0.95$$

The mass density is

$$\rho = 0.95 \times 1000 \text{ kg/m}^3$$
$$= 950 \text{ kg/m}^3$$

The kinematic viscosity is, from Eq. (2.5),

$$\nu = \frac{3.31 \times 10^{-2} \text{ N} \cdot \text{s/m}^2}{950 \text{ kg/m}^3}$$
$$= 0.000\ 034\ 8 \text{ m}^2/\text{s}$$
$$= 3.48 \times 10^{-5} \text{ m}^2/\text{s}$$

EXAMPLE 2.5

The fluid in the container of Fig. 2.3 has a total weight of 319 kN and a dynamic viscosity of 91×10^{-5} Pa·s. Determine its relative and mass density and kinematic viscosity.

SOLUTION

From Example 2.1, the volume of fluid is

$$V = 22.38 \text{ m}^3$$

Therefore

$$\gamma = \frac{319 \text{ kN}}{22.38 \text{ m}^3}$$
$$= 14.25 \text{ kN/m}^3$$

and its relative density is

$$s = \frac{14.25 \text{ kN/m}^3}{9.81 \text{ kN/m}^3}$$
$$= 1.45$$

The mass density of liquid is

$$\rho = 1000 \text{ kg/m}^3 \times 1.45$$
$$= 1450 \text{ kg/m}^3$$

and, from Eq. (2.5),

$$v = \frac{91 \times 10^{-5} \text{ N} \cdot \text{s/m}^2}{1450 \text{ kg} \cdot \text{m}^3}$$
$$= 0.000\,000\,63 \text{ m}^2/\text{s}$$
$$= 6.3 \times 10^{-7} \text{ m}^2/\text{s}$$

2.5 SUMMARY OF THE CHAPTER IN BRITISH UNITS

Density (ρ)

The density of a fluid is the mass of the fluid per unit volume. It is expressed in units of

$$\frac{\text{lb/ft}^2}{\text{s}^2}$$

Specific Weight (γ)

The specific weight of a fluid is the weight per unit volume. It is expressed in units of

$$\text{lbf/ft}^3$$

Dynamic Viscosity (μ)

The dynamic viscosity of a fluid is the measure of its resistance to internal shear stresses. The dynamic viscosity is expressed by the formula

$$\mu = \frac{\tau h}{v}$$

and its units are

$$\frac{\text{lbf/ft}^2 \times \text{ft}}{\text{ft/s}} = \text{lbf} \cdot \text{s/ft}^2$$

Kinematic Viscosity (v)

The kinematic viscosity of a fluid is the ratio of its dynamic viscosity to its density, or

$$v = \frac{\mu}{\rho}$$

It is therefore expressed in units of

$$\frac{lbf \times s/ft^2}{(lb/ft^2)/s^2} = ft^2/s$$

Surface Tension (σ)

The surface tension causes a film to form on the surface of a liquid and is an expression of the resistance to tensile stresses by this film. Its units are

$$lbf/ft$$

Vapor Pressure (p_v)

Vapor pressure is the pressure exerted by the evaporated liquid above the liquid surface. Its units are

$$lbf/in^2 = psi$$

Bulk Modulus of Elasticity (K)

The bulk modulus of elasticity is a measure of its compressibility. Its units are

$$lbf/in^2 = psi$$

EXAMPLE 2.6

Determine the total mass and weight of the fluid in the container of Fig. 2.4 if the fluid is (a) water, (b) mercury, with $s = 13.6$.

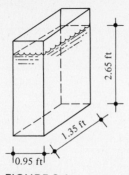

FIGURE 2.4

SOLUTION

The volume of the fluid is

$$V = 0.95 \text{ ft} \times 2.35 \text{ ft} \times 2.65 \text{ ft}$$

$$= 5.92 \text{ ft}^3$$

Therefore,

(a) The mass of the water in the container is

$$M = 5.92 \text{ ft}^3 \times 1.941 \text{ slugs/ft}^3$$

$$= 11.5 \text{ slugs}$$

and the weight of the water is

$$W = 5.92 \text{ ft}^3 \times 62.4 \text{ lb/ft}^3$$

$$= 369.4 \text{ lb}$$

$$= 369 \text{ lb}$$

(b) If the fluid is mercury the mass and weight will be 13.6 times that of water, or

$$M_{merc} = 11.5 \text{ slugs} \times 13.6$$

$$= 156.0 \text{ slugs}$$

and

$$W_{merc} = 369.0 \text{ lb} \times 13.6$$

$$= 5020.0 \text{ lb}$$

EXAMPLE 2.7

If the total weight of liquid in the container of Fig. 2.4 is 650 lb, determine the height of liquid if it is (a) water, (b) oil, with $s = 0.85$.

SOLUTION

(a) The volume V of water in the container is

$$V = \frac{650 \text{ lb}}{62.4 \text{ lb/ft}^3}$$

$$= 10.42 \text{ ft}^3$$

Consequently the liquid height h is

$$h = \frac{10.42 \text{ ft}^3}{0.95 \text{ ft} \times 1.35 \text{ ft}}$$

$$= 8.12 \text{ ft}$$

(b) The volume of oil in the container is

$$V = \frac{650 \text{ lb}}{62.4 \text{ lb/ft}^3 \times 0.85}$$

$$= 12.25 \text{ ft}^3$$

and

$$h = \frac{12.15 \text{ ft}^3}{0.95 \text{ ft} \times 1.35 \text{ ft}}$$

$$= 9.56 \text{ ft}$$

EXAMPLE 2.8

Determine the dynamic viscosity of an oil with a kinematic viscosity of 3.8×10^{-5} ft^2/s and a relative density of 0.88.

SOLUTION

From Eq. (2.5),

$$v = \frac{\mu}{\rho}$$

It follows that the dynamic viscosity μ is

$$\mu = v \times \rho$$

$$= 3.8 \times 10^{-5} \text{ ft}^2/\text{s} \times 1.941 \text{ slugs/ft}^3 \times 0.88$$

$$= 0.000\,064\,9 \text{ lb} \cdot \text{s/ft}^2$$

$$= 6.49 \times 10^{-5} \text{ lb} \cdot \text{s/ft}^2$$

EXAMPLE 2.9

A fluid has a specific weight of 61.4 lb/ft^3 and a dynamic viscosity of 0.69×10^{-3} lb $\cdot$ s/ft^2. Determine its relative and mass density and kinematic viscosity.

SOLUTION

The relative density is

$$s = \frac{61.4 \text{ lb/ft}^3}{62.4 \text{ lb/ft}^3}$$

$$= 0.98$$

The mass density is

$$\rho = 0.98 \times 1.941 \text{ slugs/ft}^3$$

$$= 1.902 \text{ slugs/ft}^3$$

The kinematic viscosity is, from Eq. (2.5),

$$\nu = \frac{0.69 \times 10^{-3} \text{ lb} \cdot \text{s/ft}^2}{1.902 \text{ slugs/ft}^3}$$

$$= 0.000\,363 \text{ ft}^2/\text{s}$$

$$= 36.3 \times 10^{-5} \text{ ft}^2/\text{s}$$

EXAMPLE 2.10

The fluid in the container in Fig. 2.4 has a total weight of 475 lb and a dynamic viscosity of 1.90×10^{-5} lb·s/ft². Determine its relative and mass density and kinematic viscosity.

SOLUTION

From Example 2.6, the volume of fluid is

$$V = 5.92 \text{ ft}^3$$

Therefore

$$\gamma = \frac{475 \text{ lb}}{5.92 \text{ ft}^3}$$

$$= 80.24 \text{ lb/ft}^3$$

and the relative density is

$$s = \frac{80.24 \text{ lb/ft}^3}{62.4 \text{ lb/ft}^3}$$

$$= 1.29$$

Hence the mass density of the liquid is

$$\rho = 1.941 \text{ slugs/ft}^3 \times 1.29$$
$$= 2.496 \text{ slugs/ft}^3$$

and, from Eq. (2.5),

$$\nu = \frac{1.90 \times 10^{-5} \text{ ft}^2/\text{s}}{2.496 \text{ slugs/ft}^3}$$
$$= 0.000\,007\,6 \text{ ft}^2/\text{s}$$
$$= 0.76 \times 10^{-5} \text{ ft}^2/\text{s}$$

SELECTED PROBLEMS IN SI UNITS

2.1   Determine the total mass and weight of the fluid in the container of Fig. P2.1, if the fluid is (a) water and (b) mercury ($s = 13.6$).

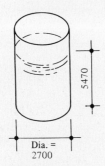

Dia. = 2700

FIGURE P2.1

2.2   Determine the total weight of the liquid in the container of Fig. P2.2 if the liquid is (a) water and (b) oil ($s = 0.83$).

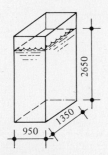

FIGURE P2.2

2.3   If the liquid in Fig. P2.2 is water and its total weight is 45.0 kN, what is the depth of water in the container?

2.4   If the liquid in the container in Fig. P2.4 has a relative density of 2.85 and $d = 1.50$ m, what are its total mass and weight?

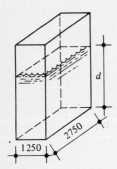

FIGURE P2.4

2.5   The mass of the liquid in Fig. P2.4 is 18900 kg and its relative density is 1.65. Determine the depth d of the liquid.

2.6   The liquid in Fig. P2.4 weighs 285 kN and its relative density is 1.06. Determine the depth of liquid in the container.

2.7   If the liquid in Fig. P2.4 weighs 125 kN and $d = 2.75$ m, what is its relative density?

SELECTED PROBLEMS IN BRITISH UNITS

2.8   Determine the total weight of the liquid in the container of Fig. P2.8, if the liquid is (a) water and (b) oil ($s = 0.83$).

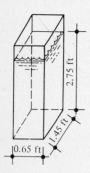

FIGURE P2.8

2.9   If the liquid in Fig. P2.8 is water and its total weight is 5025 lb, determine the depth of water in the container.

2.10 The relative density of the liquid in the container of Fig. P2.10 is 1.6 and d is 2.2 ft. Determine the total weight of the liquid.

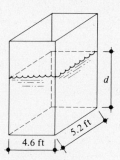

FIGURE P2.10

2.11 The liquid in Fig. P2.10 is brine with a relative density of 1.15, and its total weight is 5606 lb. Determine d.

2.12 If, in Fig. P2.10, $d = 6.4$ ft, and the liquid weighs 9000 lb, determine the relative density of the liquid.

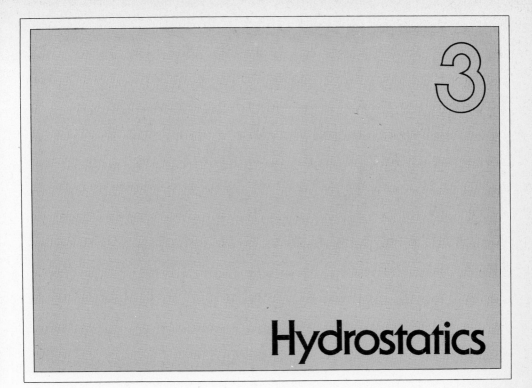

Hydrostatics

3.1 FLUIDS AT REST

Hydrostatics is the study of *fluids at rest* and their interaction with other bodies of mass. Because of the nature of a liquid at rest, the study of hydrostatics is the one branch of hydraulics in which most laws and formulas are based on theoretical considerations. Only in a relatively few instances is it necessary to rely on empirical relationships derived from observation and experimentation.

Hydrostatics is used by the civil engineer to determine pressures exerted by fluids on the surfaces that contain them, to study the stability of dams, and to measure differences in pressure between flowing fluids.

3.2 FLUID PRESSURE

Total Pressure (P)

The force exerted by a mass of fluid on a real or imaginary surface, is the *total pressure*, P, on that surface. Since P is a force, it is expressed in units of force, or newtons.

Unit Pressure (*p*)

The *unit pressure* or *intensity of pressure*, exerted by a fluid on a surface, is the force exerted by the fluid on a unit area of the surface.

If *P* is the total pressure or force acting on a surface, whose area is *A*, the unit pressure *p* will be

$$p = \frac{P}{A} \qquad (3.1)$$

and

$$P = p \times A \qquad (3.2)$$

EXAMPLE 3.1

If the container shown in Fig. 3.1 contains water, determine the total and the unit pressure on its bottom.

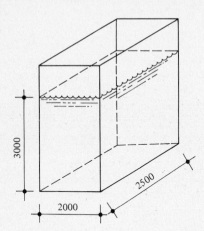

FIGURE 3.1

SOLUTION

The bottom of this container supports the total weight of the water. Hence the force or total pressure on the bottom will be

$$P = F = M \times g$$

Since the mass of the water in the container is equal to its volume times its density, ρ,

$$M = 2.0 \text{ m} \times 2.5 \text{ m} \times 3.0 \text{ m} \times 1000 \text{ kg/m}^3$$

and

$$M = 15000 \text{ kg}$$

Consequently,

$$P = 15000 \text{ kg} \times 9.81 \text{ m/s}^2$$
$$= 147\,150 \text{ N}$$

or

$$P = 147.2 \text{ kN}$$

The unit pressure on the bottom of the container will therefore be, from Eq. (3.1),

$$p = \frac{147.2 \text{ kN}}{2.0 \text{ m} \times 2.5 \text{ m}}$$
$$= 29.4 \text{ kN/m}^2$$

or

$$p = 29.4 \text{ kPa}$$

EXAMPLE 3.2

For the container in Fig. 3.2, determine the total and unit pressure on the bottom if the liquid in the container has a relative density of 1.5.

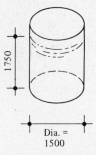

1750

Dia. =
1500

FIGURE 3.2

SOLUTION

The density of the liquid is 1.5 times that of water, or

$$\rho = 1000 \text{ kg/m}^3 \times 1.5 = 1500 \text{ kg/m}^3$$

The volume of liquid in the container is

$$V = \frac{(1.5 \text{ m})^2 \times \pi}{4} \times 1.75 \text{ m}$$

$$V = 3.09 \text{ m}^3$$

and its total mass is

$$M = 3.09 \text{ m}^3 \times 1500 \text{ kg/m}^3$$

$$M = 4635.0 \text{ kg}$$

Consequently, the total pressure on the bottom of the container is

$$P = 4635.0 \text{ kg} \times 9.81 \text{ m/s}^2$$

$$= 45470 \text{ N}$$

$$= 45.5 \text{ kN}$$

Since the unit pressure $p = P/A$, then

$$p = \frac{45.5 \text{ kN}}{(1.5 \text{ m})^2 \times \pi/4}$$

$$= 25.75 \text{ kN/m}^2$$

or

$$p = 25.8 \text{ kPa}$$

Pressure Head (h)

If a container, such as the one shown in Fig. 3.2, contains a fluid with relative density γ, then the total pressure on the bottom of the container, with an area A and a depth of liquid h, will be

$$P = A \times \gamma \times h$$

and the unit pressure p is

$$p = \frac{A \times \gamma \times h}{A}$$

and

$$p = \gamma \times h \tag{3.3}$$

or

$$h = p/\gamma \qquad (3.4)$$

and

$$\gamma = p/h \qquad (3.5)$$

Similarly, the pressure on any imaginary surface, such as *abcd* in Fig. 3.3, with an area A_1 and a distance h_1 below the fluid surface, will be

$$P_1 = A_1 \times \gamma \times h_1 \qquad (3.6)$$

and

$$p_1 = \gamma \times h_1 \qquad (3.7)$$

Equations (3.3) and (3.7) indicate that the unit pressure p on any surface in a fluid is a function of the height of the fluid above this surface and the relative density of the fluid. The term h in Eqs. (3.3) and (3.4), is called the *pressure head*.

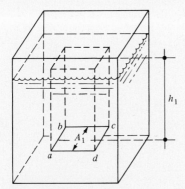

FIGURE 3.3

EXAMPLE 3.3

Determine the unit pressure in Example 3.1 using the concepts of specific weight and pressure head.

SOLUTION

The specific weight of water is

$$\gamma = 1000 \text{ kg/m}^3 \times 9.81 \text{ m/s}^2$$
$$= 9.81 \text{ kN/m}^3$$

Hence, from Eq. (3.3), the unit pressure on the bottom of the container is

$$p = 9.81 \ kN/m^3 \times 3.0 \ m$$

$$= 29.4 \ kN/m^2$$

$$= 29.4 \ kPa$$

EXAMPLE 3.4

Determine the height of water, or pressure head, required above a point 1, in a container of water, if the pressure p_1 at point 1, is 685 kPa.

SOLUTION

Since

$$p_1 = 685 \ kPa$$

$$= 685 \ kN/m^3$$

Eq. (3.4) will yield

$$h = \frac{685 \ kN/m^2}{9.81 \ kN/m^3} = 69.8 \ m$$

and the depth of water, or pressure head, is 69.8 m.

EXAMPLE 3.5

Determine the pressure head at point 1, in Example 3.4, if the pressure p_1 at point 1, is 685 kPa, and the fluid is (a) oil, with $s = 0.80$, and (b) mercury, with $s = 13.6$.

SOLUTION

(a) For oil with a relative density of 0.80,

$$\gamma = 9.81 \ kN/m^3 \times 0.80$$

$$= 7.85 \ kN/m^3$$

Therefore

$$h = \frac{685 \text{ kN/m}^2}{7.85 \text{ kN/m}^3}$$

$$= 87.3 \text{ m}$$

(b) For mercury, with a relative density of 13.6,

$$\gamma = 9.81 \text{ kN/m}^3 \times 13.6$$

$$= 133.41 \text{ kN/m}^3$$

and

$$h = \frac{685 \text{ kN/m}^2}{133.41 \text{ kN/m}^3}$$

$$= 5.13 \text{ m}$$

Pressure and Pressure Head Difference

If h_1 and h_2 in Fig. 3.4 are the respective depths of points 1 and 2 in a fluid with specific weight γ, then

$$p_1 = \gamma h_1 \tag{3.8}$$

and

$$p_2 = \gamma h_2 \tag{3.9}$$

Hence

$$p_1 - p_2 = \gamma h_1 - \gamma h_2$$

$$= \gamma(h_1 - h_2) \tag{3.10}$$

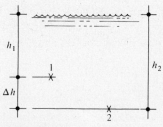

FIGURE 3.4

or generalized

$$p_1 - p_2 = \gamma \, \Delta h \qquad (3.11)$$

where Δh is the difference in pressure heads between points 1 and 2, and

$$\Delta h = \frac{p_1 - p_2}{\gamma} \qquad (3.12)$$

EXAMPLE 3.6

The container in Fig. 3.5 contains three immiscible liquids, with specific gravities $s_1 = 0.80$, $s_2 = 1.75$, and $s_3 = 13.6$, and liquid depths $d_1 = 0.30$ m, $d_2 = 0.35$ m, and $d_3 = 0.25$ m. Determine the pressures at points 1, 2, and and 3.

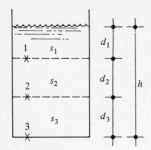

FIGURE 3.5

SOLUTION

From Eq. (3.3), at point 1,

$$p_1 = 9.81 \text{ kN/m}^3 \times 0.80 \times 0.30 \text{ m}$$

$$p_1 = 2.35 \text{ kPa}$$

At point 2,

$$p_2 = p_1 + \gamma d_2$$

$$= 2.35 \text{ kPa} + (9.81 \text{ kN/m}^3 \times 1.75 \times 0.35 \text{ m})$$

$$p_2 = 8.36 \text{ kPa}$$

At point 3,

$$p_3 = p_2 + \gamma d_3$$

$$= 8.36 \text{ kPa} + (9.81 \text{ kN/m}^3 \times 13.6 \times 0.25 \text{ m})$$

$$p_3 = 41.7 \text{ kPa}$$

EXAMPLE 3.7

If the pressures in the container of Fig. 3.5 were $p_1 = 15.50$ kPa, $p_2 = 32.70$ kPa, and $p_3 = 55.50$ kPa, determine the depths of the liquids in the container.

SOLUTION

From Eq. (3.4),

$$h_1 = d_1 = \frac{15.50}{9.81 \text{ kN/m}^3 \times 0.80}$$

$$d_1 = 1.98 \text{ m}$$

Similarly,

$$d_2 = \frac{p_2 - p_1}{\gamma}$$

$$= \frac{32.70 \text{ kPa} - 15.50 \text{ kPa}}{9.81 \text{ kN/m}^3 \times 1.75}$$

$$= 1.00 \text{ m}$$

and

$$d_3 = \frac{p_3 - p_2}{\gamma}$$

$$= \frac{55.50 \text{ kPa} - 32.70 \text{ kPa}}{9.81 \text{ kN/m}^3 \times 13.6}$$

$$= 0.17 \text{ m}$$

Hence the total depth of liquid in the container is as follows:

$$h = d_1 + d_2 + d_3$$

$$= 1.98 \text{ m} + 1.00 \text{ m} + 0.17 \text{ m}$$

$$= 3.15 \text{ m}$$

EXAMPLE 3.8

If the depths of the liquids in Fig. 3.5 are $d_1 = 0.55$ m, $d_2 = 0.35$ m, $d_3 = 0.35$ m, and the pressures are $p_1 = 20.25$ kPa, $p_2 = 43.50$ kPa, and $p_3 = 84.50$ kPa, determine the specific gravity of each of the liquids in the container.

SOLUTION

From Eq. (3.5), it follows that

$$\gamma_1 = \frac{p_1}{d_1}$$

$$= \frac{20.25 \text{ kPa}}{0.55 \text{ m}}$$

$$= 36.82 \text{ kN/m}^3$$

$$\gamma_2 = \frac{p_2 - p_1}{d_2}$$

$$= \frac{43.50 \text{ kPa} - 20.25 \text{ kPa}}{0.35 \text{ m}}$$

$$= 66.43 \text{ kN/m}^3$$

$$\gamma_3 = \frac{p_3 - p_2}{d_3}$$

$$= \frac{84.50 \text{ kPa} - 43.50 \text{ kPa}}{0.35 \text{ m}}$$

$$= 117.14 \text{ kN/m}^3$$

From the definition of specific gravity,

$$s_1 = \frac{\gamma_1}{\gamma_w}$$

$$= \frac{36.82 \text{ kN/m}^3}{9.81 \text{ kN/m}^3}$$

$$= 3.75$$

Similarly,

$$s_2 = 6.77$$

$$s_3 = 11.9$$

Transmission of Pressure

Equation (3.11) indicates that the unit pressures p_1 and p_2 are related to each other and differ from each other only by a value directly related to their difference in elevation, or depth in the liquid. Consequently, an increase in p_1 will bring about an equal increase in p_2, and vice versa.

Pascal's Law

In 1663, one year after Blaise Pascal's death, his treatise on the equilibrium of liquids was published. In it he formulated and proved a major axiom of

Hydrostatics, now known as Pascal's law, which states: *At any point in a fluid at rest, the pressure is transmitted equally in all directions.*

This law can be visualized and proved mathematically, by considering the small prism of fluid *ABCDEF* placed in a three-dimensional coordinate system in Fig. 3.6. Since the fluid is at rest, the prism must be in equilibrium, or

$$\Sigma F_x = 0 \qquad (3.13)$$

$$\Sigma F_y = 0 \qquad (3.14)$$

$$\Sigma F_z = 0 \qquad (3.15)$$

where ΣF_x, ΣF_y, and ΣF_z represent the forces acting on the prism parallel to the x, y, and z axis, respectively.

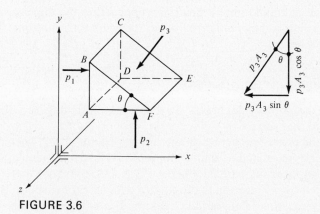

FIGURE 3.6

That Eq. (3.15) is satisfied is obvious, since the side *ABF* of the prism is parallel to and identical to the side *CDE*.

If p_1, p_2, and p_3 denote the average unit pressures on, and A_1, A_2, and A_3 are the areas of, the sides of the prism *ABCD*, *ADEF*, and *BCEF*, respectively, Eqs. (3.13) and (3.14) can be expressed as follows:

$$\Sigma F_x = p_1 A_1 - p_3 A_3 \sin \theta = 0 \qquad (3.16)$$

$$\Sigma F_y = p_2 A_2 - p_3 A_3 \cos \theta = 0 \qquad (3.17)$$

or

$$p_1 A_1 = p_3 A_3 \sin \theta \qquad (3.18)$$

$$p_2 A_2 = p_3 A_3 \cos \theta \qquad (3.19)$$

Since

$$A_3 \sin \theta = A_1 \qquad (3.20)$$

$$A_3 \cos \theta = A_2 \qquad (3.21)$$

it follows that

$$p_1 A_1 = p_3 A_1 \qquad (3.22)$$

$$p_2 A_2 = p_3 A_2 \qquad (3.23)$$

or that

$$p_1 = p_2 = p_3 \qquad (3.24)$$

If the prism is infinitesimally small and approaches the dimensions of a point, it follows that: At any point in a fluid the pressure acts with equal magnitude in all directions.

Direction of Pressure on a Surface

Pressure exerted by a fluid at rest, on a surface, can act only in a direction perpendicular to that surface.

Consider the pressure P acting on the surface AB in Fig. 3.7. The pressure or force P can be considered as the resultant of two forces: P_1 acting at right angles to AB and P_2 acting parallel to AB.

P_1 is resisted by the reaction of the surface AB and P_2 must be resisted by the friction or shear forces between the fluid and the surface AB. Since a fluid at rest cannot resist shear forces or create friction, P_2 must be equal to zero. Consequently,

$$P = P_1$$

and P must be perpendicular to the surface on which it acts.

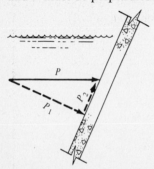

FIGURE 3.7

Atmospheric Pressure

Air is a fluid and the atmospheric air above the earth, exerts a pressure on the earth's surface. This pressure is known as *atmospheric pressure*, p_{at}, and it varies with weather conditions, the earth's latitude, and the elevation above the earth's surface at which the pressure is measured.

The standard atmospheric pressure at sea level is 101.325 kPa. For most practical applications and in this text

$$p_{at} = 101 \text{ kPa} \tag{3.25}$$

Absolute Pressure and Gage Pressure (p_a and p)

In the containers in Examples 3.1 and 3.2 atmospheric pressure acts at the surface of the liquids. The calculations for unit pressure p in those and many other previous examples did not take this atmospheric pressure into account.

From the transmission-of-pressure principle, explained on page 42, it is evident that atmospheric pressure will increase the pressures calculated in the examples by an amount equal to p_{at}.

Most gages, designed for measuring pressure, do not take into account atmospheric pressure; hence the term *gage pressure* is used when atmospheric pressure is ignored. The term *absolute pressure* is used for pressure measurements or indications that include atmospheric pressure. In this text, the term *pressure* and the symbol p will always indicate gage pressure.

From the above, the relationship between absolute, gage, and atmospheric pressure can be expressed as

$$p_a = p + p_{at} \tag{3.26}$$

EXAMPLE 3.9

For the container in Fig. 3.1, determine the absolute pressure on the bottom of the container.

SOLUTION

$$p_a = p_{at} + p$$
$$= 101 \text{ kPa} + 29.4 \text{ kPa}$$
$$= 130.0 \text{ kPa}$$

Partial Vacuum or Negative Pressure

Absolute pressures less than zero, $p_a < 0$, cannot exist. Hence, absolute pressure is always positive. Total vacuum, or $p_a = 0$, is a state that is practically impossible to attain. The latter condition is inherent in the property of gases— that they are discontinuous. Therefore, even the smallest amount of gaseous matter will continue to expand and tend to fill the container in which $p_a = 0$ is attempted.

Gage pressure is zero when $p = p_{at}$, and consequently it can have negative values. When p is negative, the condition of the fluid is called *partial vacuum*.

Figure 3.8 illustrates graphically the relationship between gage and absolute pressure.

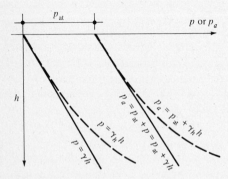

FIGURE 3.8

Graphical Representation of Pressure and Pressure Head

In the foregoing discussion the compressibility of fluids was not taken into account. Whereas gases are highly compressible, liquids are, for most practical purposes, considered to be incompressible.

If a fluid is compressible, its density will increase as the pressure increases. Consequently, since the density varies with the pressure and pressure head, so also will the specific weight vary with the pressure head, or

$$\gamma_h = f(h) \tag{3.27}$$

where γ_h is the specific weight of the fluid, corresponding to a pressure head h.

For compressible fluids Eq. (3.3) will therefore read

$$p = \gamma_h h \tag{3.28}$$

and

$$p_a = p_{at} + \gamma_h h \tag{3.29}$$

When Eqs. (3.3), (3.26), (3.28), and (3.29) are plotted on a graph, with h as the ordinate and p or p_a as the abscissa, as in Fig. 3.9, Eqs. (3.3) and (3.26) will be represented by a straight line, indicating the linear relationship between pressure and pressure head for incompressible fluids, and the curve representing Eqs. (3.28) or (3.29) will be similar to those shown in the figure.

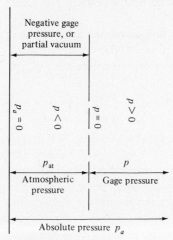

FIGURE 3.9

3.3 MEASUREMENT OF PRESSURE

The Piezometer

Consider the transparent tube a in Fig. 3.10, mounted on the container, so that fluid can freely flow from the container to the tube and vice versa. The tube a is the most elementary form of the *piezometer*.

Liquid in the tube will rise to the same level as the liquid surface in the container. Consequently, the distance h in Fig. 3.10, which can be measured in the piezometer, will represent the pressure head in the liquid at point 1 in the container.

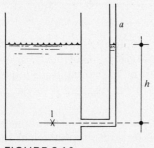

FIGURE 3.10

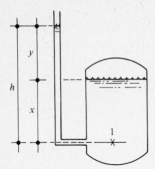

FIGURE 3.11

If the container is closed to the atmosphere, as in Fig. 3.11, and the pressure of the gas above the liquid in the container is increased, the pressure at point 1 will also increase in accordance with the principle of transmission of pressure. An increase in the pressure at point 1 is also an increase in the pressure head h at this point. Since the level of the liquid in the piezometer indicates the pressure head at point 1, it follows that the liquid level in the piezometer will rise a distance y above the liquid level in the container. The distance y will of course represent the pressure head of the gas in the liquid. Other forms of piezometers are shown in Fig. 3.12.

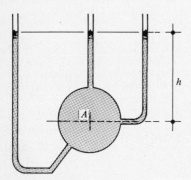

FIGURE 3.12

Piezometers have very limited applications. Obviously, a piezometer would be useless when gas pressures are to be measured, since a gas does not form a free surface; hence determination of the distance h in the piezometer tube would be impossible.

Also, for liquids with a low relative density, a large increase in h will correspond to only a small increase in pressure. As a result, piezometer tubes of impractical length would be required.

EXAMPLE 3.10

If the container in Fig. 3.11 contains water to a depth of $x = 2.20$ m above the point 1, and if the space in the container above the water is filled with air under a pressure of 175 kPa, determine (a) the pressure at point 1, (b) the height of the column of water in the piezometer, and (c) the difference in water levels between the piezometer tube and the container.

SOLUTION

(a) The pressure at point 1, p_1, will be equal to the air pressure above the water plus the water pressure at point 1; that is,

$$p_1 = 175 \text{ kPa} + \gamma h$$
$$= 175 \text{ kPa} + 9.18 \text{ kN/m}^3 \times 2.20 \text{ m}$$
$$= 196.6 \text{ kPa}$$

(b) The height of the water column, h, in the piezometer tube is equal to the pressure head at point 1:

$$h = p_1 / \gamma$$
$$= 196.6 \text{ kPa} / (9.81 \text{ kN/m}^3)$$
$$= 20.0 \text{ m}$$

(c) The difference in water levels, y, in the piezometer tube and the container is

$$y = h - x$$
$$= 20.0 \text{ m} - 2.20 \text{ m}$$
$$= 17.8 \text{ m}$$

y is also equal to the pressure head of the air above the water in the container; that is,

$$y = 175 \text{ kPa} / (9.81 \text{ kN/m}^3)$$
$$= 17.8 \text{ m}$$

The Manometer

In order to overcome the problems related to the extremely long and impractical tubes and to facilitate the measurement of gas pressures, the tubes of the piezometer are bent and filled with a liquid, the gage liquid. The gage

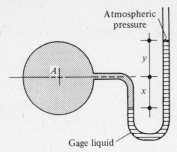

Atmospheric
pressure

FIGURE 3.13

liquid has a relative density higher than the fluid in the container and it is immiscible with it. Piezometers so constructed are called *manometers*. If the gage liquid has one surface exposed to the atmosphere, as is shown in Fig. 3.13, manometers are called *open manometers*.

The solution of manometer problems and the principles related to pressure measurements with these instruments is illustrated in the following example.

EXAMPLE 3.11

For the container in Fig. 3.14 determine the pressure at point 1 if the liquid in the container is water, the gage liquid is mercury ($s = 13.6$), $x = 0.15$ m, $y = 0.35$ m, and the air pressure in the top of the container is atmospheric.

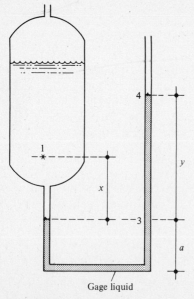

FIGURE 3.14

SOLUTION

In order to simplify the solution of manometer problems, it is convenient to identify and number points in the manometer where an interface of fluids exists or where special conditions prevail. In this example such points are point 2, where water and mercury meet, and point 4, the interface of the mercury and the atmospheric air.

With the numbering completed, as shown in Fig. 3.14, consider the pressure heads at each one of the points in turn and convert these pressure heads to equivalent water pressure head.

In this example, point 4 is a convenient point at which to commence since it is the only point at which the pressure is known. Gage pressure at point 4 is 0.0 kPa.

At point 3 the pressure head is 0.35 m of mercury more than the head at point 4, or

$$h_4 = 0.0 \text{ m}$$

$$h_3 = h_4 + (y \times s_{merc})$$

$$= 0.0 \text{ m} + (0.35 \text{ m} \times 13.6)$$

$$h_3 = 4.76 \text{ m}$$

From point 2 to 3 the pressure head first increases by a head of mercury equal to a and then decreases by the same amount when point 2 is reached. Therefore, the pressures at these points are equal, or

$$h_2 = h_3$$

At point 1 the pressure head has decreased in relation to point 2, by the distance x that point 1 is located above point 2, or

$$h_1 = h_2 - (x \times s_w)$$

$$= 4.76 \text{ m} - (0.15 \text{ m} \times 1.0)$$

$$h_1 = 4.61 \text{ m}$$

Hence

$$p_1 = 4.61 \text{ m} \times 9.81 \text{ kN/m}^3$$

$$p_1 = 45.2 \text{ kPa}$$

An alternate method for solving a simple manometer problem such as the one in this example is to start at h_1, follow along the manometer tube algebraically adding columns of liquid, converted to water pressure head, and equate this sum to the known pressure head at point 4, as follows:

$$h_1 + x - (y \times 13.6) = h_4 = 0$$

$$h_1 = 4.61 \text{ m}$$

EXAMPLE 3.12

If the container in Fig. 3.14 is closed and contains water to a depth of 1.50 m above point 1, $x = 0.25$ m and $y = 0.55$ m, determine the air pressure in the container.

SOLUTION

As before,

$$h_1 + 0.25 \text{ m} - (0.55 \text{ m} \times 13.6) = h_4 = 0$$

$$h_1 = 7.23 \text{ m}$$

Therefore, the pressure head at the water surface is

$$h_{ws} = 7.23 \text{ m} - 1.5 \text{ m}$$

$$h_{ws} = 5.73 \text{ m}$$

Hence the air pressure above the water surface is

$$p = 5.73 \text{ m} \times 9.81 \text{ kN/m}^3$$

$$p = 56.2 \text{ kPa}$$

The Differential Manometer

When the difference in pressure between two points in a hydraulic system has to be determined, this can be done by the use of *differential manometers*.

As the name implies, differential manometers do not provide a measurement of the actual pressure at a point in the system but measure the difference in pressure between two points. Typical forms and applications of differential manometers are shown in Fig. 3.15.

Calculations to determine the pressure difference, as measured by a differential manometer, can be done in a manner similar to that described for the open manometer. In this instance, too, it is convenient to identify and number strategic points in the manometer and to add algebraically columns of fluids, converted to head of water. One method is illustrated in the following example.

EXAMPLE 3.13

Determine the difference in pressure between pipes *a* and *b* in Fig. 3.16 if the liquid in pipe *a* is brine, with $s = 1.15$, the liquid in pipe *b* is oil, with $s = 0.85$, and the gage liquid is mercury, with $s = 13.6$, $x = 0.25$ m, $y = 0.20$ m, and $z = 1.10$ m.

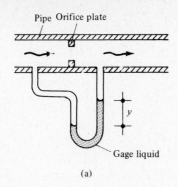

(a)

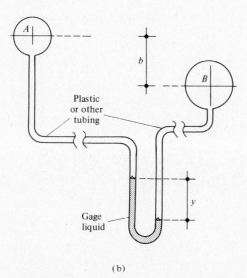

(b)

FIGURE 3.15 Manometer used to determine rate of flow by measuring pressure differential at (a) orifice plate and (b) between two remote pipes.

SOLUTION

The various points in this manometer that require consideration are numbered as shown in Fig. 3.16. Starting at point a, the center of pipe a, columns of liquid are added algebraically, after converting them to heads of water, as follows: At point a, the pressure head is h_a. At point 1,

$$h_1 = h_a + 0.20 \text{ m} \times 1.15$$
$$= h_a + 0.23 \text{ m}$$

At point 2,

$$h_2 = h_1 + 0.25 \text{ m} \times 1.15$$
$$= h_a + 0.52 \text{ m}$$

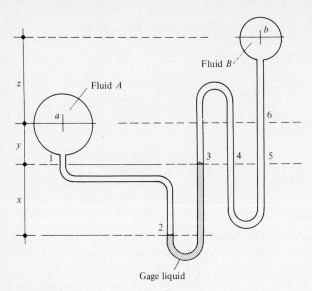

FIGURE 3.16

At point 3,

$$h_3 = h_2 - 0.25 \text{ m} \times 13.6$$
$$= h_a - 2.88 \text{ m}$$

At point 4,

$$h_4 = h_5 = h_3 = h_a - 2.88 \text{ m}$$
$$h_5 = h_a - 2.88 \text{ m}$$

At point 6,

$$h_6 = h_5 - 0.20 \text{ m} \times 0.85$$
$$= h_a - 3.05 \text{ m}$$

At point b,

$$h_b = h_6 - 1.10 \text{ m} \times 0.85$$
$$= h_a - 3.98 \text{ m}$$

Hence

$$h_a - h_b = 3.98 \text{ m}$$

and

$$p_a - p_b = 3.98 \text{ m} \times 9.81 \text{ kN/m}^3$$

$$p_a - p_b = 39.0 \text{ kPa}$$

The above calculations could have been performed in one equation, as shown in the previous example.

3.4 SUMMARY OF THE CHAPTER IN BRITISH UNITS

Total and Unit Pressure

Since pressure is a force exerted by a mass of fluid on a surface, it is expressed in units of force or weight. The unit pressure is the force exerted per unit area.

In British units the total pressure P is expressed in pounds and the unit pressure p is expressed, generally, in pounds per square inch (psi).

The specific weight γ of water at $39.2°F\,(4°C)$ is 62.427 lb/ft^3. For most practical applications and in this text, unless otherwise noted,

$$\gamma_w = 62.4 \text{ lb/ft}^3 \tag{3.30}$$

will be used as the specific weight of water.

EXAMPLE 3.14

Determine the total and unit pressure on the bottom of the container in Fig. 3.17 if the liquid in the container is water.

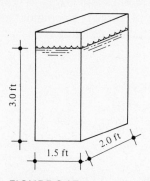

FIGURE 3.17

SOLUTION

The total pressure P is the weight of the water in the container; hence

$$P = W = \text{volume} \times \text{specific weight}$$

Therefore

$$P = 1.5 \text{ ft} \times 2.0 \text{ ft} \times 3.0 \text{ ft} \times 62.4 \text{ lb/ft}^3$$

$$P = 561.6 \text{ lb}$$

From Eq. (3.1),

$$p = \frac{561.6 \text{ lb}}{1.5 \text{ ft} \times 2.0 \text{ ft}}$$

$$= 187.2 \text{ lb/ft}^2$$

and

$$p = \frac{187.2 \text{ lb/ft}^2}{144 \text{ in}^2/\text{ft}^2}$$

$$= 1.30 \text{ psi}$$

Pressure Head

The pressure head is expressed in feet of water. It is important to note that Eqs. (3.3) and (3.4) will, in the commonly used units of feet for head and pounds per square inch for pressure, require conversion for the inconsistency in units.

If γ is expressed in pounds per cubic foot and h in feet, p in Eq. (3.3) will be expressed in pounds per square foot. In order to obtain the pressure in pounds per square inch, a further division by 144 in^2/ft^2 is required.

A convenient manner to avoid difficulties and errors in problems related to pressure and pressure head is to convert directly to pounds per square inch, and vice versa, by means of conversion factors as follows:

$$p = \gamma h$$

$$= \frac{62.4 \text{ lb/ft}^3 \times h}{144 \text{ in}^2/\text{ft}^2}$$

$$p = 0.433 \text{ psi/ft} \times h \tag{3.31}$$

or

$$1 \text{ ft of head of water} = 0.433 \text{ psi} \tag{3.32}$$

and

$$h = \frac{p}{\gamma}$$

$$= \frac{p \times 144 \ \text{in}^2/\text{ft}^2}{62.4 \ \text{lb/ft}^3}$$

$$h = p \ \text{psi} \times 2.31 \ \text{ft/psi} \qquad (3.33)$$

or

$$1 \ \text{psi} = 2.31 \ \text{ft of head of water} \qquad (3.34)$$

EXAMPLE 3.15

Determine the height of water, or pressure head, required above a point 1, in a container of water, if the pressure at point 1, $p_1 = 65.00$ psi.

SOLUTION

From Eq. (3.4),

$$h = \frac{p}{\gamma}$$

$$= \frac{65 \ \text{psi} \times 144 \ \text{in}^2/\text{ft}^2}{62.4 \ \text{lb/ft}^3}$$

$$h = 150 \ \text{ft}$$

Or, more simply, from Eq. (3.34),

$$h = 65 \ \text{psi} \times 2.31 \ \text{ft/psi}$$

$$h = 150 \ \text{ft}$$

EXAMPLE 3.16

Determine the pressure head in the previous example if the liquid is (a) oil, with $s = 0.80$, and (b) mercury, with $s = 13.6$.

SOLUTION

For a liquid with $s = 1$,

$$\gamma = s \times \gamma_w$$

$$h = \frac{p}{s \times \gamma_w}$$

where γ_w is the specific weight of water.

Therefore the pressure head for a fluid other than water relates to the pressure head of water h_w as follows:

$$h = \frac{h_w}{s}$$

Consequently,

(a) If the liquid is oil, with a specific gravity of 0.80,

$$h_{oil} = \frac{150 \text{ ft}}{0.80}$$

$$h_{oil} = 187.5 \text{ ft}$$

(b) If the liquid is mercury, with a specific gravity of 13.6,

$$h_{merc} = \frac{150 \text{ ft}}{13.6}$$

$$h_{merc} = 11.02 \text{ ft}$$

Atmospheric Pressure

Under normal conditions at sea level, standard atmospheric pressure is 2116 lb/ft^2 or 14.69 psi. For most civil engineering applications and in this text, the standard value used is as follows:

$$p_{at} = 14.7 \text{ psi} \tag{3.35}$$

EXAMPLE 3.17

Determine the heights to which the liquid will rise in the piezometers of Fig. 3.18. Assume that the gage liquid in each instance is the liquid in the container in which the piezometer is located.

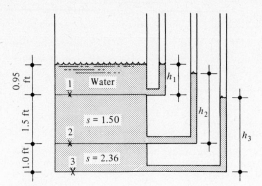

FIGURE 3.18

SOLUTION

The pressure at each interface can be calculated by means of Eqs. (3.3), (3.32), and (3.34), as follows:

$$p_1 = 0.95 \text{ ft} \times 2.31 \text{ psi/ft}$$
$$= 2.19 \text{ psi}$$
$$p_2 = 2.19 \text{ psi} + (1.50 \text{ ft} \times 1.50) \times 2.31 \text{ psi/ft}$$
$$= 7.39 \text{ psi}$$
$$p_3 = 7.39 \text{ psi} + (1.0 \text{ ft} \times 2.36) \times 2.31 \text{ psi/ft}$$
$$= 12.84 \text{ psi}$$

At each piezometer the liquid column will be the pressure head expressed in terms of feet of head of the actual liquid in the piezometer, or

$$h_1 = 2.19 \text{ psi} \times 0.433 \text{ ft/psi}$$
$$= 0.95 \text{ ft}$$
$$h_2 = 7.39 \text{ psi} \times (0.433 \text{ psi/ft/1.5})$$
$$= 2.13 \text{ ft}$$
$$h_3 = 12.84 \text{ psi} \times (0.433 \text{ psi/ft/2.36})$$
$$= 2.36 \text{ ft}$$

In the above calculations it must be remembered that if 0.433 ft of head of water represents 1 psi, it will only take 0.433/1.5 feet of head of fluid with $s = 1.50$ for 1 psi.

EXAMPLE 3.18

If the pressure in pipe 1 of Fig. 3.19 is 75 psi, determine the pressure at pipe 2. Liquids A and B are water, the gage liquid is mercury and $y=0.75$ ft and $z=0.86$ ft.

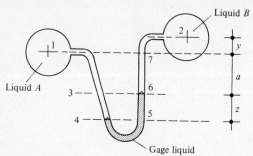

FIGURE 3.19

SOLUTION

Commencing at point 1 and adding columns of fluid converted to feet of head of water will yield

$$75 \text{ psi} \times 2.31 \text{ ft/psi} +a +0.86 \text{ ft} -0.86 \text{ ft} \times 13.6 -a -0.75 \text{ ft} = h_2$$

or

$$h_2 = 161.7 \text{ ft}$$

and

$$p_2 = 70 \text{ psi}$$

SELECTED PROBLEMS IN SI UNITS

3.1   For the manometer of Fig. P3.1, determine the pressure at point 1 if (a) liquid A is water, liquid B is mercury, $z=180$ mm, and $y=300$ mm; (b) liquid $A=$ brine, liquid B is oil, $z=0.75$ m, and $y=0.45$ m.*

3.2   For the manometer in Fig. P3.1, determine the distance y if liquid A is brine, liquid B is oil, $z=0.75$ m, and the pressure at point 1 is 12 kPa.

*Values of s for the fluids in these problems are as follows: mercury, 13.6; oil, 0.80; brine, 1.5; bromoform, 2.50.

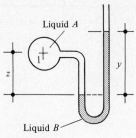

Liquid A

Liquid B

FIGURE P3.1

3.3 For the manometer in Fig. P3.3, determine the relative density of the fluid B if liquid A is water, $z = 0.25$ m, $y = 0.76$ m, and the pressure at the point 1 is 99 kPa.

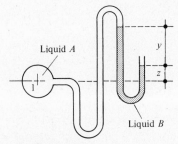

Liquid A

Liquid B

FIGURE P3.3

3.4 For the manometer in Fig. P3.3, determine the pressure at the point 1 if liquid A is water, liquid B is bromoform, $z = 115$ mm, and $y = 236$ mm.

3.5 For the manometer in Fig. P3.3, determine the distance y if liquid A is water, liquid B is bromoform, $z = 65$ mm, and the pressure at the point 1 is 8.6 kPa.

3.6 Determine the difference in pressure between points a and b in Fig. P3.6, if liquid A is water, liquid B is mercury, $y = 725$ mm, and $z = 235$ mm.

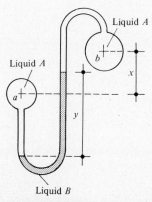

Liquid A

Liquid A

Liquid B

FIGURE P3.6

3.7 Determine the difference in pressure between points m and n in Fig. P3.7 when
(a) liquid A is water, liquid B is mercury, $y = 0.565$ m, $x = 1.675$ m, and $z = 0.125$ m;
(b) liquid A is brine, liquid B is mercury, $y = 0.753$ m, $x = 2.050$ m, and $z = 0.250$ m;
(c) liquid A is oil, liquid B is bromoform, $y = 14$ cm, $x = 20$ cm, and $z = 8$ cm; (d) liquid
A is oil, liquid B is mercury, $y = 360$ mm, $x = 180$ mm; (e) liquid A is oil, liquid B is
bromoform, $y = 210$ mm, $x = 180$ mm. (*Note:* The dimension z is not needed for
this question. Why?)

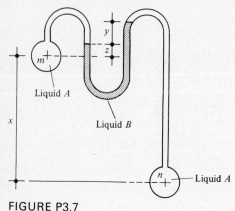

FIGURE P3.7

3.8 Determine the pressure at point n for the manometer of Fig. P3.7, if liquid A is oil,
liquid B is bromoform, $y = 240$ mm, $x = 120$ mm, and the pressure at m is 10 kPa.

3.9 Determine the difference in pressure between points m and n in Fig. P3.9 if liquid A
is water, liquid B is oil, $x = 57.6$ cm, and $y = 15.6$ cm.

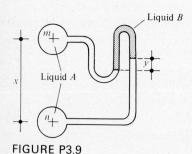

FIGURE P3.9

3.10 For the manometer in Fig. P3.9, determine the pressure at the point m if the pressure
at n is 100 kPa, liquid A is water, liquid B is oil, $x = 57.6$ cm, and $y = 30.6$ cm.

4
Civil Engineering Applications of Hydrostatics

4.1 PRESSURE ON PLANE SURFACES

Pressure on Vertical Surfaces

A vertical wall such as the one shown in Fig. 4.1 is subjected to horizontal hydrostatic pressure, varying from $p = 0$ at the water surface to $p = \gamma h$ at the bottom of the wall.

If the fluid causing the pressure is incompressible, p will increase at a constant rate from 0 to γh. The unit pressures on a section of the wall, with a length of one unit, can therefore be represented by the pressure diagram of Fig. 4.2.

The total pressure P on such a wall, of unit length, will be equal to the average unit pressure times the area on which it acts, or

$$P = \frac{\gamma h}{2} \times h \times 1 \text{ m}$$

Hence the total pressure per unit length of wall will be

$$P = \frac{\gamma h^2}{2} \tag{4.1}$$

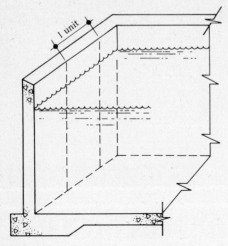

FIGURE 4.1

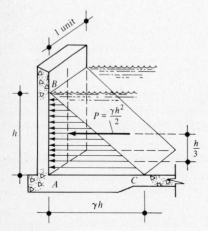

FIGURE 4.2

From the study of statics it is known that the resultant pressure, for a pressure diagram such as the one in Fig. 4.2, acts through the centroid of the pressure diagram. In this instance, therefore, P will act at a distance $h/3$ from the bottom of the wall.

The value of P, as can be noted from studying Fig. 4.2, is equal to the area of the triangle ABC, representing the pressure diagram.

EXAMPLE 4.1

If the depth of water in Fig. 4.1 is 4.5 m, determine (a) the total pressure per unit length of wall, (b) the point of application of this pressure, and (c) the total pressure on the wall if its length is 14.0 m.

SOLUTION

(a) From Eq. (4.1), the total pressure per unit length of wall will be

$$P = \frac{\gamma h^2}{2}$$

$$= \frac{9.81 \text{ kN/m}^3 \times (4.5 \text{ m})^2}{2}$$

$$P = 99.3 \text{ kN/m}$$

(b) P will act at $h/3$ from the bottom of the wall, or at

$$\bar{y} = \frac{4.5 \text{ m}}{3} = 1.5 \text{ m}$$

(c) The total pressure on the wall will be

$$\Sigma P = 99.3 \text{ kN/m} \times 14.0 \text{ m} = 1390 \text{ kN}$$

Pressure on Inclined Surfaces

The pressure of any fluid on a surface acts at right angles to that surface. Therefore, the triangle ABC in Fig. 4.3, with its base forming a right angle with the wall, represents the pressure diagram on the wall, inclined at an angle α with the horizontal.

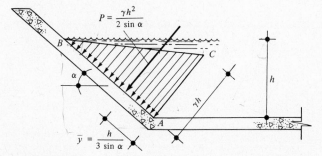

FIGURE 4.3

In this instance, the unit pressure at the bottom of the wall is again $p = \gamma h$. The total pressure per unit length of wall is equal to the average unit pressure multiplied by the area on which it acts.

For the inclined wall, the area on which this average unit pressure acts is

$$A = \frac{h}{\sin \alpha} \times 1 \text{ m}$$

and, consequently,

$$P = \frac{\gamma h}{2} \times \frac{h}{\sin \alpha} \times 1$$

$$P = \frac{\gamma h^2}{2 \sin \alpha} \tag{4.2}$$

P will, again, act through the centroid of the pressure diagram or at a distance

$$\bar{y} = \frac{h}{3 \sin \alpha}$$

from the bottom of the wall.

EXAMPLE 4.2

If in Fig. 4.3 the depth of water is 4.5 m and the wall forms an angle of 60° with the horizontal, determine (a) the pressure per unit length of wall and (b) the point of application of this pressure.

SOLUTION

(a)

$$P = \frac{\gamma h^2}{2 \sin \alpha}$$

$$= \frac{9.81 \text{ kN/m}^3 \times (4.5)^2}{2 \sin 60°}$$

$$P = 115 \text{ kN/m}$$

(b) The point of application of P is at

$$\bar{y} = \frac{4.5 \text{ m}}{3 \sin 60°}$$

$$= 1.73 \text{ m}$$

from the bottom of the wall, measured along the sloping face of the wall, or at

$$\bar{y}' = \frac{4.5 \text{ m}}{3} = 1.5 \text{ m}$$

from the bottom of the wall, measured vertically.

The pressure P will, of course, act at right angles to the face of the wall.

EXAMPLE 4.3

A concrete gravity dam, with a vertical upstream face, is 30.0 m high. Determine the total pressure per lineal meter on this dam and its point of application, when the water level is 3.0 m below the top of the dam.

SOLUTION

Figure 4.4 represents the pressure diagram for this dam. From this figure,

$$P = \frac{\gamma h^2}{2}$$

$$= \frac{9.81 \text{ kN/m}^3 \times (27.0 \text{ m})^2}{2}$$

$$P = 3580 \text{ kN/m}$$

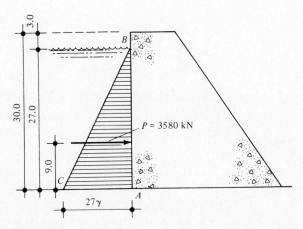

FIGURE 4.4

P will act at

$$\bar{y} = \frac{27.0 \text{ m}}{3}$$

$$= 9.0 \text{ m}$$

from the bottom of the dam.

EXAMPLE 4.4

If the dam in Example 4.3 has an upstream face, with a slope of 1 horizontal to 6 vertical, determine the resultant pressure P and its point of application.

SOLUTION

Figure 4.5 represents the pressure diagram for this case. The angle of the upstream face with the horizontal is determined from

$$\tan \alpha = \frac{6}{1} \quad \text{and} \quad \alpha = 80.5°$$

Hence the pressure per linear meter of dam is as follows:

$$P = \frac{\gamma h^2}{2 \sin \alpha}$$

$$= \frac{9.81 \text{ kN/m}^3 \times (27.0 \text{ m})^2}{2 \sin 80.5°}$$

$$P = 3630 \text{ kN/m}$$

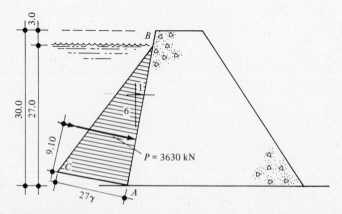

FIGURE 4.5

P will act at right angles to the face of the dam at a distance $\bar{y}$ from the bottom of the dam. Hence

$$\bar{y} = \frac{27.0 \text{ m}}{3 \times \sin 80.5°}$$

$$= 9.1 \text{ m}$$

measured along the face of the dam.

EXAMPLE 4.5

A schematic representation of an attic water storage tank, connected to a hot-water tank in a cottage is shown in Fig. 4.6. Determine the pressure diagram on the hot-water tank walls.

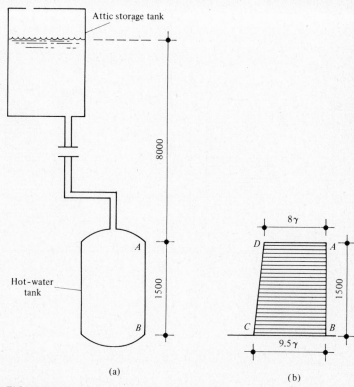

Attic storage tank

8000

Hot-water tank

1500

A

B

8γ

D A

1500

C B

9.5γ

(a)

(b)

FIGURE 4.6

SOLUTION

The unit pressure on the walls, at the top of the tank is

$$p_{top} = 9.81 \text{ kN/m}^3 \times 8.0 \text{ m}$$
$$p_{top} = 78 \text{ kPa}$$

The unit pressure at the bottom of the tank is

$$p_{btm} = 9.81 \text{ kN/m} \times 9.5 \text{ m}$$
$$p_{btm} = 93 \text{ kPa}$$

The pressure diagram is shown in Fig. 4.6.

EXAMPLE 4.6

Figure 4.7a represents one side of a rectangular concrete water storage tank. With the water level as shown, determine the resultant pressure P per lineal meter of the wall section AB and the point of application of P.

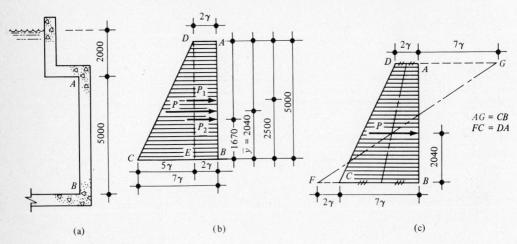

(a)   (b)   (c)

FIGURE 4.7

SOLUTION

The unit pressure at the top of the wall, or at point A, is

$$p_A = 9.81 \text{ kN/m}^3 \times 2.0 \text{ m}$$
$$p_A = 19.6 \text{ kPa}$$

The unit pressure at the bottom of the wall, or at point B, is

$$p_B = 9.81 \text{ kN/m}^3 \times 7.0 \text{ m}$$

$$p_B = 68.7 \text{ kPa}$$

The trapezoid $ABCD$ in Fig. 4.7b represents the pressure diagram on this wall. Hence the resultant pressure P per unit width of wall is

$$P = \frac{19.6 \text{ kPa} + 68.7 \text{ kPa}}{2} \times 5.0\text{m} \times 1.0\text{m}$$

$$P = 221 \text{ kN/m}$$

The pressure diagram $ABCD$ can be broken down into the triangle CDE and the rectangle $ABCE$, as shown in Fig. 4.7b. Each one of these two portions can be represented by its resultants and the total pressure exerted by them will be the same as that resulting from P.

For the rectangle $ABCE$ the resultant pressure per linear meter is

$$P_1 = 9.81 \text{ kN/m}^3 \times 2.0 \text{ m} \times 5.0 \text{ m}$$

$$P_1 = 98 \text{ kN/m}$$

For the triangle CDE the resultant pressure per linear meter is

$$P_2 = \frac{9.81 \text{ kN/m}^3 \times (5.0 \text{ m})^2}{2}$$

$$P_2 = 123 \text{ kN/m}$$

P_1 acts at a distance of 5.0 m/2 = 2.5 m and P_2 at 5.0 m/3 = 1.67 m from the bottom of the wall. Also, the total pressure is

$$P = P_1 + P_2$$

or

$$P = 98 + 123$$

$$= 221 \text{ kN}$$

The location of the resultant pressure P can be found by application of the law of statics, which states that the sum of the moments of a number of forces acting on a body must equal the moment of the resultant of these forces.

Therefore, denoting by $\bar{y}$ the moment arm of P and taking moments about the point B, the following statement must be true for P to be the resultant of P_1 and P_2.

$$P_1 \times 2.5 \text{ m} + P_2 \times 1.67 \text{ m} = P \times \bar{y}$$

or

$$98 \text{ kN} \times 2.5 \text{ m} + 123 \text{ kN} \times 1.67 \text{ m} = 221 \text{ kN} \times \bar{y}$$

and

$$\bar{y} = 2.04 \text{ m}$$

The location of the point of application of P could also have been found graphically, as demonstrated in Fig. 4.7c.

4.2   GENERAL CASE OF PRESSURE ON PLANE SURFACES

Total Pressure on Plane Surfaces

The total pressure exerted by a fluid on areas with rectangular or square configurations and with one side parallel with the water surface can be readily found by methods explained previously.

The general solution for determining resultant pressures applicable to all plane surfaces is more complex and requires a different treatment.

Figures 4.8a and b show an irregularly shaped plane submerged in a liquid in such a way that it forms an angle α with the liquid surface. The distances h, $\bar{h}$, and h_p indicate the head of liquid to an elementary strip of the plane, Δ_A, to the center of gravity, and to the center of pressure, respectively; h', $\bar{h}'$ and h'_p, are the distances from these same points to the liquid surface, but measured along the plane.

At the elementary area ΔA, the unit pressure is

$$p = \gamma h = \gamma h' \sin \alpha$$

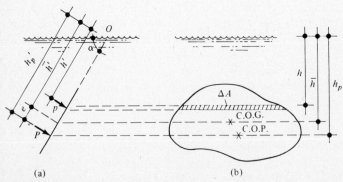

(a)   (b)

FIGURE 4.8

and the total pressure on this strip, with area ΔA, is

$$\Delta P = \gamma h \times \Delta A = \gamma h' \sin \alpha \times \Delta A \tag{4.3}$$

Hence the total pressure on the plane could be expressed as the summation of an infinite number of such elementary strips, making up the total area of the plane.

If the areas ΔA are chosen sufficiently small, the distances h and h_1 can be considered to be distances to their center of gravity. Hence it follows that the summation of the moments of these ΔA areas will also be the moment of the entire area A with the moment arm to its center of gravity. Consequently, the total pressure P will be

$$P = \gamma \sin \alpha \bar{h}' A \tag{4.4}$$

or

$$P = \gamma \bar{h} A \tag{4.5}$$

In other words: *The total pressure on a plane surface submerged in a fluid is equal to the product of the area of the plane, the specific weight of the fluid, and the head on the center of gravity of the plane.*

Center of Pressure on Plane Surfaces

The point at which the resultant pressure P in Fig. 4.8 acts is called the center of pressure. Depending on the configuration of the plane, the location of the center of pressure deviates from the center of gravity—horizontally as well as vertically.

However for planes with a regular configuration that are symmetrical about a vertical centerline, the center of pressure will always fall on this line. This is the case in most civil engineering applications, where the areas used almost always are made up of triangles, rectangles, squares, trapezoids, circles, or other regular geometric figures.

To determine the vertical location of the center of pressure, consider the moments of the resultant pressure P acting at the center of pressure, and the summation of the moments of the pressures on the elementary areas, ΔA, about the axis O.

Keeping in mind that the moment of inertia of the plane about the axis O is

$$I_O = Ak^2 + A\bar{h}'^2$$

where k is the radius of gyration of the plane, leads to the following expression for the distance to the center of pressure.

$$h'_p = \frac{k^2 + \bar{h}'^2}{\bar{h}'} \tag{4.6}$$

The vertical distance e between the center of gravity and the center of pressure can be found as follows:

$$e = h'_p - \bar{h}'$$

$$= \frac{k^2 + \bar{h}'^2}{\bar{h}'} - \bar{h}' \tag{4.7}$$

or

$$e = \frac{k^2}{\bar{h}'} \tag{4.8}$$

The solution of problems related to pressures on such surfaces is facilitated by reference to Table 4.1, showing properties of commonly encountered geometric figures.

EXAMPLE 4.7

A vertical circular gate in the low flow pipe of a dam has its centerline at a distance of 6.35 m below the water surface. The gate is 1.50 m in diameter. Determine the total pressure on this gate and the point where it acts.

SOLUTION

From Eq. (4.5)

$$P = 9.81 \text{ kN/m}^3 \times 6.35 \text{ m} \times \frac{(1.5 \text{ m})^2 \times \pi}{4}$$

$$= 110.0 \text{ kN}$$

and from Eq. (4.8) and Table 4.1

$$e = \frac{(1.5/4)^2}{6.35}$$

$$= 0.022 \text{ m}$$

Consequently, the center of pressure is 22 mm below the center of the gate.

EXAMPLE 4.8

The gate in Fig. 4.9 is hinged as shown. Determine the force F required to open this gate.

TABLE 4.1   Properties of Selected Geometric Figures

Square:

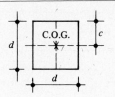

$$A = d^2 \qquad\qquad s = \frac{d^3}{6}$$

$$c = \frac{d}{2}$$

$$I = \frac{d^4}{12} \qquad\qquad k = \frac{d}{12^{1/2}}$$

Rectangle:

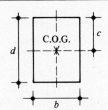

$$A = bd \qquad\qquad s = \frac{bd^2}{6}$$

$$c = \frac{d}{2}$$

$$I = \frac{bd^3}{12} \qquad\qquad k = \frac{d}{12^{1/2}}$$

Triangle:

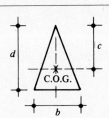

$$A = \frac{bd}{2} \qquad\qquad s = \frac{bd^2}{24}$$

$$c = \frac{2d}{3}$$

$$I = \frac{bd^3}{36} \qquad\qquad k = \frac{d}{18^{1/2}}$$

Circle:

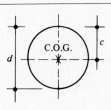

$$A = \frac{\pi d^2}{4} \qquad\qquad s = \frac{\pi d^3}{32}$$

$$c = \frac{d}{2}$$

$$I = \frac{\pi d^4}{64} \qquad\qquad k = \frac{d}{4}$$

Semicircle:

$$A = \frac{\pi d^2}{8} \qquad\qquad s = \frac{d^3(9\pi^2 - 64)}{192(3\pi - 4)}$$

$$c = \frac{d}{2}\left(1 - \frac{4}{3\pi}\right)$$

$$I = \frac{d^4}{16}\left(\frac{\pi}{8} - \frac{8}{9\pi}\right) \qquad\qquad k = \frac{d(9\pi^2 - 64)^{1/2}}{12\pi}$$

Legend: C.O.G = Center of gravity; A = area; c = distance to C.O.G.; I = moment of inertia; s = section modulus; k = radius of gyration.

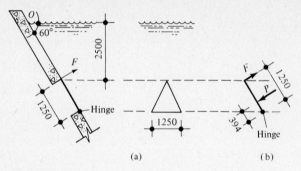

(a)   (b)

FIGURE 4.9

SOLUTION

To determine the total force, the area and the distances from the center of gravity to the point O must be known. The center of gravity is a distance

$$c = \frac{2 \times 1.25}{3}$$

$$= 0.833 \text{ m}$$

from the top of the gate. Consequently,

$$\bar{h}' = \frac{2.5}{\sin 60°} + 0.833$$

$$= 3.72 \text{ m}$$

and, from Eq. (4.4),

$$P = 9.81 \text{ kN/m}^3 \times \sin 60° \times 3.72 \text{ m} \times \frac{1.25 \text{ m} \times 1.25 \text{ m}}{2}$$

$$= 24.7 \text{ kN}$$

From Table 4.1,

$$k = \frac{1.25 \text{ m}}{18^{1/2}}$$

$$= 0.295$$

and hence

$$e = \frac{0.295^2}{3.72}$$

$$= 0.023$$

Figure 4.9b shows a free-body diagram of the gate. The moment arm of P with respect to the hinge is

$$\bar{y} = 1.25 - [(2/3)1.25 + 0.023]$$

$$= 0.394$$

Therefore, taking moments about the hinge yields

$$F \times 1.25 \text{ m} = P \times 0.394 \text{ m}$$

Hence

$$F = \frac{24.7 \text{ kN} \times 0.394 \text{ m}}{1.25 \text{ m}}$$

$$= 7.8 \text{ kN}$$

EXAMPLE 4.9

The gate on the low flow pipe of the dam in Fig. 4.10 is semicircular. Determine the force F required to keep this gate closed when the water level is as shown.

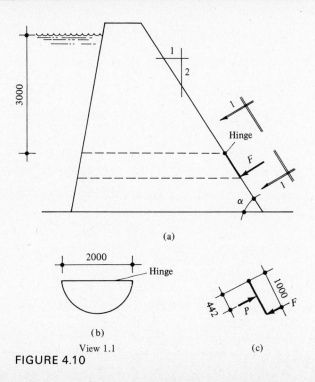

(a)

(b)

View 1.1

(c)

FIGURE 4.10

SOLUTION

From the geometry of the figure,

$$\tan\left(\alpha = \frac{2}{1}\right)$$

and

$$\alpha = 63.4°$$

From Table 4.1 the location of the center of gravity of this gate is a distance

$$c = \frac{2.0 \text{ m}}{2}\left(1 - \frac{4}{3\pi}\right)$$

$$= 0.576 \text{ m}$$

from the bottom of the gate, or a distance

$$c' = 1.0 \text{ m} - 0.576 \text{ m}$$

$$= 0.424 \text{ m}$$

from the hinge. Therefore

$$\bar{h}' = \frac{3.0 \text{ m}}{\sin 63.4°} + 0.424 \text{ m}$$

$$= 3.78 \text{ m}$$

and, from Eq. (4.4),

$$P = 9.81 \text{ kN/m}^3 \times \sin 63.4° \times 3.78 \text{ m} \times \frac{\pi \times (2.0 \text{ m})^2}{8}$$

$$= 52.1 \text{ kN}$$

Also, from Table 4.1,

$$k = \frac{2.0 \text{ m}(9\pi^2 - 64)^{1/2}}{12\pi}$$

$$= 0.264 \text{ m}$$

Hence, from Eq. (4.8),

$$e = \frac{0.264^2}{3.78}$$

$$= 0.018 \text{ m}$$

The free-body diagram of this gate in Fig. 4.10b indicates that

$$F \times 1.0 = P \times (0.424 + 0.018)$$

$$F = \frac{52.1 \text{ kN} \times 0.442 \text{ m}}{1.0 \text{ m}}$$

$$= 23.0 \text{ kN}$$

4.3   PRESSURE ON CURVED SURFACES

General Case of Pressure on Curved Surfaces

Hydrostatic pressures on curved surfaces occur in civil engineering in such applications as gates on dams or sides of culverts, among others. In most instances these surfaces will be of a regular configuration, such as circular, elliptical, parabolic, and hyperbolic sections.

For the sake of generality, however, the discussion of pressures on curved surfaces herein is based on a surface with any curvature. For example, in Fig. 4.11 consider the prism of liquid $ABCA'B'C'$, having a width equal to one unit and bounded by the curved surface AC, the water surface AB, and the imaginary plane BC.

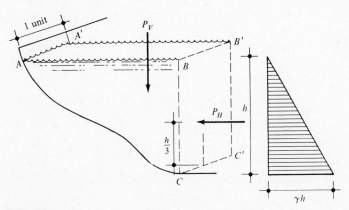

FIGURE 4.11

One force, acting on the surface AC, is the force of gravity on the prism of liquid. This force is equal to the volume of liquid in the prism, multiplied by the specific weight of the liquid and will act vertically downward. This force will be referred to as P_V.

The only other force acting on the surface AC is the pressure exerted by the liquid. This pressure is zero at the liquid surface and equal to γh at the point C. This total pressure will act horizontally and can be represented by the pressure diagram in Fig. 4.11. The resultant horizontal force P_H due to hydrostatic pressure acts at a point $2h/3$ below the liquid surface.

The resultant of the pressures P_V and P_H is the total pressure P exerted by the liquid on the surface AC and

$$P = (P_V{}^2 + P_H{}^2)^{1/2} \qquad (4.9)$$

From earlier discussions it is evident that the total pressure P must act at right angles to the surface, and this fact can be used to determine its location, as will be demonstrated in the following examples.

EXAMPLE 4.10

Determine the horizontal, vertical, and total pressures acting on the Tainter gate in Fig. 4.12a.

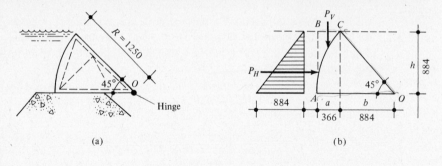

(a)   (b)

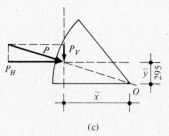

(c)

FIGURE 4.12

SOLUTION

Figure 4.12b shows the geometric arrangement and dimensions of this gate. They were calculated as follows:

$$h = 1.25 \text{ m} \times \sin 45°$$

$$= 0.884 \text{ m}$$

$$b = 1.25 \text{ m} \times \cos 45°$$

$$= 0.884 \text{ m}$$

$$a = 1.25 \text{ m} - 0.884 \text{ m}$$

$$= 0.366 \text{ m}$$

The horizontal pressure P_H on a 1.0-m section of this gate is shown in the pressure diagram in Fig. 4.12b and is calculated as follows:

$$P_H = \frac{0.884\gamma \times 0.884 \text{ m}}{2}$$

$$= 0.39\gamma$$

$$= 3.8 \text{ kN}$$

acting at a point a distance

$$\bar{y} = 0.884/3$$

$$= 0.295 \text{ m}$$

from the hinge.

The vertical pressure P_v is due to the weight of the wedge of water ABC in Fig. 4.12b. The force P_v can be calculated by determining the area of the trapezoid $ABCO$ and subtracting from it the area of the circular sector ACO; that is,

$$A_{trap} = \frac{0.366 + 1.25}{2} \times 0.884$$

$$= 0.714 \text{ m}^2$$

$$A_{sector} = \frac{2.5^2 \times \pi}{4} \times \frac{45°}{360°}$$

$$= 0.614 \text{ m}^2$$

Consequently, the area of the water producing P_v can be found as follows:

$$A_{ABC} = 0.714 \text{ m}^2 - 0.614 \text{ m}^2$$

$$= 0.100 \text{ m}^2$$

If the gate is considered to be 1.0 m in length, then

$$P_v = 0.100 \text{ m}^2 \times 1.0 \times \gamma$$

$$= 0.1\gamma$$

$$= 0.98 \text{ kN}$$

P_H is now completely defined by its value, direction, and point of application. P_V is defined only from the point of view of value and direction. Its point of application will be through the center of gravity of the area ABC. This center of gravity is not readily found.

However, since the resultant force P must act at right angles to the surface of the gate, it must also act through the center of the circle or point O. Using this fact leads to the free-body diagram of Fig. 4.12c. From similar triangles on this figure, the moment arm of P_V in relation to O, $\bar{x}$, can be calculated as follows,

$$\frac{\bar{x}}{P_H} = \frac{\bar{y}}{P_V}$$

or

$$\bar{x} = \frac{0.295 \times 0.39 \gamma}{0.1 \gamma}$$

$$= 1.15 \text{ m}$$

The total force P is, from Fig. 4.12c,

$$P = (P_H^2 + P_V^2)^{1/2}$$

$$= (3.8^2 + 0.98^2)^{1/2}$$

$$= 3.92 \text{ kN}$$

An interesting fact about this gate is that the resultant force P acts through the center of the circle at the hinge about which the gate turns. Consequently, the hydrostatic pressure has a moment arm and moment about this hinge equal to zero. Opening such a gate therefore involves overcoming only the moment of the weight of the gate. Thus hydrostatic pressure does not need to be overcome in operating this gate.

EXAMPLE 4.11

Figure 4.13 is a schematic drawing of a tainter gate with the upstream water elevation 0.75 m above the top of the gate and water spilling over as shown. Determine the total force acting on this gate and the horizontal and vertical components of this force. The radius of the gate is 3.5 m.

SOLUTION

From the geometrics of the figure, the following dimensions are found.

$$b = 3.5 \times \cos 55°$$

$$= 2.01 \text{ m}$$

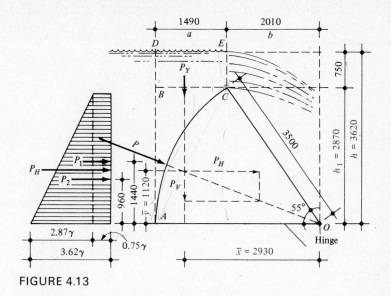

FIGURE 4.13

$$a = 3.5 - 2.01$$
$$= 1.49 \text{ m}$$
$$h_1 = 3.5 \times \sin 55°$$
$$= 2.87 \text{ m}$$
$$h = 2.87 + 0.75$$
$$= 3.62 \text{ m}$$

Consequently, the horizontal pressure diagram will be trapezoidal with the top unit pressure 0.75γ and the bottom unit pressure 3.62γ. This diagram can be divided into the rectangle and triangle as shown in Fig. 4.13. If P_1 and P_2 are the resultant horizontal forces for the rectangular and triangular portion, respectively, then

$$P_1 = 0.75\gamma \times 2.87$$
$$= 2.15\gamma \text{ per meter of gate}$$

and

$$P_2 = \frac{2.87\gamma}{2} \times 2.87$$
$$= 4.12\gamma \text{ per meter of gate}$$

Their respective moment arms in relation to the hinge O are

$$\bar{y}_1 = \frac{2.87}{2}$$
$$= 1.44 \text{ m}$$

$$\bar{y}_2 = \frac{2.87}{3}$$

$$= 0.96 \text{ m}$$

The net resultant horizontal force P_H is

$$P_H = 2.15\gamma + 4.12\gamma$$

$$= 6.27\gamma \text{ per meter of gate}$$

$$= 61.5 \text{ kN/m}$$

Its moment arm $\bar{y}$ is found by equating moments about the hinge, as follows:

$$6.27\gamma \times \bar{y} = (2.15\gamma \times 1.44) + (4.12 \times 0.96)$$

or

$$\bar{y} = 1.12 \text{ m}$$

P_1, P_2, and P_h are shown on Fig. 4.13.

Calculations of the vertical force P_v can be done in a manner similar to that in the previous example. P_v will be equal to the weight of the water with a volume equal to the area $ADEC$ by one meter for a unit section of the gate. Therefore

$$P_v = (1.49 \times 0.75)\gamma + \left(\frac{1.49 + 3.5}{2} \times 2.87 - \frac{7.0^2 \times \pi}{4} \times \frac{55}{360}\right) \times \gamma$$

$$= 2.40\gamma \text{ per meter of gate}$$

$$= 23.5 \text{kN/m}$$

Its point of application and moment arm $\bar{x}$ are found from the geometrical relationships of the free-body diagram in Fig. 4.13, remembering that the resultant force P must pass through O. Hence

$$\frac{\bar{x}}{6.27\gamma} = \frac{1.12}{2.40\gamma}$$

or

$$\bar{x} = 2.93 \text{ m}$$

and

$$P = (61.5^2 + 23.5^2)^{1/2}$$

$$= 6.58 \text{ kN per meter of gate}$$

Pressure on Cylindrical Surfaces

Cylindrical surfaces are a common occurrence in civil engineering and other applications. Storage tanks and pipes are almost always circular and are in everyday use.

Figure 4.14 represents a cylindrical tank with a vertical longitudinal axis. At any point along this axis the unit pressure on the tank wall will be

$$p = \gamma h$$

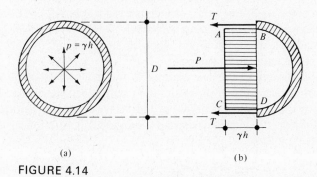

(a)   (b)

FIGURE 4.14

This pressure will be the same at any point in the horizontal plane at the depth h. Consequently, this pressure will also be exerted on the tank walls.

For these reasons the total horizontal pressure per unit length on one-half of the tank wall can be represented by the pressure diagram $ABCD$, as shown in Fig. 4.14b. This total pressure is

$$P = \gamma h \times d \times 1.0 \qquad (4.10)$$

where d is the diameter of the tank.

The forces resisting this total pressure are the tensile forces T in the tank wall. These forces act tangentially to the tank wall and

$$T = \frac{\gamma h \times d}{2} \qquad (4.11)$$

If the cylinder or pipe has a horizontal longitudinal centerline, the above reasoning regarding wall pressures and tensile forces does not apply exactly. It is obvious that in this instance the pressure at the bottom of the tank or pipe would be larger than the pressure at the top by

$$\Delta p = \gamma d \qquad (4.12)$$

Practical applications of this situation arise in the case of pipes carrying liquids under pressure. In almost all these types of installations, however, the pipe diameter d is only a small fraction of the total head h under which the pipe will operate. Therefore, unless exceptional circumstances prevail, it is generally assumed that the pressure in the horizontal pipe is uniform throughout its cross section and equal to that produced by the head on the pipe's centerline.

In the case of civil engineering and other applications of this principle to pipes under pressure, other serious considerations must be remembered. Some of these are the fluctuation in pressures causing repeated loading of the pipe wall, short bursts of very high pressure due to water hammer problems, and corrosion of the inside and the outside surfaces of the pipe wall. For this reason, often the design of the wall thickness of such pipes is governed by considerations other than the tensile stresses, T, discussed here. Most of these designs are governed by rigid specifications set by appropriate governing authorities or industrial and professional organizations.

A discussion similar to the one leading to the determination of internal pressures in a cylindrical tank or pipes and the resulting tensile stresses in their walls will lead to the conclusion that a cylindrical container submerged in a liquid, will be subject to compressive stresses on the wall, as follows:

$$C = \frac{\gamma h d}{2} \qquad (4.13)$$

EXAMPLE 4.12

A cylindrical water standpipe for a small municipality has a diameter of 6.0 m and a maximum water level of 20.0 m above its bottom, as shown in Fig. 4.15. Determine (a) the maximum theoretical tensile stresses in the tank walls at the bottom of the tank; (b) the minimum wall thickness at the bottom of the tank if the tank is built of steel, with a maximum allowable tensile stress of 125 MPa.

SOLUTION

(a) The unit pressure at the bottom of the tank is

$$p = 9.81 \times 20.0$$

$$= 196 \text{ kN/m}^2$$

Consequently, the maximum tensile stress in the tank wall is

$$T = \frac{196 \times 20.0}{2}$$

$$= 589 \text{ kN/m}$$

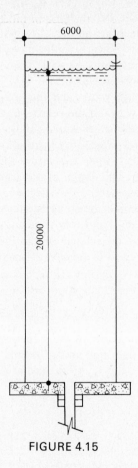

6000

20000

FIGURE 4.15

(b) At the bottom of the tank the required steel plate thickness t is

$$t = \frac{589 \text{ kN/m}}{125 \text{ MPa}}$$

$$= 0.005 \text{ m} = 5.0 \text{ mm}$$

Obviously, a thicker plate should be used to allow for corrosion.

4.4   GRAVITY DAMS

Most reservoir outlet structures and other type of dams are classified as gravity dams. These structures rely entirely on their mass to remain in equilibrium and to resist sliding or overturning movement resulting from hydrostatic pressures or other forces acting on them. The two most common types of gravity dams built today are the mass concrete dam and the buttress dam.

Mass concrete gravity dams generally have a nearly uniform cross section throughout their entire length, whereas buttress dams consist of heavy gravity buttresses joined by thin overflow sections or spillways.

Gravity dams are constructed to hold back water, and the forces acting on them are the hydrostatic pressure of the water in the reservoir; the uplift pressure, or hydrostatic pressure, exerted by the water seeping in the foundations; earthquake forces; and the ice pressures created by a frozen reservoir surface. The resultant of all these forces will tend either to overturn the dam or to make it slide along its foundation, or both. The design of the gravity dam relies on the mass of the dam to resist this movement.

The design of large gravity dams is a complex civil engineering problem requiring expert engineering skills and experience and is beyond the scope of this book. Consequently, the discussions herein are related only to the general principles involved—particularly as they apply to the design of small gravity dams.

Hydrostatic Pressure on Gravity Dams

The upstream face of a dam, such as the one shown in Fig. 4.16, is in contact with the water. Considering a section of the dam one unit in length, the unit hydrostatic pressures on the face of the dam will vary from zero at the water surface to

$$p = \gamma h$$

at the bottom of the dam.

The total pressure on the face of the dam is represented by the pressure

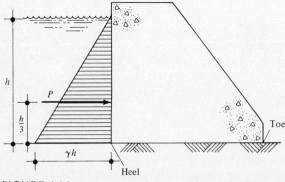

FIGURE 4.16

diagram in Fig. 4.16, and the resultant pressure has a value of

$$P = \frac{\gamma h^2}{2}$$

and P will act at a point $h/3$ above the heel of the dam.

It is evident that P in this instance represents the total hydrostatic pressure exerted by the upstream water on a section of the dam only one unit in length. It is general practice to investigate dams in this manner. Of course, in cases where the cross section varies considerably along the length of the dam more than one such section might need to be investigated.

In the event, as shown in Fig. 4.17, that the water level exceeds the top of the dam, as would be the case for an overflow or spillway section, the pressure diagram will assume the form of the trapezoid $ABCD$. Because of overflow condition, some of the water near the top will have an upward velocity and therefore will not contribute to the hydrostatic pressure on the dam. The actual pressure diagram will more or less approximate that shown by broken line in Fig. 4.17. The actual reduction in pressure due to this phenomenon is generally very small in comparison with the overall hydrostatic pressure on the dam. Most designers therefore assume that the pressure diagram is the full trapezoid, as shown. The actual shape of the pressure diagram in these cases can be approximated by means of a flow-net analysis, as discussed later in this text.

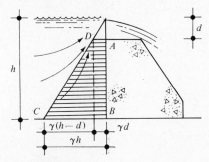

FIGURE 4.17

EXAMPLE 4.13

Determine the total horizontal and vertical components of the hydrostatic pressure on a section, one meter in length, of the dam in Fig. 4.18.

SOLUTION

It will be noted that this dam is subject to hydrostatic pressure not only of the water on the upstream side but also of the tailwater downstream of the dam.

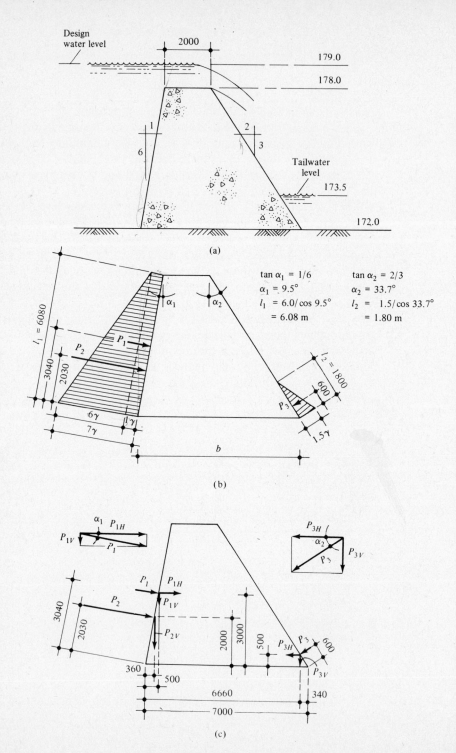

(a)

$\tan \alpha_1 = 1/6$ $\tan \alpha_2 = 2/3$

$\alpha_1 = 9.5°$ $\alpha_2 = 33.7°$

$l_1 = 6.0/\cos 9.5°$ $l_2 = 1.5/\cos 33.7°$

     $= 6.08$ m $= 1.80$ m

(b)

(c)

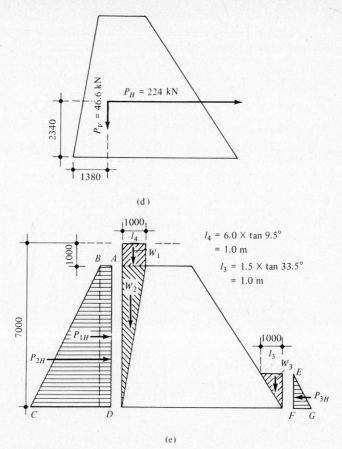

FIGURE 4.18

The pressure diagrams for this condition are shown in Fig. 4.18*a*. It should be noted that since the faces of the dam are not vertical, the pressure diagrams are not horizontal but at right angles to the faces of the dam. Also, in order to simplify the calculations, forces have been expressed including the term γ rather than converting to numerical values. Actual numerical values of the forces will be shown at intermediate stages, as appropriate.

As shown in Fig. 4.18*b*, the resultant forces P_1, P_2, and P_3 can be calculated as follows:

$$P_1 = 1.0\gamma \times 6.08 \text{ m}$$

$$= 6.08\gamma$$

$$P_2 = \frac{6.0\gamma}{2} \times 6.08 \text{ m}$$

$$= 18.24\gamma$$

$$P_3 = \frac{1.5\gamma}{2} \times 1.80$$

$$= 1.35\gamma$$

The points of application for these forces are as follows: For P_1, the distance from the heel is

$$\frac{6.08 \text{ m}}{2} = 3.04 \text{ m}$$

For P_2, the distance from the heel is

$$\frac{6.08 \text{ m}}{3} = 2.03 \text{ m}$$

For P_3, the distance from the toe is

$$\frac{1.80 \text{ m}}{3} = 0.60 \text{ m}$$

In order to determine the total horizontal and vertical hydrostatic pressures on the dam, consider the free-body diagram in Fig. 4.18c. Then,

$$P_{1H} = 6.08\gamma \times \cos 9.5° = 6.0\gamma$$

$$P_{1V} = 6.08\gamma \times \sin 9.5° = 1.0\gamma$$

$$P_{2H} = 18.24\gamma \times \cos 9.5° = 18.0\gamma$$

$$P_{2V} = 18.24\gamma \times \sin 9.5° = 3.0\gamma$$

$$P_{3H} = 1.35\gamma \times \cos 33.7° = 1.12\gamma$$

$$P_{3V} = 1.35\gamma \times \sin 33.7° = 0.75\gamma$$

The horizontal and vertical hydrostatic pressures will therefore be

$$P_H = 6.0\gamma + 18.0\gamma - 1.12\gamma = 22.88\gamma = 224 \text{ kN}$$

$$P_V = 1.0\gamma + 3.0\gamma + 0.75\gamma = 4.75\gamma = 46.6 \text{ kN}$$

The points of applications of P_H and P_V can be found by equating moments of the various components with the moments of the resultants. If the moments about the heel of the dam are considered, then the moment arms of the components can be computed.
    For

$$P_{1H}: \quad 3.04 \times \cos 9.5° = 3.0 \text{ m}$$

$$P_{2H}: \quad 2.03 \times \cos 9.5° = 2.0 \text{ m}$$

$$P_{3H}: \quad 0.60 \times \cos 33.7° = 0.5 \text{ m}$$

$$P_{1V}: \quad 3.04 \times \sin 9.5° \ = 0.5 \text{ m}$$

$$P_{2V}: \quad 2.03 \times \sin 9.5° \ = 0.36 \text{ m}$$

$$P_{3V}: \quad 0.60 \times \sin 33.7° \ = 0.34 \text{ m (from the toe of the dam)}$$

The base of the dam has a dimension b of

$$b = 2.0 \text{ m} + \frac{6.0 \text{ m}}{6} + \frac{6.0 \text{ m}}{1.5} = 7.0 \text{ m}$$

Therefore the moment arm of P_{3V} about the heel of the dam is

$$7.0 \text{ m} - 0.34 \text{ m} = 6.66 \text{ m}$$

Consequently, the moment arms $\bar{y}$ for P_H and $\bar{x}$ for P_V about the heel of the dam are

$$\bar{y} = \frac{(6.0\gamma \times 3.0 + 18.0\gamma \times 2.0 - 1.12\gamma \times 0.5)}{22.88\gamma}$$

$$= 2.34 \text{ m}$$

and

$$\bar{x} = \frac{(1.0\gamma \times 0.5 + 3.0\gamma \times 0.36 + 0.75\gamma \times 6.66)}{4.75\gamma}$$

$$= 1.38 \text{ m}$$

Figure 4.18d shows the dam and the resultant hydrostatic forces.

The above calculations could have been carried out in a somewhat simpler manner by considering the pressure diagram as illustrated in Fig. 4.18e.

The pressure diagrams ABCD and EFG neglect the fact that the faces of the dam are not vertical. Therefore, allowance must be made for the vertical forces W_1, W_2, and W_3, equal to the weight of the prisms of water acted upon by the force of gravity.

It will be noted that, as could be expected in this instance, the horizontal forces P_1, P_2 and P_3 correspond to the horizontal components calculated earlier and that the gravitational forces W_1, W_2, and W_3 are equivalent to the vertical components calculated above.

Uplift Pressures on Gravity Dams

In almost all cases, water will penetrate the foundation material on which the dam rests. At the heel and the toe of the dam, this water will be subjected to hydrostatic pressures created by the head of water at the upstream and downstream faces of the dam, respectively.

This hydrostatic pressure, according to Pascal's law, will act equally in all directions. Consequently, it will also act upward on the base of the dam, with an intensity equal to the horizontal hydrostatic pressures at these points. Figures 4.19a and b show theoretical uplift pressure diagrams on the base of gravity dams. In Fig. 4.19a is shown the case where no water exists at the toe of the dam, or without tailwater. When tailwater exists, the pressure diagram theoretically is that as shown in Fig. 4.19b.

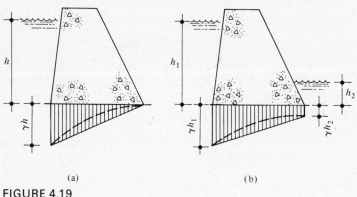

(a)   (b)

FIGURE 4.19

Experimental investigations of the actual uplift pressures on numerous gravity dams have shown that the reduction in pressure, from heel to toe, is not a linear (straight-line) relationship but more or less follows the curve represented by the broken lines in Fig. 4.19.

This nonlinear relationship is primarily due to the fact that the water under the dam is not at rest, or in a static condition, but is actually flowing. As will be illustrated in this text, flowing water loses energy in overcoming the resistance to flow offered by the material in contact with the water, in this case the foundation material. The actual reduction of the uplift pressures and the shape of the pressure diagram will be a reflection of the resistance to flow offered by the foundation materials.

If no means are employed to reduce the uplift pressures, the uplift pressure diagrams for dams on rock foundations are almost always considered to follow a linear pattern. For dams on soil foundations, the pressure diagram is also often assumed to be linear, thus introducing a certain safety factor. Reasonably accurate estimates of the actual pressure diagram to be expected in soil foundations can be made by means of a flow-net analysis, which will be discussed later in this text.

The effects of uplift pressures on gravity dams can be reduced by means of drainage structures, grouting of the foundation material, cutoff walls, and other means.

Drainage structures designed to remove the water from underneath the base of the dam can be very effective in reducing uplift pressures, since they increase the velocity of the flowing water under the dam. This higher velocity will result in increased pressure losses. Drains under a dam can, however, become clogged easily and their actual conditions or effectiveness during the lifetime of the dam are not easily monitored. Therefore, unless such actual continuous monitoring is available and it is certain that the effectiveness of the drains will not be reduced, the use of these devices for reducing uplift pressures is not recommended. It should also be remembered that drains removing water from beneath the dam will remove water from the reservoir as well, and thus increase the water loss.

Grouting of the foundations, cutoff walls, and impervious blankets located upstream are shown schematically in Fig. 4.20. The effectiveness of these methods in reducing uplift pressure lies in the fact that water will travel a considerably longer distance to reach the base of the dam. Hence, there will be an increased pressure loss. Their effectiveness is readily seen since these methods do not require continuous monitoring and actually diminish water loss.

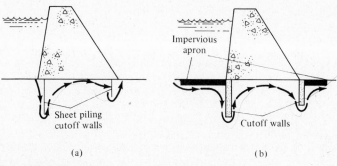

(a)   (b)

FIGURE 4.20

EXAMPLE 4.14

For the dam in Example 4.13, determine the resultant uplift force, assuming the uplift pressure diagram to vary linearly along the base.

SOLUTION

The uplift pressure diagram is shown in Fig. 4.21. Since the diagram is trapezoidal, it can, as shown earlier, be broken down into a rectangular and a triangular section. If U_1 and U_2 represent the resultant forces due to the triangular and rectangular sections, respectively, then

$$U_1 = \frac{5.5\gamma}{2} \times 7.0 = 19.25\gamma$$

$$U_2 = 1.5\gamma \times 7.0 = 10.5\gamma$$

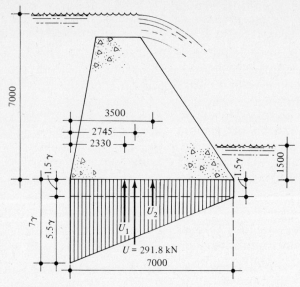

FIGURE 4.21

and the total uplift force U will be the sum of U_1 and U_2, or

$$U = 19.25\gamma + 10.5\gamma = 29.75\gamma$$

$$U = 291.8 \text{ kN}$$

The moment arm of each section of the pressure diagram and the total resultant uplift force can be found in a manner similar to that employed for the horizontal hydrostatic pressures. On this basis the moment arm for U will be found to be 2.75 m from the heel of the dam.

Ice Pressure on Gravity Dams

In climates where reservoir water surfaces will freeze over, extensive additional horizontal forces can be exerted on the face of the dam, caused by the expansion of the ice sheet with fluctuating temperatures. The intensity of the pressure produced by the ice depends on a number of factors, the more important of which will be discussed below.

TEMPERATURE VARIATIONS
Rapid variations in temperature of the ice sheet will induce sudden expansion of the ice surface and could produce large pressures on the face of a dam. Relatively little information of the behavior of ice under these conditions is available.

Investigations by the U.S. Bureau of Reclamation indicate that once the ice sheet approaches a thickness of 500 mm, temperature variations do not exist at the bottom of the sheet and hence ice pressures can be considered negligible.

Ice temperature variations are caused primarily by the air temperature above the sheet and absorption of solar energy. A layer of snow with a thickness of 150 mm has sufficient insulating value to remove these effects and the consequent expansion of the ice sheet.

Figure 4.22 represents estimates of ice forces produced by various ice sheet thicknesses and ice temperatures. These curves were developed in 1947 by E. Rose, and subsequent extensive field and laboratory tests by the U.S. Bureau of Reclamation have found close agreement between actual ice pressures and those calculated from Fig. 4.22.

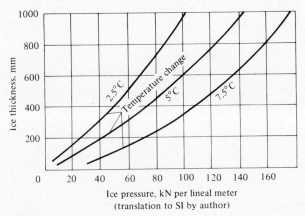

FIGURE 4.22

LATERAL RESTRAINT OF THE ICE SHEET

The ice sheet when expanding will act somewhat similarly to a structural column or compression member. As with a structural member, therefore, length-to-thickness ratio and end support conditions will have some effect on the total pressure the sheet can develop before it will buckle or break.

Consequently, the slope of the reservoir shores and the size of the reservoir, particularly the length at right angles to the longitudinal axis of the dam will greatly affect the ice pressures.

WATER-LEVEL FLUCTUATIONS

An ice sheet resting on a stationary water surface will, in a sense, have a certain amount of lateral restraint. As in the case of structural compression members, such lateral restraint will increase the total pressure that the sheet will be able to develop. On the other hand, fluctuating water levels in the reservoir will tend to break the ice sheet during its formation. This will result in uneven thicknesses of the sheet, and earlier failure and buckling in the thinner sections will occur.

Silt Pressures on Gravity Dams

A reservoir, regardless of its purpose, will act as a large settling basin. Silt and soil are carried into the reservoir by the streams discharging into it. When the waters of the streams reach the reservoir, there is a considerable reduction in the velocity of flow, resulting in silt and other sedimentary materials settling to the bottom of the reservoir. Obviously the heavier sediment will settle out early, while the finer silt particles will stay suspended in the water longer. The result of this process is that near the face of the dam the sediment consists of the finest silt particles carried by the water. These particles have a density higher than that of water and could conceivably exert pressures on the upstream face of the dam in excess of those expected due to hydrostatic pressure alone.

The actual amount of the pressure produced by this silt accumulation is uncertain. As the layers of silt deposits develop, the lower layers become compacted and stop acting as a fluid. The well-compacted layers of silt will become essentially impervious and act as a drainage blanket, thus reducing the uplift forces somewhat.

Forces due to silt loads are therefore seldom considered in the design of small dams unless special conditions exist where these forces could be considered significant. In such cases their effects should be carefully studied on the basis of soil mechanics and possibly model tests.

Resistance Against Sliding

The horizontal component of the resultant of all the external forces and hydrostatic, ice, and silt pressures acting on the dam, will tend to make the dam slide on its foundation. The only resistance provided to prevent this sliding is the friction forces developed between the base of the dam and the foundation material.

These friction forces will increase as the vertical forces on the dam increase. They will also depend on the type of material in the foundation in contact with the base of the dam.

The following two methods are in general use, in the analysis of small dams, to ensure safety against sliding. Each of these methods is based on calculating a coefficient that, when not exceeded, will provide safety for the dam against sliding.

SAFE FRICTION COEFFICIENT (C_f)

This method relies on the friction coefficient between the foundation materials and the base of the dam. This coefficient, C_f, is related to the static friction coefficient of the materials. The safe friction coefficient is calculated as follows:

$$C_f = \frac{R_H}{R_V}$$ (4.14)

where R_H and R_V are the resultants of all the horizontal and vertical forces on the dam, respectively, including forces due to the mass of the dam, uplift forces, and ice pressures. Values of the safe friction coefficient C_f are shown in Table 4.2.

TABLE 4.2   Safe Friction Coefficient, C_f

Foundation Material	C_f
Sound rock	0.70–0.80
Gravel and coarse sand	0.40
Sand	0.30
Shale	0.30

SAFETY FACTOR (F_s)

The safety factor against sliding is defined as the ratio of the coefficient of static friction, f, to the tangent of the angle θ made by the resultant of all the forces acting on the dam and a line perpendicular to the base of the dam. A study of Fig. 4.23 indicates that

$$F_s = \frac{f}{\tan \theta}$$ (4.15)

or

$$F_s = f \times \frac{R_V}{R_H}$$ (4.16)

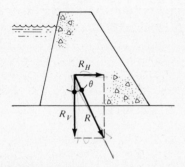

FIGURE 4.23

TABLE 4.3   Minimum Safety Factor, F_s

Foundation Material	f	F_s
Rock	0.80	1.0–1.15
Gravel	0.40	2.5
Sand and shale	0.30	2.5
Silt or clay		2.5*

*Tests are required to determine actual value.

Values of the static friction factor, f, and the safety factor, F_s, are shown in Table 4.3.

Most gravity dams are constructed with their bases partially below the surface of the ground, or contain cutoff trenches or keys in their bases, as shown in Fig. 4.24. The volume of earth downstream of the dam and above the base will, to some extent, aid in preventing the dam from moving horizontally. To include this in the design for sliding safety, however, must be done only with great care. Only when the exact nature of the shear forces in the earth can be determined and when no danger of soil condition changes, due to seepage and other factors, can occur should these forces be relied on to provide safety against sliding.

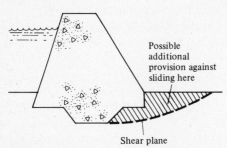

Possible additional provision against sliding here

Shear plane

FIGURE 4.24

EXAMPLE 4.15

For the dam in Example 4.13 determine whether the design is safe against sliding if (a) it is constructed on sound rock and (b) it is constructed on sand. Assume the density of concrete to be $\rho_c = 2400$ kg/m³ or $\gamma_c = 23.5$ kN/m³.

SOLUTION

In order to determine the values of C_f and F_s for this dam it is necessary to find the horizontal and vertical components of all the forces acting on the dam. The hydrostatic pressure and uplift forces, were calculated in Examples 4.14 and 4.15, respectively. Assuming ice pressures and forces due to earthquake and wave action are negligible, the only other forces to be considered are those due to the force of gravity acting on the mass of the dam.

For ease of calculation, the dam cross section has been divided in triangles and a rectangle, as shown in Fig. 4.25a. Consequently, the forces due to gravity that will have to be taken into account are G_1, G_2, and G_3.

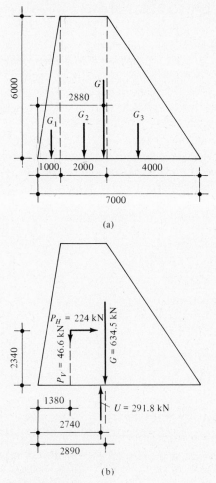

(a)

(b)

FIGURE 4.25

These forces and their moment arms about the heel of the dam are as follows:

$$G_1 = 1.0 \times \frac{6.0}{2} \times 23.5 = 70.5 \text{ kN at } 0.67 \text{ m}$$

$$G_2 = 2.0 \times 6.0 \times 23.5 = 282.0 \text{ kN at } 2.0 \text{ m}$$

$$G_3 = 4.0 \times \frac{6.0}{2} \times 23.5 = 282.0 \text{ kN at } 4.33 \text{ m}$$

The total force due to gravity, G, is the sum of the above three forces, and its moment arm, $\bar{x}$, can be found by equating the moment of G with the sum of the moments of G_1, G_2, and G_3, or

$$G = 70.5 + 282.0 + 282.0$$

$$= 634.5 \text{ kN}$$

and

$$\bar{x} = \frac{70.5 \times 0.67 + 282.0 \times 2.0 + 282.0 \times 4.33}{634.5}$$

$$\bar{x} = 2.89 \text{ m}$$

The resultant forces—representing the hydrostatic pressure, the uplift force, and the force of gravity of the dam—are shown on the free-body diagram in Fig. 4.25b. Hence

$$R_H = P_H = 224 \text{ kN} \qquad \text{(from Example 4.13)}$$

$$R_V = P_V - U + G$$

$$= 46.6 - 291.8 + 634.5$$

$$= 389.3 \text{ kN}$$

From Tables 4.2 and 4.3:
   (a) On sound rock foundation,

$$C_f = \frac{224}{389.3} = 0.58 < 0.75$$

and

$$F_s = 0.8 \times \frac{389.3}{224} = 1.40 > 1.15$$

Consequently, the dam would be safe against sliding on a sound rock foundation.
   (b) On sand foundation,

$$C_f = 0.58 > 0.40$$

and

$$F_s = 0.3 \times \frac{389.3}{224.0} = 0.52 < 2.5$$

The dam would obviously be unsafe from a sliding point of view if constructed on a sand foundation.

Resistance Against Overturning

The horizontal forces acting on a dam will tend to make the dam turn about the toe. Resistance against this overturning moment is provided by the mass of the concrete in the dam. From the study of statics it is known that a body will overturn when the resultant of all the forces acting on this body acts along a line that falls outside of its base. Examples of these conditions are shown in Fig. 4.26.

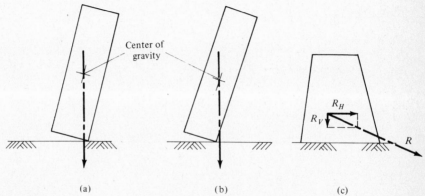

FIGURE 4.26   The object in (a) will right itself.   Objects in (b) and (c) will fall or overturn.

The design of concrete gravity dams takes this condition into account. Most common design criteria demand that the resultant force acts through a line that will intersect the base within the middle third of its length, as shown in Fig. 4.27. With the resultant force acting at this location, no negative stresses will develop in the concrete or at the interface between the concrete base and the foundation material at the upstream face of the dam.

When the resultant force intersects the base outside of the middle third of the base length, negative stresses will develop at the upstream face of the dam. Although these negative stresses do not result in overturning of the structure, they will create tension in the materials at the face where they occur. Concrete, for all intents and purposes, cannot resist tensile stresses without a certain amount of cracking. In a wet environment, especially where freezing and thawing cycles occur, such cracks could lead to serious deterioration of the concrete. Good practice therefore consists of checking the design at strategic points in the dam to ensure no such tensile stresses develop.

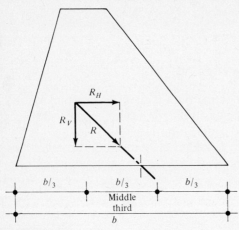

FIGURE 4.27

EXAMPLE 4.16
Determine whether the dam in Example 4.13 is safe against overturning.
Assume that no ice pressures or forces due to earthquakes or wave action
are to be considered.

SOLUTION

In order to determine the safety against overturning for this dam,
it is necessary to ascertain whether the resultant of all forces acting on it
will intersect the base in its middle third. The free-body diagram of Fig.
4.25b shows the forces contributing to the overturning moment of the dam.

The moment arm of R_H is equal to the moment arm of P_H since the
latter is the only horizontal force acting on the dam. Consequently,

$$\bar{y} = 2.34 \text{ m}$$

The moment arm of R_V can again be found by comparing the moments
of P_V, U, and G with the moment of R_V, or

$$\bar{x} = \frac{46.6 \times 1.38 - 291.8 \times 2.74 + 634.5 \times 2.89}{389.3}$$

$$\bar{x} = 2.82 \text{ m}$$

The horizontal and vertical resultants, R_H and R_V, are shown in Fig.
4.28. If R_H, R_V, and the dam are drawn to scale, as is the case in this
figure, it is evident that R intersects the base of the dam within the
middle third of its length.

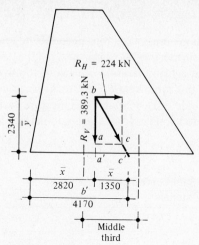

FIGURE 4.28

The location of the intersection of R with the base of the dam can also be found as follows:

The triangle abc is similar to the triangle $a'bc'$; therefore,

$$\frac{\bar{x}'}{R_H} = \frac{\bar{y}}{R_V}$$

or

$$\frac{\bar{x}'}{224} = \frac{2.34}{389.3}$$

and

$$\bar{x}' = 1.35 \text{ m}$$

Hence

$$b' = 2.82 \text{ m} + 1.35 \text{ m}$$
$$= 4.17 \text{ m}$$

and

$$\frac{7.0}{3} < 4.17 < 7.0 \times \frac{2}{3}$$

R acts through the middle third of the base of the dam, and thus the dam is safe against overturning and no negative stresses will develop at the upstream face.

Foundation Stresses for Gravity Dams

The design of gravity dams must take into account the stresses produced in the foundation materials supporting the dam.

Table 4.4 lists allowable bearing stresses of various foundation materials. Since the values shown in the table relate only to average conditions, they should be considered as a guide only. For any specific application care should be taken to undertake detailed soils investigation and studies for determining the actual values to be used.

TABLE 4.4   Safe Load Bearing Values of
Foundation Materials

Material	kPa	kip/ft²
Fine sand and silt	75	1.5
Medium sand	75	1.5
Fine gravel	120	2.5
Gravel and sand	120–140	2.5–3.0
Coarse gravel	120–240	2.5–5.0
Boulders, gravel, and sand	120	2.5
Medium clay	100	2.0
Hardpan	240	5.0
Good rock	2400	50.0
Fissured rock	850	17.0

EXAMPLE 4.17

Determine the foundation stresses for the dam in Example 4.13.

SOLUTION

Figure 4.29 shows the base of the dam and the assumed shape of the foundation stress diagram. This diagram should not be confused with the uplift pressure diagram. It actually represents the assumed distribution of the pressures exerted by the foundation materials in order to keep the dam in equilibrium.

Since R intercepts the base at a point 4.17 m from the heel, R_H and R_V can be applied at this point as well. R_H will be resisted by the sliding friction developed between the base of the dam and the foundation material, as discussed and investigated on page 100. The force R_V therefore is the only force to be resisted by the foundation material.

If F_1 and F_2 represent the resultants of the triangular and rectangular sections of the foundation stress diagram, respectively, then

$$F_1 + F_2 = R_V = 389.3 \text{ kN} \qquad \text{(Example 4.15)}$$

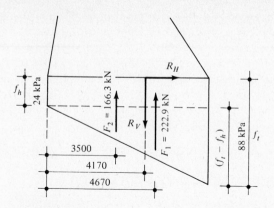

FIGURE 4.29

or

$$F_1 = 389.3 - F_2$$

Furthermore, the sum of the moments of F_1 and F_2 must be equal to the moment of R_V, or

$$F_1 \times 4.67 + F_2 \times 3.5 = 389.3 \times 4.17$$

Substituting the value for F_1 found above into the last equation yields

$$(389.3 - F_2) \times 4.67 + F_2 \times 3.5 = 389.3 \times 4.17$$

or

$$F_2 = 166.4 \text{ kN}$$

and

$$F_1 = 222.9 \text{ kN}$$

A careful study of Fig. 4.29 indicates that the following relationships exist between F_1, F_2, f_t, and f_h. The latter two symbols represent the unit foundation stresses at the toe and the heel of the dam, respectively.

$$F_1 = \frac{f_t - f_h}{2} \times b$$

and

$$F_2 = f_h \times b$$

where b is the length of the base of the dam. In this instance $b = 7.0$ m. Substituting the values of F_1 and F_2 found above, and solving for f_t and f_h, yields

$$f_t = 87.1 \text{ kPa or } 87 \text{ kPa}$$

and

$$F_h = 23.8 \text{ kPa or } 24 \text{ kPa}$$

Other Considerations in Gravity Dam Design

The foregoing discussion is only a very generalized outline of the principles involved in the design of gravity dams. The design of dams is a very specialized area in the civil engineering field and should not be attempted without detailed soils and field investigations, laboratory and model testing, and, above all, the advice from experts in this field.

Other than the items mentioned in the previous paragraphs, the following considerations and parameters should be carefully studied and checked.

EARTHQUAKE FORCES

Earthquakes can appreciably increase hydrostatic pressures and reduce bearing capacities of soils. Allowances for these conditions, if they are prevalent in the area under study, must be made.

The forces and problems created by earthquakes are beyond the scope of this text and reference should be made to local building regulations and other numerous publications relating to this subject. Some of these publications are listed in the list of references.

CONCRETE STRESSES

Although internal concrete tensile stresses should be avoided, they could occur in the upstream face of the dam—particularly at the horizontal plane, where the cross section of the dam changes. Care must be taken that these tensile stresses are well within the allowable values for the concrete used and the construction practices employed.

Another problem area where overstressing of the concrete may occur is at the toe of the dam. Generally at this point the foundation stresses are the highest, and these must be resisted by the concrete in the toe of the dam. Since this is generally the point in the dam where the concrete cross section is the smallest, high tensile and/or compressive stresses can be expected.

SOIL MOISTURE CONDITIONS

Excessive seepage under the dam could result in certain foundation materials being carried away, thereby creating a condition known as "piping." When piping occurs, it becomes progressively more pronounced and could in a relatively short time destroy the entire foundation of the dam. Seepage also tends to create an upward flow of water at the toe of the dam. This condition could destroy the bearing capacity of the soil at that point. Further discussions on seepage and counteractive measures are included in future chapters in this text.

WAVE ACTION

Although they are not generally very significant in the design of small dams, wave action and the forces created by them on the upstream face of the dam require consideration for large reservoirs.

CONSTRUCTION PRACTICES

Many specialized construction practices are to be employed in the building of gravity dams. Among these are considerations of concrete quality, as well as placing and curing techniques. A great deal of literature is available on this subject and careful reference should be made to it. As already mentioned, particular care is required at the construction joints to ensure that individual sections of the dam meet overturning and sliding criteria.

CAVITATION FORCES

The overflow or spillway sections of almost all dams are concrete gravity sections. In many instances flow conditions could be such that the water will separate, on the downstream side, from the concrete section, as illustrated in a later chapter. Unless special provisions are made, this condition will create a vacuum between the downstream face of the dam and the lower nappe of the water stream. This condition is called cavitation. Forces resulting from cavitation not only can damage the downstream face of the dam by spalling off concrete, but also can add to the overturning moment of the dam.

Cavitation can be avoided either by providing for air to enter the space between the water stream and the surface of the dam or by proper spillway design.

ADDITIONAL FOUNDATION PRESSURES

Aside from the foundation stresses discussed in this chapter, additional pressures can result from the water falling downstream of the dam or flowing over the downstream face. For higher dams with large flows, these forces can induce considerable additional stresses in the foundation materials and the concrete at the toe of the dam.

A complete design of a gravity dam should also include a check of the foundation stresses when the reservoir is empty and no uplift or hydrostatic pressures are present. In some instances the weight of the concrete could produce foundation stresses in excess of the allowable.

4.5 SUMMARY OF THE CHAPTER IN BRITISH UNITS

Theoretical Considerations

The general discussions and the theoretical development of formulas and principles related to pressure on surfaces given in the earlier parts of this chapter apply equally in the British System of units. None of the formulas presented require conversion from SI to British, or vice versa.

Allowable stresses and other tabulated or graphed values related to the design of dams are, where required, shown in both systems of units.

In order to illustrate this chapter and its relation to the British System of units the following examples are presented here. The explanation of the various steps in solving problems related to the following examples have not been carried out to the same detailed extent as earlier in the chapter. It is assumed that the examples and the methods presented earlier have been worked through and are familiar.

Selected Examples in British Units

EXAMPLE 4.18

If the depth of water in Fig. 4.1 is 4.5 ft, determine (a) the total pressure per unit length of wall, (b) the point of application of this pressure, and (c) the total pressure on this wall, if its length is 14.0 ft.

SOLUTION

(a) From the pressure diagram and Eq. (4.1) the total pressure per unit length will be

$$P = \frac{62.4 \text{ lb/ft}^3 \times (4.5 \text{ ft})^2}{2}$$

$$= 632.0 \text{ lb per linear foot}$$

$$= 0.63 \text{ kip per linear foot}$$

where 1.0 kip is 1000 lb.

(b) P will act through the centroid of the pressure diagram, or a distance of $h/3$ from the bottom of the wall, or

$$\bar{y} = \frac{4.5 \text{ ft}}{3}$$

$$= 1.5 \text{ ft}$$

(c) The total pressure on the wall will be

$$\Sigma P = 632 \text{ lb/ft} \times 14.0 \text{ ft}$$

$$= 8850$$

$$= 8.85 \text{ kip}$$

EXAMPLE 4.19

In Fig. 4.3, if the depth of water is 4.5 ft and the wall forms an angle of 60° with the horizontal, determine (a) the pressure per unit length of wall and (b) the point of application of this pressure.

SOLUTION

(a) From the pressure diagram and Eq. (4.2),

$$P = \frac{62.4 \text{ lb/ft}^3 \times (4.5 \text{ ft})^2}{2 \times \sin 60°}$$

$$= 730 \text{ lb per linear foot}$$

(b) The point of application of P is at

$$\bar{y} = \frac{4.5 \text{ ft}}{3 \times \sin 60°}$$

$$= 1.73 \text{ ft}$$

from the bottom of the wall, measured along the sloping wall face, or at

$$\bar{y}' = \frac{4.5 \text{ ft}}{3}$$

$$= 1.5 \text{ ft}$$

from the bottom of the wall, measured vertically. Of course, the pressure P will act at right angles to the face of the wall.

EXAMPLE 4.20

A concrete gravity dam with a vertical upstream face is 100.0 ft high. Determine the total pressure per linear foot on this dam and its point of application when the water level is 10.0 ft below the top of the dam.

SOLUTION

Figure 4.30 represents the pressure diagram for this dam. From this diagram,

$$P = \frac{62.4 \text{ lb/ft}^3 \times (90.0 \text{ ft})^2}{2}$$

$$= 253 \text{ kip per linear foot}$$

P will act at

$$\bar{y} = \frac{90.0 \text{ ft}}{3}$$

$$= 30.0 \text{ ft}$$

from the bottom of the dam.

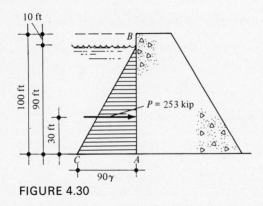

FIGURE 4.30

EXAMPLE 4.21

If the dam in Example 4.20 has an upstream face, with a slope of 1 horizontal to 6 vertical, determine the resultant hydrostatic pressure P and its point of application.

SOLUTION

Figure 4.31 represents the pressure diagram in this case. The angle α is found as follows:

$$\tan \alpha = 6/1$$

$$\alpha = 80.5°$$

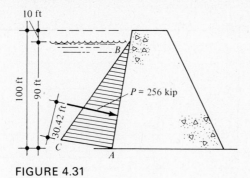

FIGURE 4.31

Hence

$$P = \frac{62.4 \text{ lb/ft}^3 \times (90.0 \text{ ft})^2}{2 \times \sin 80.5°}$$

$$= 256 \text{ kip}$$

P will act at

$$\bar{y} = \frac{90.0 \text{ ft}}{3 \times \sin 80.5°}$$

$$= 30.4 \text{ m}$$

from the bottom of the dam, measured along the face of the dam.

EXAMPLE 4.22

The concrete wall in Fig. 4.32a represents the side of a water storage reservoir. In this wall is a rectangular gate AB, which is 4.5 ft high and 5.5 ft wide. A cable at the bottom of the gate leads via a pulley to a winch. Determine the force required to be exerted by the winch to open this gate.

SOLUTION

The pressure diagram on this gate is shown in Fig. 4.32b. This diagram is constructed remembering that the pressures at the top and the bottom of the gate are, respectively, 2.5γ and 7.0γ, since the pressure heads at these points are 2.5 ft and 7.0 ft.

The resulting pressure diagram is a trapezoid, and the total pressure on this gate will act through the centroid of this diagram. Mathematical location of this centroid is complex, and therefore the trapezoidal pressure diagram is divided into the rectangular and triangular sections as shown.

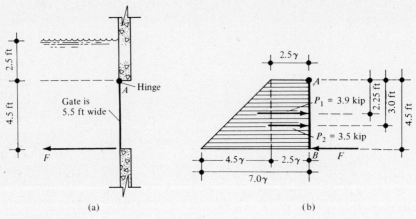

(a)   (b)

FIGURE 4.32

If P_1 and P_2 represent the total pressures corresponding to the rectangular and triangular sections of the pressure diagram, respectively, the following values result.

$$P_1 = 2.5\gamma \times 4.5 \text{ ft} \times 5.5 \text{ ft}$$

$$= 61.9\gamma$$

$$= 3.9 \text{ kip}$$

and

$$P_2 = \frac{4.5\gamma}{2} \times 4.5 \text{ ft} \times 5.5 \text{ ft}$$

$$= 55.7\gamma$$

$$= 3.5 \text{ kip}$$

P_1 and P_2 act through the centroid of their respective sections of the pressure diagram. Hence their moments about the hinge are as shown in Fig. 4.32b, or

$$\bar{a}_1 = 2.25 \text{ ft}$$

$$\bar{a}_2 = 3.0 \text{ ft}$$

Comparing the moment of the force F required to open the gate with the moments of P_1 and P_2 yields

$$F \times 4.5 \text{ ft} = 61.9\gamma \times 2.25 \text{ ft} + 55.7\gamma \times 3.0 \text{ ft}$$

or

$$F = 68.1\gamma$$

$$= 4.2 \text{ kip}$$

EXAMPLE 4.23

The seawall in Fig. 4.33a has a gate 3.5 ft square, which is hinged as shown. With the water levels as indicated in the figure, determine the force F required to open this gate. Assume the relative density of seawater to be 1.04.

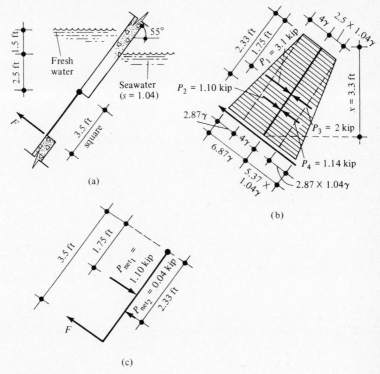

FIGURE 4.33

SOLUTION

The pressure diagram for this gate is shown in Fig. 4.33b. The values for the unit pressures at the bottom of the gate are found, bearing in mind that the head of water at this point is

$$h_{left} = 4.0 \text{ ft} + 3.5 \text{ ft} \times \sin 55°$$
$$= 6.87 \text{ ft}$$

and

$$h_{right} = 2.5 \text{ ft} + 3.5 \text{ ft} \times \sin 55°$$
$$= 5.37 \text{ ft}$$

Hence the forces created by the water pressures and their moment arms for each of the four portions of the pressure diagram are as follows:

$$P_1 = 4.0\gamma \times (3.5 \text{ ft})^2$$
$$= 49\gamma$$
$$= 3.1 \text{ kip, acting at 1.75 ft from the hinge}$$
$$P_2 = \frac{2.87\gamma}{2} \times (3.5)^2 \text{ ft}$$
$$= 17.6\gamma$$
$$= 1.10 \text{ kip, acting at 2.33 ft from the hinge}$$
$$P_3 = 2.5\gamma \times 1.04 \times 3.5^2 \text{ ft}$$
$$= 31.8\gamma$$
$$= 2.0 \text{ kip, acting at 1.75 ft from the hinge}$$
$$P_4 = \frac{2.87\gamma \times 1.04}{2} \times 3.5^2 \text{ ft}$$
$$= 18.3\gamma$$
$$= 1.14 \text{ kip, acting at 2.33 ft from the hinge}$$

The resulting free-body diagram, showing the net pressures on this gate, is given in Fig. 4.33c. The net pressures are

$$P_{net_1} = 49\gamma - 31.8\gamma$$
$$= 17.2\gamma$$
$$= 1.1 \text{ kip, acting at 1.75 ft from the hinge}$$
$$P_{net_2} = 17.6\gamma - 18.3\gamma$$
$$= -0.7\gamma$$
$$= -0.04 \text{ kip, acting at 2.33 ft from the hinge}$$

Therefore, comparing moments of F and the net hydrostatic pressures leads to

$$F \times 3.5 \text{ ft} = 17.2\gamma \times 1.75 \text{ ft} - 0.7\gamma \times 2.33 \text{ ft}$$

or

$$F = 8.1\gamma$$
$$= 0.5 \text{ kip}$$

EXAMPLE 4.24

The tainter gate in Fig. 4.34 is subjected to water pressures as shown. Determine the total hydrostatic pressure exerted per lineal foot of gate.

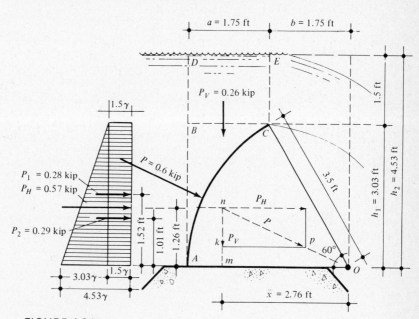

FIGURE 4.34

SOLUTION

Figure 4.33 shows the dimensions of the gate, which were calculated as follows:

$$h_1 = 3.5 \text{ ft} \times \sin 60°$$
$$= 3.0 \text{ ft}$$
$$h_2 = 3.03 \text{ ft} + 1.5 \text{ ft}$$
$$= 4.53 \text{ ft}$$

$$b = 3.5 \text{ ft} \times \cos 60°$$
$$= 1.75 \text{ ft}$$
$$a = 3.5 \text{ ft} - 1.75 \text{ ft}$$
$$= 1.75 \text{ ft}$$

From the resulting pressure diagram, also shown in Fig. 4.34, the horizontal pressures P_1 and P_2 are calculated to be

$$P_1 = 1.5\gamma \times 3.03 \text{ ft}$$
$$= 4.55\gamma, \text{ acting at 1.52 ft from the hinge}$$
$$P_2 = \frac{3.03\gamma}{2} \times 3.03 \text{ ft}$$
$$= 4.59\gamma, \text{ acting at 1.01 ft from the hinge}$$

Hence the total horizontal pressure on the gate is

$$P_H = 4.55\gamma + 4.59\gamma$$
$$= 9.14\gamma$$

and its moment arm about the hinge is

$$\bar{y} = \frac{4.55\gamma \times 1.52 \text{ ft} + 4.59\gamma \times 1.01 \text{ ft}}{9.14\gamma}$$
$$= 1.26 \text{ ft}$$

The vertical force P_V is due to the weight of the prism *ABDEC* of water, 1 ft wide, above the gate. The value of this force is found as follows:

$$\text{Area}_{ABCO} = \frac{1.75 \text{ ft} + 3.5 \text{ ft}}{2} \times 3.03 \text{ ft}$$
$$= 7.95 \text{ ft}^2$$
$$\text{Area}_{ACO} = \frac{(7.0 \text{ ft})^2 \times \pi}{4} \times \frac{60°}{360°}$$
$$= 6.41 \text{ ft}^2$$
$$\text{Area}_{ABC} = A_{ABCO} - A_{ACO}$$
$$= 7.95 \text{ ft}^2 - 6.41 \text{ ft}^2$$
$$= 1.54 \text{ ft}^2$$

and

$$\text{Area}_{ABCDE} = 1.54 \text{ ft}^2 + 1.5 \text{ ft} \times 1.75 \text{ ft}$$
$$= 4.17 \text{ ft}^2$$

and

$$V_{ABCDE} = 4.17 \text{ ft}^2 \times 1 \text{ ft}$$

$$= 4.17 \text{ ft}^3$$

Hence

$$P_V = 4.17\gamma$$

The value of the total resultant force on the gate is found as follows:

$$P = [(9.14\gamma)^2 + (4.17\gamma)^2]^{1/2}$$

$$= 10.0\gamma$$

$$= 0.62 \text{ kip}$$

The vector diagrams representing these forces are shown in Fig. 4.34. Since the resultant force P acts at right angles to the circular face of the gate, it follows that it must also act through the hinge at O. Using this fact and subsequently comparing the similar triangles knp and mnO, leads to the following relationship, determining the moment arm of P_V.

$$\frac{\bar{x}}{P_H} = \frac{1.26 \text{ ft}}{P_V}$$

or

$$\bar{x} = \frac{1.26 \text{ ft} \times 9.14\gamma}{4.17\gamma}$$

$$= 2.76 \text{ ft}$$

EXAMPLE 4.25

Investigate the dam in Fig. 4.35 for stability against overturning and sliding, and find the foundation stresses. Assume the specific weight of concrete to be 150 lb/ft³.

SOLUTION

The forces acting on this dam are the hydrostatic pressures P_1 and P_2, the uplift forces U_1 and U_2, and the gravitational forces G_1, G_2, and G_3. Their respective locations and values per lineal foot of dam are shown in Fig. 4.35a and have been calculated as follows:

$$P_1 = \frac{21.2\gamma}{2} \times 21.2 \text{ ft}$$

$$= 224.7\gamma$$

$$= 14.0 \text{ kip, acting at 7.07 ft from the heel}$$

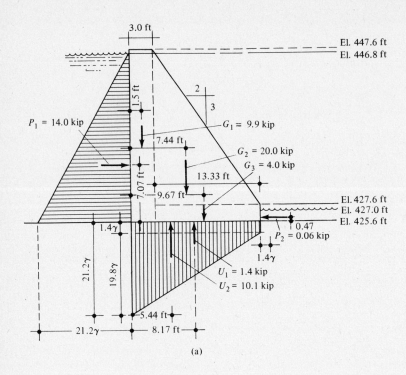

(a)

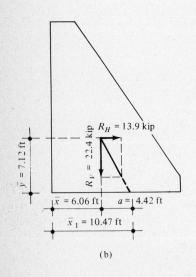

(b)

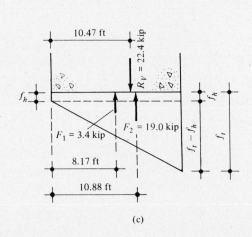

(c)

FIGURE 4.35

$$P_2 = \frac{1.4\gamma}{2} \times 1.4 \text{ ft}$$

$$= 1.0\gamma$$

$$= 0.06 \text{ kip, acting at 0.47 ft from the heel}$$

$$U_1 = 1.4\gamma \times 16.33 \text{ ft}$$

$$= 22.9\gamma$$

$$= 1.4 \text{ kip, acting at 8.17 ft from the heel}$$

$$U_2 = \frac{19.8\gamma}{2} \times 16.33 \text{ ft}$$

$$= 161.7\gamma$$

$$= 10.1 \text{ kip, acting at 5.44 ft from the heel}$$

$$G_1 = 3.0 \text{ ft} \times 22.0 \text{ ft} \times 150 \text{ lb/ft}^3$$

$$= 9.9 \text{ kip, acting at 1.5 ft from the heel}$$

$$G_2 = \frac{20.0 \text{ ft} \times 13.33 \text{ ft}}{2} \times 150 \text{ lb/ft}^3$$

$$= 20.0 \text{ kip, acting at 7.44 ft from the heel}$$

$$G_3 = 2.0 \text{ ft} \times 13.33 \text{ ft} \times 150 \text{ lb/ft}^3$$

$$= 4.0 \text{ kip, acting at 9.67 ft from the heel}$$

The resultant horizontal and vertical components are then

$$R_H = 14.0 \text{ kip} - 0.06 \text{ kip}$$

$$= 13.9 \text{ kip, acting at 7.12 ft from the heel}$$

and

$$R_V = 9.9 \text{ kip} + 20.0 \text{ kip} + 4.0 \text{ kip} - 1.4 \text{ kip} - 10.1 \text{ kip}$$

$$= 22.4 \text{ kip, acting at 6.06 ft from the heel}$$

These forces are shown on the free-body diagram in Fig. 4.35b.
In order to determine the distance $\bar{x}$, where the resultant R intersects the base of the dam, the distance a must be found. This distance can be determined from the similar triangles, shown in Fig. 4.34b, as follows

$$a = \frac{13.9 \times 7.12 \text{ ft}}{22.4}$$

$$= 4.42 \text{ ft}$$

Hence

$$\bar{x} = 6.06 \text{ ft} + 4.42 \text{ ft}$$

$$= 10.47 \text{ ft}$$

The middle third of the base of this dam lies between points 5.44 ft and 10.88 ft from the heel of the dam. Consequently, the resultant force R falls within the middle third of the dam base and the dam can be considered safe against overturning.

The friction coefficient for this dam is

$$C_f = \frac{13.9 \text{ kip}}{22.4 \text{ kip}}$$

$$= 0.62$$

From values read from Table 4.2, it is evident that this dam would be safe against sliding when built on a rock foundation. On a sand foundation, the friction coefficient exceeds the safe friction coefficient value and the dam is unsafe and liable to fail due to sliding.

Figure 4.35c shows the relationship between R_V and the foundation stresses. From this relationship the following equations can be written.

$$F_1 = f_h \times 16.33 \text{ ft, acting at 8.17 ft from the heel}$$

$$F_2 = \frac{f_t - f_h}{2} \times 16.33 \text{ ft, acting at 10.88 ft from the heel}$$

$$F_1 + F_2 = R_V$$

$$= 22.4 \text{ kip}$$

Hence

$$F_1 = 22.4 \text{ kip} - F_2$$

and comparing moments of F_1 and F_2 with R_V,

$$F_1 \times 8.17 \text{ ft} + F_2 \times 10.88 = R_V \times 10.47 \text{ ft}$$

Substituting values for F_1 and R_V in the above yields

$$(22.4 \text{ kip} - F_2) \times 8.17 \text{ ft} + F_2 \times 10.80 \text{ ft} = 22.4 \text{ kip} \times 10.47 \text{ ft}$$

From this relationship,

$$F_2 = 19.0 \text{ kip}$$

and

$$F_1 = 22.4 \text{ kip} - 19.0 \text{ kip}$$

$$= 3.4 \text{ kip}$$

Hence

$$f_h = \frac{3.4 \text{ kip}}{16.33 \text{ ft}^2}$$

$$= 0.21 \text{ kip/ft}^2$$

$$f_t - f_h = \frac{2 \times 19.0 \text{ kip}}{16.33 \text{ ft}^2}$$

$$= 2.33 \text{ kip/ft}^2$$

and

$$f_t = 2.33 \text{ kip/ft}^2 + 0.21 \text{ kip/ft}^2$$

$$= 2.54 \text{ kip/ft}^2$$

Therefore the maximum foundation stress is

$$f_t = 2.5 \text{ kip/ft}^2$$

which would be satisfactory on some gravel and sand foundations, as well as on hardpan and rock foundations (Table 4.4).

SELECTED PROBLEMS IN SI UNITS

4.1   The wall in Fig. P4.1 has a length of 12.0 m. Determine the total pressure exerted by the water on this wall.

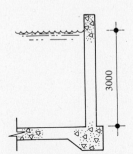

3000

FIGURE P4.1

4.2   A vertical gate 3.0 m high and 1.25 m wide has water standing on one side to a depth of 1.5 m above the top of the gate. Determine the magnitude and resultant total pressure on this gate.

4.3   The gate in Fig. P4.3, is square. Determine the force F required to open it if (a) $z = 2.0$ m, $y = 3.0$ m, $x = 3.5$ m; (b) $z = 1.5$ m, $y = 2.0$ m, $x = 1.5$ m; (c) $z = 1.5$ m, $y = 4.0$ m, $x = 4.0$ m; (d) $z = 3.0$ m, $y = 3.5$ m, $x = 2.0$ m.

4.4   Determine the force F required to keep the gate in Fig. P4.4 in the closed position. The gate is 2.0 m wide.

4.5   Determine the force required to open the square gate in Fig. P4.5.

4.6   Determine the total pressure P on the 1.0-m-wide gate in Fig. P4.6.

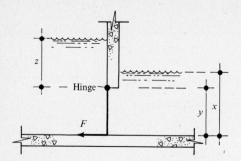

FIGURE P4.3

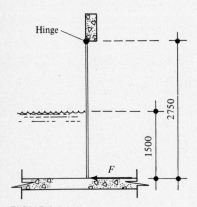

FIGURE P4.4

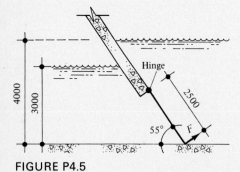

FIGURE P4.5

4.7   Figure P4.7 represents the longitudinal section of a rectangular channel that is 1.4 m wide. The gate must open when the water level in the channel reaches an elevation of 1.0 m above the bottom. Determine the required weight W of the counterweight.

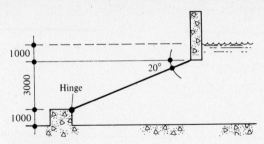

FIGURE P4.6

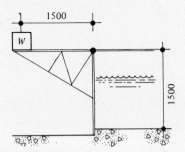

FIGURE P4.7

4.8   The gate in Fig. P4.8 is 6.0 m wide and is opened by pulling on two cables, one attached on each top corner of the gate. Determine the force required in each cable.

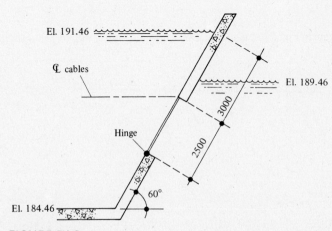

FIGURE P4.8

4.9   The gate in Fig. P4.9 is square. Determine the force F required to open it.

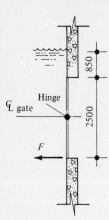

FIGURE P4.9

4.10   The gate in Fig. P4.10 is 4.15 m long. Determine the force F required to open it.

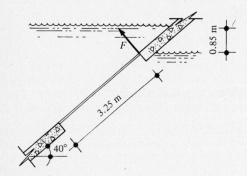

FIGURE P4.10

4.11   The gate in Fig. P4.11 is square. Determine the force F required to open it.

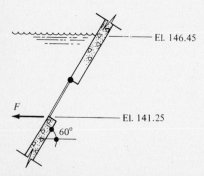

FIGURE P4.11

4.12 The gate in Fig. P4.12 is a flash gate, designed to open (rotate about the hinge) when the water level reaches a certain distance h above the top of the sill. Determine the height of water required to open this gate.

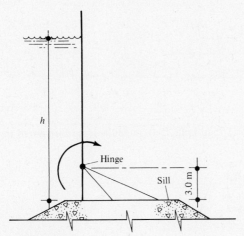

FIGURE P4.12

4.13 For the gates in Problem 4.3 determine the force F required to open the gates if the water on the right side of the gates is seawater ($s = 1.04$).

4.14 Determine the force F required to open the square gate in Fig. P4.14.

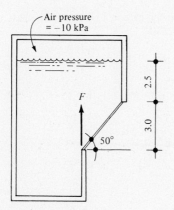

FIGURE P4.14

4.15 For the gate in Fig. P4.15 define the vertical and horizontal pressure forces and their points of application.

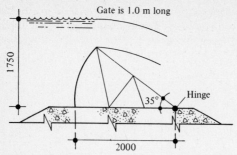

FIGURE P4.15

4.16   The gate in Fig. P4.16 is 1.75 m long. Determine the force F required to open it.

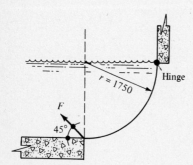

FIGURE P4.16

4.17   Determine the force F required to open the gate in Fig. P4.17. The gate is 3.55 m long and the weight W of the gate is 225 kN.

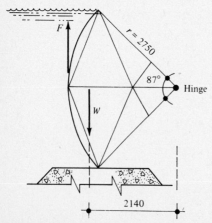

FIGURE P4.17

4.18 Is the dam in Fig. P4.18 safe against sliding and overturning? (Provide detailed sketches, showing all pressure diagrams and resultant forces.)

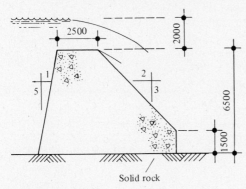

FIGURE P4.18

4.19 Perform a stability and sliding safety analysis for the gravity dam in Fig. P4.19. Determine the unit stresses at the heel and the toe, as well as at points *A* and *B*. Show all pressure diagrams and resultant forces. Include a short discussion of your findings, together with your recommendations regarding possible improvements.

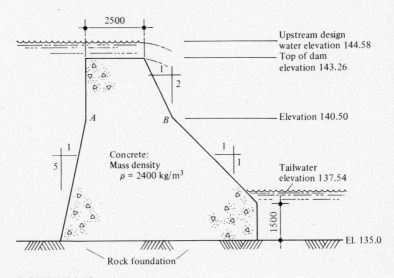

FIGURE P4.19

5

Buoyancy and Relative Equilibrium of Liquids

5.1 BUOYANCY

Archimedes' Principle

In the third century B.C., a Greek scientist and mathematician at the court of the Egyptian pharaoh, formulated the principle related to buoyant forces. His observations and conclusions are as valid today as they were 2000 years ago. Archimedes' principle, stated in today's terms and language reads as follows: "Any body submerged in a fluid is subject to an upward buoyant force equal to the gravitational force that would be exerted upon the volume of fluid being displaced."

It is this principle that explains phenomena such as the flotation of ships and wood in water, for instance. The validity of the principle can be visualized by observation of the prism of mass $ABCDEFGH$ in Fig. 5.1.

This prism is subject to hydrostatic pressures because of its submergence in the fluid. The horizontal pressures on the sides of the prism are all equal and in opposite directions. Therefore the horizontal forces are in equilibrium. Vertically, the prism is subjected to the hydrostatic pressures on the top and on the bottom surface. Using the notation of Fig. 5.1 and Eq. (3.2), these pressures can be expressed as

$$P_{\text{btm}} = \gamma h_1 \times ab \qquad (5.1)$$

131

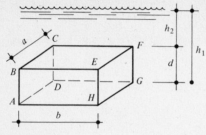

FIGURE 5.1

and

$$P_{top} = \gamma h_2 \times ab \tag{5.2}$$

The prism is therefore subject to a pressure difference of

$$\Delta P = \gamma(h_2 - h_1) \times ab \tag{5.3}$$

From Fig. 5.1 it can be observed that

$$h_2 - h_1 = d \tag{5.4}$$

Substituting these values in Eq. (5.3) yields

$$\Delta P = \gamma abd \tag{5.5}$$

or

$$B = \gamma abd \tag{5.6}$$

where B is the unbalanced vertical pressure, or buoyant force, γ is the specific weight of the fluid, and abd is the volume of the prism. This discussion, presented for simplicity as pertaining to a regular-shape body, can be easily extended to a body of any shape.

Although the prism is subjected to only one resultant hydrostatic pressure, it is also acted upon by gravity and has a gravitational force G that counteracts the buoyant force B, as shown in the free-body diagram of Fig. 5.2. A study of the figure indicates that there are three possible conditions that could exist with relation to the action of G and B.

1. If B is larger than G, the prism is acted upon by a resultant force in an upward direction and the prism will rise upward in the fluid.
2. If B is equal to G, the prism is in equilibrium and it will float at some position as shown in Fig. 5.1. From Eqs. (5.5) and (5.6) it is evident that

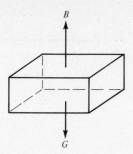

FIGURE 5.2

the values of B and G are related only to the volume of the prism and the relative densities of the prism's materials and the fluid respectively. Hence neither B nor G is dependent on the depth at which the prism is located and the prism will float at a constant depth.

3. If B is smaller than G, the resultant force on the prism is downward and the prism will sink.

EXAMPLE 5.1

The prism in Fig. 5.1 is made of concrete with a mass density of 2400 kg/m³. If $a=1.0$ m, $b=2.0$ m, and $d=1.5$ m, determine the force required to keep the prism stationary in the water.

SOLUTION

The prism is acted upon by two forces: the buoyancy force B and the force of gravity G. The latter is

$$G = 1.0 \times 2.0 \times 1.5 \times 2400 \times 9.81$$

$$= 70.6 \text{ kN}$$

and from Eq. (5.6),

$$B = 9.81 \times 1.0 \times 2.0 \times 1.5$$

$$= 29.4 \text{ kN}$$

The force G is larger than B, and therefore the prism would sink, unless supported by the force F. This force is equal to the difference between G and B, or

$$F = 70.6 - 29.4$$

$$= 41.2 \text{ kN}$$

EXAMPLE 5.2

A 300-mm-square timber with a relative density of 0.48, floats in the water as shown in Fig. 5.3. Determine the portion of the timber that will not be submerged.

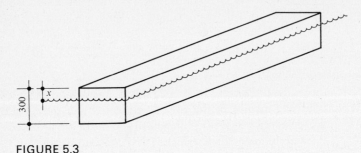

FIGURE 5.3

SOLUTION

Consider a section of the timber 1.0 m long and floating as shown in Fig. 5.3. If the portion above the water surface has a height equal to x, then the submerged portion is $(0.3 \text{ m} - x)$.

The buoyant force B therefore relates to the amount of water displaced by the latter portion of the timber, or

$$B = (0.3 - x) \times 0.3 \times 1.0 \times 9.81$$
$$= 0.88 - 2.94x$$

The force of gravity G acting on a 1.0-m-long section of the timber

$$G = 0.3 \times 0.3 \times 1.0 \times 9.81 \times 0.48$$
$$= 0.42 \text{ kN}$$

The timber is floating in equilibrium on the surface. Therefore B must equal G, or

$$0.88 - 2.94x = 0.42$$

and

$$x = 0.156 \text{ m}$$
$$= 156 \text{ mm}$$

Consequently, the portion of the prism that is submerged is 144 mm high.

Stability of Floating Bodies

When a floating body is tilted as shown in Fig. 5.4a, it will react in one of three possible ways:

1. It will remain in the position shown in Fig. 5.4a.
2. It will right itself to take on the position in Fig. 5.4b.
3. It will capsize and take on the position shown in Fig. 5.4c.

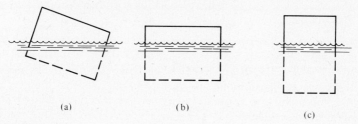

(a)   (b)   (c)

FIGURE 5.4

The final position assumed by this body will depend on the action of all the forces acting on it. These forces, shown in Fig. 5.5, consist of the force of gravity G and the hydrostatic forces or pressures P_1 and P_2 on the left and right sides, respectively, and P_3 on the bottom. As they are shown in the figure, P_2 and P_3 tend to impart a counterclockwise rotation to the body, whereas P_1 acts in a clockwise direction. The actual rotation that will take place is of course dependent upon the force B, the final resultant of the three hydrostatic pressures. This force B as discussed earlier, is the buoyant force and will act through the center of gravity of the submerged section of the body.

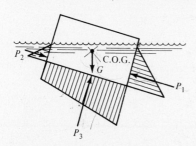

FIGURE 5.5

EXAMPLE 5.3

A barge weighing 56.86 kN per linear meter is located in water in the position shown in Fig. 5.6. Determine the position this barge will assume when floating in the water.

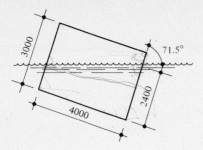

FIGURE 5.6

SOLUTION

The pressure diagrams of all the hydrostatic pressures exerted on the barge per lineal meter are shown in Fig. 5.7. From the geometry of the figure and the barge's location in the water the following values are readily calculated.

$$h = 2.4 \times \sin 71.5° = 2.28 \text{ m}$$

$$h' = 2.28 - 4.00 \times \cos 71.5° = 1.01 \text{ m}$$

$$P_1 = \frac{2.28\gamma \times 2.40}{2}$$

$$= 2.74\gamma, \text{ acting at } 0.80 \text{ m from } A$$

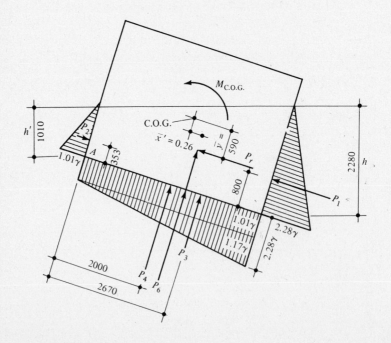

FIGURE 5.7

$$P_2 = \frac{1.01\gamma \times 1.06}{2}$$

$$= 0.54\gamma, \text{ acting at } 0.353 \text{ m from } A$$

$$P_3 = \frac{(2.28-1.01)\gamma \times 4.0}{2}$$

$$= 2.54\gamma, \text{ acting at } 2.67 \text{ m from } A$$

$$P_4 = 1.01\gamma \times 4.0$$

$$= 4.04\gamma, \text{ acting at } 2.0 \text{ m from } A$$

The forces P_1 and P_2 can be represented by their resultant force P_r as follows:

$$P_r = 2.74\gamma - 0.54\gamma$$

$$= 2.20\gamma$$

The moment arm $\bar{y}$ of P_r is

$$\bar{y} = \frac{2.74\gamma \times 0.80 - 0.54\gamma \times 0.353}{2.20\gamma}$$

$$= 0.91 \text{ m from } A$$

In a similar manner the resultant P_b of P_3 and P_4 can be found; that is

$$P_b = 2.54\gamma + 4.04\gamma$$

$$= 6.58\gamma$$

The moment arm $\bar{x}$ of P_b is

$$\bar{x} = \frac{2.54\gamma \times 2.67 + 4.04\gamma \times 2.00}{6.58\gamma}$$

$$= 2.26 \text{ m}$$

The locations of P_r and P_b are shown in Fig. 5.7. Also indicated on this figure are the moment arms $\bar{y}'$ and $\bar{x}'$ of these forces with respect to the center of gravity of the barge, or

$$\bar{y}' = 0.594 \text{ m}$$

$$\bar{x}' = 0.26 \text{ m}$$

Hence, the moment of the forces acting on this body with respect to its center of gravity is

$$M_{cog} = 2.20\gamma \times 0.594 - 6.58\gamma \times 0.26$$

$$= -0.41\gamma$$

$$= -4.0 \text{ kN} \cdot \text{m}$$

Since this moment is negative, it indicates a counterclockwise direction, as shown by the arrow in Fig. 5.7. Therefore the barge will turn in that direction and assume a position with its long side parallel to the water surface.

In order to determine the final position and the depth d at which the barge will float, a procedure similar to that used in previous examples will yield

$$B = 4.0 \times d \times 9.81$$

$$= 39.24d$$

B is equal to the force of gravity of the barge, or 58.86 kN, and

$$d = \frac{58.86}{39.24}$$

$$= 1.50 \text{ m}$$

Metacenter and Metacentric Height

As is evident from the previous example and Fig. 5.5, the forces acting on floating bodies can be resolved into a couple, with the force of gravity G acting through the center of gravity of the body and B the buoyant force. Whether such a body turns clockwise or counterclockwise depends on the relative positions of G and B. If B is to the right of G, the body will turn counterclockwise, and vice versa.

The buoyant force B acts in a vertical direction and will intersect the centerline of the body at the point M. This point is called the *metacenter*, and the distance from it to the center of gravity is the metacentric height. The location of the metacenter and the value of the metacentric height provide a great deal of information as to the stability of the floating body, as indicated in Fig. 5.8.

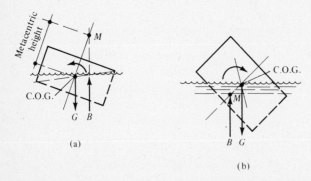

(a)

(b)

FIGURE 5.8

When the metacenter and the center of gravity are in the same vertical line, the body is in equilibrium and is said to be stable. The stability resulting from this position can be readily visualized by considering that the moment arm of the G and B couple of forces is zero. Therefore there can be no turning moment acting on the body. When, as in Fig. 5.8a the metacenter is located above the center of gravity, the moment of G and B is in a direction, that will right the body or in a direction to reduce the heel—that is, the angle of the body with the vertical. If the heel of the body were in a direction opposite to the one shown in Fig. 5.8a, the force B would be located to the left of G and the righting moment would be in the other direction. Obviously, the larger the moment of the couple, B and G, the larger the righting force. A large righting force will tend to make the body rotate at a higher rate and result in a less stable condition. The closer B is to G, the more stable the body will become. As B approaches G, the metacentric height increases until it reaches infinity just at the point where B and G become colinear.

With M below the center of gravity, the moment of the B and G couple will produce a force that will make the body capsize, as shown in Fig. 5.8b. The force providing for the capsizing will become increasingly large because the moment arm of the couple B and G increases as the angle of heel increases.

The location of the metacenter is a very important consideration in the stability of ships. It is often necessary to take on ballast in order to provide sufficient distance between the metacenter and the center of gravity under all conditions of heel, particularly during stormy weather conditions. For small angles of heel the metacentric height of shallow-draft and high ships, such as war and sailing ships, is approximately 1.0 m, whereas other ships have metacentric heights of 300 to 600 mm.

EXAMPLE 5.4

For the barge in Example 5.3, determine the metacentric height.

SOLUTION

Figure 5.9 shows the location of the forces B and G, on the barge. The buoyant force B intersects the centerline of the barge at the point M, the metacenter. The value of the distance m can be readily calculated by considering the two similar triangles ABC and MDA. Hence

$$\frac{m}{AC} = \frac{0.260}{CB}$$

or

$$m = \frac{0.260 \times 6.58\gamma}{2.20\gamma}$$

$$= 0.78 \text{ m}$$

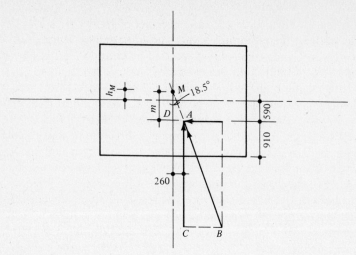

FIGURE 5.9

Alternatively, m could have been found directly from triangle MDA, as follows:

$$m = \frac{0.26}{\tan 18.5°}$$

$$= 0.78$$

Therefore, from the geometry of Fig. 5.9, the metacentric height h_M is

$$h_M = 0.91 + 0.78 - 1.50$$

$$= 0.18 \text{ m}$$

$$= 18 \text{ mm}$$

5.2 RELATIVE EQUILIBRIUM OF FLUIDS

Relative Equilibrium

Relative equilibrium for a fluid refers to the condition where the fluid is at rest with one body and not with another. For instance, a tank truck transporting a liquid will have that liquid at rest or in equilibrium with respect to the truck's body. The liquid itself, however, will not be at rest relative to the earth. Similarly, on a lake with water essentially at rest relative to the earth there is no equilibrium with the sides of a boat traveling on the lake's surface.

The study of relative equilibrium enables the engineer to determine speeds

with which vehicles carrying liquids can be accelerated. It also helps in determining the drag exerted on a body by stationary or moving fluids. Of particular
interest to the civil engineer is the study of the relative equilibrium of liquids in
spinning vessels and centrifugal pumps.

In this chapter two cases of relative equilibrium are considered: the
behavior of a liquid under constant linear acceleration and the effect of rotation
of a container holding a liquid on the equilibrium of that liquid.

Relative Equilibrium Under Constant Horizontal Acceleration

As long as the container in Fig. 5.10 is stationary, the liquid surface will be
horizontal, as indicated by the line AB. When this container is accelerated to
the left, the liquid surface will take on the shape indicated by the line $A'B'$. The
liquid surface will remain in this position until the acceleration ceases. Obviously this will happen only when the container travels at constant velocity or
when it is again stopped. The liquid surface during constant acceleration will
be a plane surface and will form an angle θ with the horizontal. The above
statements and the value of θ can be deduced by examining the following
considerations.

Consider a small mass M located anywhere at the surface of the liquid.
This mass is moving at the same velocity and with the same acceleration as the
other particles of the liquid surrounding it. It is therefore subjected to a force,
creating this motion. If the acceleration of the container and the liquid is a,
this force P will be, from Newton's laws of motion,

$$P = Ma \qquad (5.7)$$

The force P is also the resultant of all the forces acting on the mass, M.
The forces acting on M are the force of gravity G and the pressure of particles
of the fluid immediately adjacent to the mass M. The force of gravity is

$$G = Mg \qquad (5.8)$$

Figure 5.11a shows the mass M, subjected to the pressures of the adjacent
liquid. It follows from a careful examination that the pressures from left and

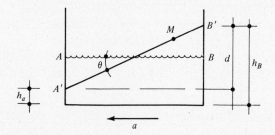

FIGURE 5.10

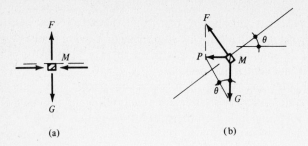

(a)   (b)

FIGURE 5.11

right are equal and will cancel, leaving only the pressures from below the mass, to make up the force F. This force therefore must act at right angles to the liquid surface as shown in the free-body diagram of M in Fig. 5.11b. The forces P, G, and M are shown in this figure by their vectors. From this diagram it follows that

$$\tan \theta = \frac{P}{G} \tag{5.9}$$

where θ is the angle of the water surface with the horizontal. Substituting values of P and G from Eqs. (5.7) and (5.8) into Eq. (5.9) yields

$$\tan \theta = \frac{Ma}{Mg} \tag{5.10}$$

and thus

$$\tan \theta = \frac{a}{g} \tag{5.11}$$

Equation (5.11) states that the slope of the liquid surface, $\tan \theta$, is equal to the ratio of the horizontal acceleration and the acceleration due to gravitational forces.

Since the mass M was located arbitrarily on the surface of the liquid, it follows that its location has no effect on the value of θ. Therefore, θ, will have the same value at any point on the liquid surface, or the line $A'B'$ in Fig. 5.10 will be a straight line and the surface will be a plane surface.

Further, since only a and g are terms determining the slope of the liquid surface, it follows that the surface will take the same slope regardless of the density of the liquid.

A third important observation indicates that when the container is moving at a constant velocity (that is, when the acceleration a is zero), θ will be zero and the water surface will again be horizontal.

EXAMPLE 5.5

The container in Fig. 5.10 is rectangular and has a length of 1.70 m. When at rest, the depth of the liquid in this container is 1.0 m. Determine the depth of liquid at A and B if the container is accelerated to the left with a constant acceleration of 2.50 m/s².

SOLUTION

From Eq. (5.11),

$$\tan \theta = \frac{2.50 \text{ m/s}^2}{9.81 \text{ m/s}^2}$$

$$= 0.255$$

and therefore

$$\theta = 14.30°$$

A study of Fig. 5.10 and the water surface $A'B'$ indicates that

$$AA' = BB'$$

Therefore the vertical distance $A'B'$, or d, is

$$d = \tan 14.30° \times 1.70 \text{ m}$$

$$= 0.433 \text{ m}$$

Hence, the rise and drop in water levels at A and B respectively are

$$AA' = BB' = \frac{0.433 \text{ m}}{2}$$

$$= 0.217 \text{ m}$$

and the corresponding water depths at A and B are

$$h_A = 1.0 \text{ m} - 0.217 \text{ m}$$

$$= 0.78 \text{ m}$$

and

$$h_B = 1.0 \text{ m} + 0.217 \text{ m}$$

$$= 1.22 \text{ m}$$

EXAMPLE 5.6

If the container in Example 5.5 has sides with a height of 1.40 m, determine the maximum horizontal acceleration it can be subjected to without spilling any liquid.

SOLUTION

When the container is accelerated until liquid will spill out, the liquid surface will have the form shown in Fig. 5.12, and the vertical distance d will be

$$d = 2 \times (1.40 \text{ m} - 1.0 \text{ m})$$

$$= 0.80 \text{ m}$$

Therefore, from the geometry of Fig. 5.12,

$$\tan \theta = \frac{0.80}{1.70}$$

$$= \quad 0.4706$$

and from Eq. (5.11) the acceleration at this point is

$$a = 0.4706 \times g$$

$$= 4.62 \text{ m/s}^2$$

The maximum allowable acceleration before spilling liquid out of this container is therefore 4.62 m/s². Any further acceleration would increase the slope of the liquid surface and cause the liquid to spill.

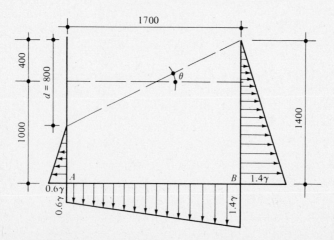

FIGURE 5.12

Hydrostatic Pressures Under Constant Horizontal Acceleration

When a liquid is subjected to constant horizontal acceleration, the liquid is in equilibrium with the container. Therefore the elementary prism AB shown in Fig. 5.13 will also be in equilibrium and all the forces acting on it will have a resultant force of zero. The hydrostatic pressures acting on this prism from the left and the right side are, of course, equal and in opposite direction, and hence their resultant will be zero. The only other forces acting on this prism are then the force of gravity G and the pressure P acting on the bottom of the prism.

If the cross-sectional area of the prism is ΔA and its height is h, and the liquid has a specific weight γ, then

$$G = \Delta A \times \gamma \times h \tag{5.12}$$

The pressure on the bottom of the prism is the product of the unit pressure p_B at B, times the cross-sectional area ΔA. The unit pressure at B is

$$p_B = \gamma h \tag{5.13}$$

and hence

$$P = \Delta A \times \gamma h \tag{5.14}$$

The resultant force of G and P must equal zero, and therefore the unit pressure at B is identical to the pressure at B found for liquids at rest. It must be remembered however, that points of equal pressure in this situation will not lie on a horizontal plane but on a plane parallel to the liquid surface.

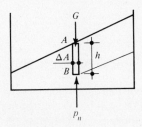

FIGURE 5.13

EXAMPLE 5.7

When the liquid in the container in Example 5.6 has reached the acceleration corresponding to the water surface slope shown in Fig. 5.12, what will be the pressures at points A and B if (a) the liquid is water, (b) the liquid is mercury ($s=13.6$).

SOLUTION

The liquid levels at A and B can be read from Fig. 5.12 and are

$$h_A = 1.40 \text{ m}$$

and

$$h_B = 0.60 \text{ m}$$

Therefore, if the liquid is water, the pressures at these points will be

$$P_A = 1.40 \text{ m} \times 9.81 \text{ kN/m}^3$$
$$= 13.7 \text{ kPa}$$

and

$$p_B = 0.60 \text{ m} \times 9.81 \text{ kN/m}^3$$
$$= 5.9 \text{ kPa}$$

The corresponding pressure diagrams of pressures acting on the sides and bottom of the container are shown in Fig. 5.12.

A procedure similar to the above will lead to finding the pressures at A and B when the liquid is mercury. They are

$$p_A = 186 \text{ kPa}$$

and

$$P_B = 80 \text{ kPa}$$

Relative Equilibrium Under Constant Vertical Acceleration

When a container such as the one shown in Fig. 5.14 is accelerated upward, with a constant rate of acceleration a, there will be no change in the water surface. However, there will be some effect from this acceleration on the pressure within the liquid.

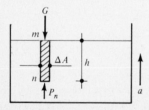

FIGURE 5.14

Consider the elementary prism *mn* shown in Fig. 5.13. The force F that will produce the acceleration is the resultant of all the forces acting on this prism. These forces are the force of gravity G and the hydrostatic pressures P_m and P_n at the top and the bottom of the prism, respectively. The horizontal pressures on the prism will be equal and in opposite direction and will produce no resultant pressure. The pressure at the top of the prism is zero. Therefore the only forces remaining acting on the prism are

$$G = \Delta a \times \gamma h \tag{5.15}$$

and

$$P_n = \Delta a \times p_n \tag{5.16}$$

where p_n = unit pressure at point n in the liquid
$\quad\quad \Delta a$ = cross-sectional area of the prism
$\quad\quad \gamma$ = specific weight of the liquid
$\quad\quad h$ = height of the prism

Hence the resultant force F is

$$F = \Delta a \times p_n - \Delta a \times \gamma h$$

$$= \Delta a(p_n - \gamma h) \tag{5.17}$$

Since F is also the force creating the vertical acceleration a, it must be equal to the product of the mass of the liquid and a. The mass of the prism, where ρ is the density of the liquid, is

$$M = \Delta a \times \rho \times h \tag{5.18}$$

and remembering that $\rho = \gamma/g$,

$$F = Ma = \Delta a \times \frac{\gamma}{g} \times h \times a \tag{5.19}$$

Equating the right sides of Eqs. (5.17) and (5.19) and simplifying yields

$$p_n - \gamma h = \gamma h \times \frac{a}{g} \tag{5.20}$$

or

$$p_n = \gamma h + \gamma h \frac{a}{g} \tag{5.21}$$

$$= \gamma h \left(1 + \frac{a}{g} \right) \qquad (5.22)$$

Equations (5.21) and (5.22) indicate that the pressure in a liquid subjected to vertical acceleration is equal to the normal hydrostatic pressure, increased by a proportion equal to the ratio of the horizontal acceleration and the acceleration due to gravity. In the case of a falling container or with acceleration in a downward direction, a becomes negative and the pressure in the liquid will be reduced correspondingly. A free-falling container moving at a downward acceleration equal to g will have zero pressures in the liquid.

EXAMPLE 5.8

If the container in Fig. 5.14 is accelerated upward at a constant rate of 2.50 m/s², what will be the pressure diagrams on the sides and bottom of this container? The liquid is brine and has a relative density s of 1.4 and is 0.85 m deep.

SOLUTION

Since the acceleration is upward only, the liquid surface will remain horizontal. The pressure on the bottom of the container is, from Eq. (5.22),

$$p_{btm} = 9.81 \text{ kN/m}^3 \times 1.4 \times 0.85 \text{ m} \times \left(1 + \frac{2.50 \text{ m/s}^2}{9.81 \text{ m/s}^2} \right)$$

$$= 14.6 \text{ kPa}$$

At the top of the liquid surface, h is equal to zero and so is the pressure. The resulting pressure diagrams are shown in Fig. 5.15.

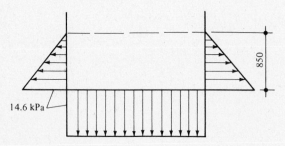

14.6 kPa

850

FIGURE 5.15

EXAMPLE 5.9

The container in Fig. 5.16a is pulled upward along the inclined track with a constant accceleration. Determine the maximum acceleration this container may undergo before the water in it will spill out and construct pressure diagrams for the sides and the bottom of the container.

SOLUTION

As indicated by the vector diagram of Fig. 5.16a, the acceleration is both horizontal and vertical. The vertical and horizontal components, a_V and a_H, respectively, of the acceleration are

$$a_V = a \times \sin 30°$$

and

$$a_H = a \times \cos 30°$$

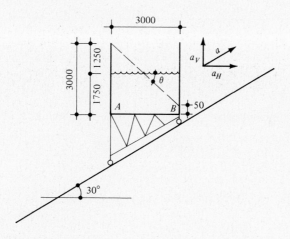

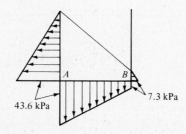

FIGURE 5.16

The maximum slope the liquid surface can assume prior to spilling, can be calculated from the geometry of the water surface as follows:

$$\tan \theta = \frac{2.50 \text{ m}}{3.0 \text{ m}}$$

$$= 0.833$$

Also, from Eq. (5.11),

$$\tan \theta = \frac{a_H}{g}$$

Hence, equating the right sides of the above equations,

$$0.833 = \frac{a_H}{9.81}$$

or

$$a_H = 8.17 \text{ m/s}^2$$

Therefore

$$a = \frac{8.17 \text{ m/s}^2}{\cos 30°}$$

$$= 9.44 \text{ m/s}^2$$

Consequently the vertical component of the acceleration is

$$a_V = 9.44 \text{ m/s}^2 \times \sin 30°$$

$$= 4.72 \text{ m/s}^2$$

From Eq. (5.22) the pressures at points A and B will be

$$p_A = 3.0 \text{ m} \times 9.81 \text{ kN/m}^3 \times \left(1 + \frac{4.72 \text{ m/s}^2}{9.81 \text{ m/s}^2}\right)$$

$$= 43.6 \text{ kPa}$$

and

$$p_B = 0.50 \text{ m} \times 9.81 \text{ kN/m}^3 \times \left(1 + \frac{4.72 \text{ m/s}^2}{9.81 \text{ m/s}^2}\right)$$

$$= 7.3 \text{ kPa}$$

The corresponding pressure diagrams are shown in Fig. 5.16b.

Relative Equilibrium in a Rotating Container

A cylindrical container originally containing a liquid with its surface at AB is spun about its vertical axis with a constant rate of rotation. The centrifugal force created by this movement will make the liquid surface take on the shape of the curve $A'CB'$ shown in Fig. 5.17. There is a constant relationship, for any point on the liquid surface, between h, its height above C, and its distance r from the centerline. Reasoning similar to the discussions related to the liquid surface with horizontal acceleration shows that this relationship can be mathematically expressed as

$$h = \frac{\omega^2 \times r^2}{2g} \tag{5.23}$$

where h and r are values mentioned above and ω is the angular velocity of rotation in radians per unit of time. Since ω and g are constant, Eq. (5.23) represents the equation of a parabola, and hence the shape of the liquid surface will be parabolic.

One rotation of the container is equivalent to 2π radians. It follows therefore that a point on the circumference of a circle with radius r will travel a distance equal to πr during an angular displacement of 2π radians or one revolution. Therefore Eq. (5.23) can be expressed as

$$h = \frac{v^2}{2g} \tag{5.24}$$

where v is the velocity of a point on the surface of the liquid.

The liquid surface is caused by rotation at constant velocity. Since there is no acceleration, vertically or horizontally, pressures in the liquid will be

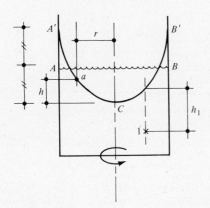

FIGURE 5.17

directly related to the pressure heads resulting from the shape of the liquid surface. That is,

$$p_1 = \gamma h_1 \tag{5.25}$$

A pressure diagram for a fluid in a rotating cylindrical container is shown in Fig. 5.18*b*.

It should also be noted that Eqs. (5.23) and (5.24) do not contain any term related to the density of the liquid. Therefore the shape of the liquid surface is independent of the type of liquid in the container.

When the rotation of the container is allowed to be sufficiently rapid, the liquid surface will rise above the top of the container, and liquid will spill out. If at the same time liquid is allowed to enter at the bottom of the container, an elementary form of a centrifugal pump is created. The relationship expressed in Eqs. (5.23) and (5.24) is in fact the basic principle underlying the operation of centrifugal pumps.

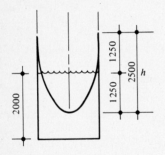 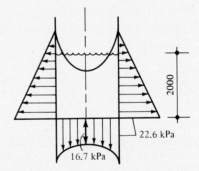

FIGURE 5.18

EXAMPLE 5.10

A cylindrical container 1.0 m in diameter has water in it to a depth of 2.0 m and is spun about its vertical axis with a constant number of rotations per minute. If the total height of this cylinder is 3.25 m, determine the maximum rotations (or revolutions) per minute (RPM_{max}) this cylinder can undergo before water will spill out of the top.

SOLUTION

From Fig. 5.18*a* it can be determined that the maximum allowable *h* is 2.50 m. Therefore, from Eq. (5.24),

$$2.50 \text{ m} = \frac{v^2}{2g}$$

or

$$v = (2g \times 2.50 \text{ m})^{1/2}$$

$$= 7.0 \text{ m/s}$$

One rotation of a point on the circumference of the cylinder, or one revolution, is

$$1 \text{ rev} = \pi \times 1.0 \text{ m}$$

$$= 3.14 \text{ m}$$

Therefore the cylinder will complete one revolution (rev) in

$$t = \frac{3.14 \text{ m/rev}}{7.0 \text{ m/s}}$$

$$t = 0.45 \text{ s/rev}$$

or

$$RPM_{\text{max}} = \frac{60 \text{ s/min}}{0.45 \text{ s/rev}}$$

$$= 134 \text{ rev/min}$$

EXAMPLE 5.11

The cylinder in the previous example is spun at a rate of 40 rpm. Determine the pressure diagrams on the bottom and sides of the cylinder.

SOLUTION

From previous calculations the circumference of the cylinder was found to be 3.14 m. Hence the velocity of a point on the circumference can be calculated as follows:

$$v = \frac{3.14 \text{ m/rev} \times 40 \text{ RPM}}{60 \text{ s/min}}$$

$$= 2.09 \text{ m/s}$$

Therefore, from Eq. (5.24),

$$h = \frac{(4.19 \text{ m/s})^2}{2g}$$

$$= 0.60 \text{ m}$$

and

$$\frac{h}{2} = 0.30 \text{ m}$$

Consequently, the water levels at the outside and the middle of the cylinder will be

$$h_A = 2.0 \text{ m} + 0.30 \text{ m}$$

$$= 2.30 \text{ m}$$

and

$$h_C = 2.0 \text{ m} - 0.30 \text{ m}$$

$$= 1.70 \text{ m}$$

Corresponding pressures are then

$$p_A = p_B = 22.6 \text{ kPa}$$

and

$$p_C = 16.7 \text{ kPa}$$

Pressure diagrams for this cylinder are shown in Fig. 5.18b.

EXAMPLE 5.12

A cylindrical vessel, when rotated at the rate of 150 rpm, has its water depth at the circumference of the cylinder change in depth from 0.50 m to 1.0 m. Determine the diameter of the cylinder.

SOLUTION

From Fig. 5.19.

$$h = 1.0 \text{ m}$$

therefore, from Eq. (5.24),

$$v = (2g \times 1.0 \text{ m})^{1/2}$$

$$= 4.43 \text{ m/s}$$

The velocity of a point on the circumference is also given by the

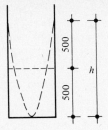

FIGURE 5.19

rotations, as follows, if d is the cylinder's diameter.

$$v = \frac{\pi d \times 150 \text{ RPM}}{60 \text{ s/min}}$$

Hence, substituting $v = 4.43$ m/s, the result is

$$d = 0.564 \text{ m}$$

$$= 564 \text{ mm}$$

5.3   SUMMARY OF THE CHAPTER IN BRITISH UNITS

Theoretical Considerations

The general discussions and the theoretical development of the formulas and principles related to buoyancy and relative equilibrium, of the earlier parts of this chapter apply equally to the British System of units. None of the formulas presented require conversion from SI to British, or vice versa.

In order to illustrate this chapter and its relation to the British System of units the following examples are presented here. The explanations of the various steps in solving these examples have not been given in the same detail as earlier in the chapter. It is assumed that the examples and methods for solving them have been worked through and are familiar.

Selected Examples in British Units

EXAMPLE 5.13

The prism in Fig. 5.1 is made of concrete, with a unit weight of 150 lb/ft³. If $a = 2.50$ ft, $b = 3.75$ ft, and $d = 2.0$ ft, determine the force required to keep the prism stationary in the water.

SOLUTION

The force of gravity G on the prism is

$$G = 2.5 \text{ ft} \times 3.75 \text{ ft} \times 2.0 \text{ ft} \times 150 \text{ lb/ft}^3$$

$$= 2810 \text{ lb}$$

The bouyant force B on the prism is

$$B = 2.5 \text{ ft} \times 3.75 \text{ ft} \times 2.0 \text{ ft} \times 62.4 \text{ lb/ft}^3$$

$$= 1170 \text{ lb}$$

The gravitational force G is larger than the bouyant force B, and unless an upward force F is applied, the prism will sink. This force F must be

$$F = 2810 - 1170$$

$$= 1640 \text{ lb}$$

EXAMPLE 5.14

Figure 5.20 is a graph representing the area of a segment of a circle as it relates to a percentage of its diameter. With the aid of this figure determine the height to which a 12-inch-diameter cedar log will protrude above the surface of the water in which it floats. The unit weight of cedar is 45 lb/ft³.

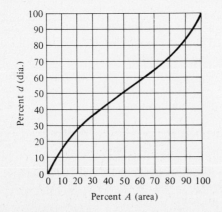

FIGURE 5.20

SOLUTION

Consider a section of the log 1 ft in length. The weight of such a section is

$$W = \frac{1.0^2 \times \pi}{4} \times 1.0 \times 45 \text{ lb/ft}^3$$

$$= 17.4 \text{ lb/ft}$$

Since the log is in equilibrium and floating, it follows that the bouyant force B must equal W.

If the portion of the log's cross section that is submerged is denoted by A_s, it follows that

$$B = A_s \times 62.4 \text{ lb/ft}^3$$

or

$$A_s \times 62.4 = 17.4$$

and

$$A_s = 0.28 \text{ ft}^2$$

Hence, the area submerged as a percentage of the total cross-sectional area of the log is found to be

$$A = \frac{0.28}{0.78} \times 100$$

$$= 36\%$$

From Fig. 5.20, this corresponds to 38% of the log's diameter. Hence, the log is submerged a distance d_s, of

$$d_s = 12 \text{ in. } \times 0.38$$

$$= 4.56 \text{ in. or } 4\frac{9}{16} \text{ in.}$$

Therefore the log will protrude above the water surface a distance of $7\frac{7}{16}$ in.

EXAMPLE 5.15

The container in Fig. 5.10 is 6.5 ft long and 4.0 ft deep and, when at rest, has water in it to a depth of 2.5 ft.

(a) Find the liquid depth at A and B when the container is accelerated to the left with a constant acceleration of 3.5 ft/s².

(b) Determine the maximum acceleration this container can be subjected to, without water spilling over its sides.
(c) At the latter rate of acceleration, determine the pressures at A and B and sketch the pressure diagrams on the sides and bottom of the container.

SOLUTION

(a) From Eq. (5.11),

$$\tan \theta = \frac{3.5 \text{ ft/s}^2}{32.2 \text{ ft/s}^2}$$

$$= 0.109$$

and

$$\theta = 6.2°$$

Hence d in Fig. 5.21 will be as follows:

$$d = \tan 6.2° \times 6.5 \text{ ft}$$

$$= 0.38 \text{ ft}$$

and

$$h_A = 2.5 \text{ ft} - \frac{0.38 \text{ ft}}{2}$$

$$= 2.31 \text{ ft}$$

$$h_B = 2.5 \text{ ft} + \frac{0.38 \text{ ft}}{2}$$

$$= 2.69 \text{ ft}$$

(b) When the water level reaches the top of the container at side B,

$$d = 3.0 \text{ ft}$$

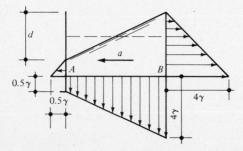

FIGURE 5.21

and

$$\tan \theta = \frac{3.0 \text{ ft}}{6.5 \text{ ft}}$$

$$= 0.46$$

From Eq. (5.11) it follows that the acceleration a is

$$a = 32.2 \text{ ft/s}^2 \times 0.46$$

$$= 14.8 \text{ ft/s}^2$$

(c) The resulting pressure diagrams are shown in Fig. 5.21. In accordance with Eq. (3.31), the pressures at A and B are

$$p_A = 0.5 \text{ ft} \times 0.433 \text{ psi/ft}$$

$$= 0.2 \text{ psi}$$

and, similarly,

$$p_B = 1.7 \text{ psi}$$

EXAMPLE 5.18

A cylindrical container 3.5 ft in diameter has water in it to a depth of 48 in. This cylinder is spun on its vertical axis, with a constant rotation of 75 rpm. Determine the water depth at the center and at the walls of this cylinder. Calculate also the pressures at these points.

SOLUTION

The velocity of a point on the wall of this cylinder is

$$v = 3.5 \text{ ft} \times \pi \times 75 \text{ rpm}$$

$$= 825 \text{ ft/min}$$

$$= 13.8 \text{ ft/s}$$

and, from Eq. (5.24),

$$h = \frac{13.8^2}{32.2 \times 2}$$

$$= 2.97 \text{ ft}$$

Hence the depth of the liquid at the center d_c and at the wall, d_w are

$$d_c = 4 - \frac{2.97}{2}$$

$$= 2.52 \text{ ft}$$

and, similarly,

$$d_w = 5.49 \text{ ft}$$

The respective pressures are

$$p_c = 1.1 \text{ psi}$$

and

$$p_w = 3.4 \text{ psi}$$

SELECTED PROBLEMS IN SI UNITS

5.1 For the gate in Fig. P5.1 find the force F required to keep the gate in the closed position, (a) $r = 1.75$ m, $\theta = 45°$, $h_1 = 1.5$ m, $h_2 = 1.25$ m; (b) $r = 1.45$ m, $\theta = 25°$, $h_1 = 2.5$ m, $h_2 = 3.0$ m; (c) $r = 0.75$ m, $\theta = 40°$, $h_1 = 0.0$ m, $h_2 = 3.0$ m.

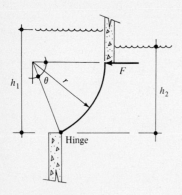

FIGURE P5.1

5.2 Determine the total load per square meter that can be supported by the floating dock, shown in cross section in Fig. P5.2, so that the dock will have a freeboard of 75 mm.

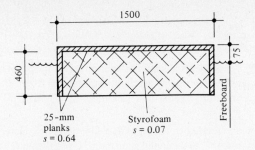

FIGURE P5.2

5.3 A caisson in the form of a prismatic hollow box has outside dimensions of 4.0 m long, 2.0 m wide, and 2.0 m high. The caisson is constructed of concrete with all walls and floor thicknesses equal to 150 mm. It is floated out to sea, to be sunk in water that is 1.75 m deep. Determine the volume of sand ballast required to sink this caisson, if $\rho = 2200 \text{ kg/m}^3$ for concrete and 1800 kg/m^3 for sand, and $s = 1.06$ for seawater.

5.4 The cylindrical object in Fig. P5.4 has a relative density of $s = 0.76$. The object is floated in the water in the container. Determine the depth to which this cylindrical object will be submerged and the rise in water level in the container.

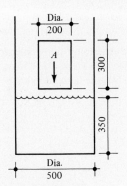

FIGURE P5.4

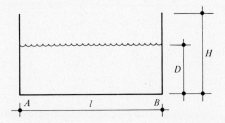

FIGURE P5.5

5.5   The container in Fig. P5.5 is rectangular and is accelerated to the right at a rate of a m/s². Determine the pressures at points A and B, if (a) $L = 1.5$ m, $D = 1.5$ m, $a = 2.0$ m/s²; (b) $L = 3.0$ m, $D = 0.5$ m, $a = 2.0$ m/s²; (c) $L = 1.75$ m, $D = 1.0$ m, $a = 4.5$ m/s².

5.6   The container in Fig. P5.5 is accelerated until the water in it is at the point of spilling out at the left side. Determine the rate of acceleration if (a) $L = 2.0$ m, $D = 0.75$ m, $H = 1.50$ m; (b) $L = 1.75$ m, $D = 2.50$ m, $H = 4.25$ m; (c) $L = 1.75$ m, $D = 1.25$ m, $H = 3.50$ m.

5.7   The tube in Fig. P5.7 is accelerated to the right at a rate of a m/s². Determine the pressures at points A, B, and C if (a) the liquid is mercury, $L = 250$ mm, $D = 125$ mm, $a = 2.5$ m/s²; (b) the liquid is water, $L = 1.2$ m, $D = 0.5$ m, $a = 5.6$ m/s²; (c) the liquid is mercury, $L = 1.0$ m, $D = 1.0$ m, $a = 3.5$ m/s².

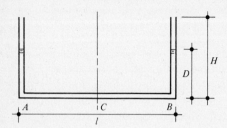

FIGURE P5.7

5.8   For the container in Fig. P5.8, determine the maximum allowable acceleration parallel to the track before liquid will spill out, if (a) $L = 1.5$ m, $D = 0.95$ m, $H = 1.5$ m and (b) $L = 3.5$ m, $D = 1.50$ m, $H = 3.5$ m, where D is the liquid depth when the container is horizontal.

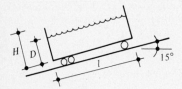

FIGURE P5.8

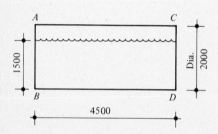

FIGURE P5.9

5.9   The cylindrical tank in Fig. P5.9 is filled with water to a depth of 1.5 m. It is accelerated to the left, with an acceleration of 3.5 m/s². Determine the total pressure on the
end CD of this tank and the location of the center of pressure.

5.10 The tube in Fig. P5.7 is spun about the point C, at a rate of 0.35 rpm. If the liquid in
the tube is water, $L = 3.0$ m, and $D = 1.25$ m, determine the pressures at points
A, B, and C.

5.11 The liquid in the tube of Fig. P5.7 is mercury and the tube is spun about the vertical
axis through the point A, at a rate of 7.0 rpm. If $L = 2.0$ m and $D = 1.0$ m, determine
the pressures at points A, B, and C.

5.12 A cylindrical container, 1.75 m in diameter and 3.0 m high, has water in it to a depth
of 0.75 m. Determine the maximum speed at which this container can be spun
without spilling water. Draw the pressure diagrams on the walls and the bottom of
the container.

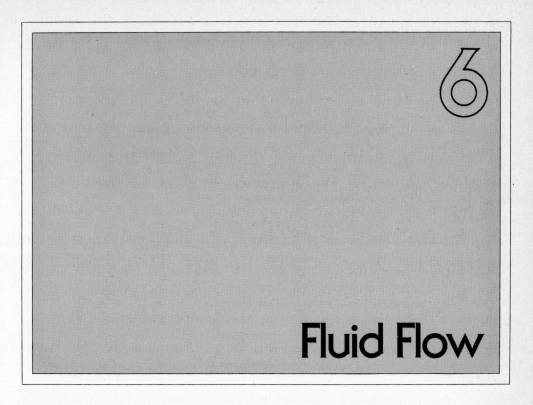

Fluid Flow

6.1 FLUIDS IN MOTION

The behavior of a flowing fluid, or a *fluid in motion*, is entirely different from that of a fluid at rest. When a fluid is at rest, its interaction with the bodies with which it is in contact is entirely predictable and subject to precise and definite mathematical relationships. Essentially, only the density of the fluid has to be determined, experimentally. From that point on, the mathematical relationships are sufficient to permit accurate predictions of the behavior of a fluid at rest.

A flowing fluid, on the other hand, is subjected to—and influenced by— an extremely complex system of forces and phenomena. Even today, many of the phenomena related to the flow of fluids cannot be expressed in simple theoretical formulas or rational relationships.

Although certain aspects of fluid-flow relationships can be approximated using rational expressions, they almost always require modifications to ensure that they represent accurately what happens in reality. These modifications are generally based on experimental data, gathered from observation of certain types of fluid in motion. The results of such observations are not always applicable to all fluids or flow conditions that might be encountered. Obviously, then, application of the results of such experiments to fluid flow in general without further modification would lead to serious error. Historically, the study of fluid flow has been preoccupied with finding, defining, and formulating empirical

165

relationships that will enable the engineer to predict, with reasonable accuracy, the behavior of fluids in motion.

The study of fluid flow enables the civil engineer to determine the size of a water supply main needed to supply a neighborhood, to determine the distance from the river's bank where floodwaters will reach, how far apart catch basins should be spaced along the curb, and many other applications commonly encountered in daily life.

6.2 CONCEPTS AND DEFINITIONS

Essentially, the only fluids with which the civil engineer is concerned are water or fluids whose properties relate very closely to those of water. Storm and sanitary sewage, for instance, have a viscosity and density almost identical to that of water. Consequently, most of the discussions related to fluid flow will be in terms of applicability to the flow of water. In the event that fluids other than water are discussed, this fact will be carefully noted.

Liquids are considered to be incompressible, as discussed earlier. This fact has an important bearing on fluid-flow conditions. The behavior of compressible fluids, such as air, in many instances is entirely different from that of water.

In order to facilitate discussion of the various important flow phenomena a number of terms and definitions are introduced here.

Pressure Flow and Gravity Flow

Water flowing along the gutter in a rainstorm seemingly moves along on its own force or initiative. Actually, this is not strictly true. The water in the gutter flows downhill and is pulled downward by gravitational attraction of the earth. In other words, gravity is the major direct cause of this flow.

On the other hand, water delivered to the nineteenth floor of a high-rise apartment building is pushed up there through the action of a pump. The pump in fact, creates pressure in the supply line, sufficient to overcome the gravitational pull and allow the water to flow up to the nineteenth floor.

Pressure flow can therefore be described as flow that occurs when a fluid is forced to flow by pressure supplied to the fluid by mechanical means. Since pressure can exist only in a closed conduit, and if the conduit is filled entirely by the fluid, pressure flow is often referred to as *pipe flow*.

Gravity flow can be defined as the flow of a liquid in a conduit, where the only major force creating flow is the force of gravity. Gravity flow always occurs when the liquid surface is open to the atmosphere. Typical gravity flow conditions are the flow of rain water in a gutter, flow of water in a river. Often gravity flow is also referred to as *open-channel flow*.

Various forms of open-channel and pipe flow conditions are shown in Fig. 6.1. It should be pointed out that in the case of Fig. 6.1b, sewer flow, open-channel flow conditions exist, even though the flow takes place in a pipe.

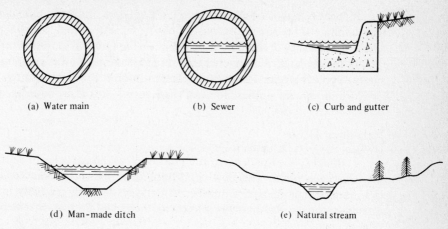

(a) Water main   (b) Sewer   (c) Curb and gutter

(d) Man-made ditch   (e) Natural stream

FIGURE 6.1

Path Lines and Stream Tubes

In the study of flow phenomena it is often advantageous to consider the behavior of one specific particle of water and its interaction with the other particles in the stream. Each particle of water follows a certain path of its own, the configuration of which depends on the conditions of flow and the action of other particles and stationary objects on it.

Figure 6.2 represents a schematic form of the path a particle could take in a flowing fluid stream. Velocity vectors of the particle at points 1, 2,... in its path are shown as V_1, V_2,.... Such a line is referred to as a *path line*. A *stream tube* is a group or bundle of such path lines flowing generally together and in the same direction.

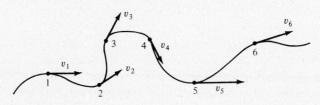

FIGURE 6.2

Laminar Flow

Laminar flow exists, in a stream tube, when the path lines of the various particles of the liquid do not intersect. Typical visual forms of laminar flow are the stream of smoke rising from a cigarette or a chimney in still air.

Laminar-flow phenomena are, by their nature, relatively simple. However, true laminar flow essentially exists only in the laboratory or in applications where the velocity of the fluid is very low or the fluid's viscosity is very high.

In the practical field of civil engineering, laminar flow is encountered only rarely. Typical laminar-flow conditions of interest are flow through soils and in hydraulic-model studies. Some of these topics are discussed later in this text.

Turbulent Flow

Turbulent flow exists when the path lines of the particles in a stream tube intersect each other. The flow conditions in this case become very complex. It is understandable that a number of particles flowing generally in the same direction, constantly bouncing into each other, must present extremely complex phenomena.

Turbulent-flow conditions can be seen in the earlier mentioned smoke stream, when it gets broken up by air currents. Extremely turbulent flow exists at rapids in a river.

Hydraulic Radius (R)

The term *hydraulic radius* is often used in formulas and therefore must be defined. The hydraulic radius R is an element introduced into most fluid formulas to incorporate the relationship between the area of flow and the area of contact of the fluid with the walls of the conduit. By definition, the hydraulic radius is equal to the area of the cross section of flowing fluid A, divided by the length of perimeter wetted by the fluid—the *wetted perimeter*, P; that is,

$$R = \frac{A}{P} \tag{6.1}$$

Since A is area and P is length, the units of R will be length.

EXAMPLE 6.1

Determine the hydraulic radius for a circular pipe, with a diameter of 500 mm if (a) the pipe is flowing full, (b) the pipe is flowing half full, and (c) the depth of water in the pipe is 150 mm.

SOLUTION

(a) When the pipe is full,

$$A = \frac{\pi \times 0.5^2}{4}$$

$$= 0.196 \text{ m}^2$$

and

$$P = \pi \times 0.5$$

$$= 1.571 \text{ m}$$

Consequently, from Eq. (6.1),

$$R = \frac{0.196 \text{ m}^2}{1.571 \text{ m}}$$

$$= 0.125 \text{ m}$$

(b) When the pipe is half full, the values for A and P will be exactly half of those calculated in a, and

$$R = \frac{0.196/2}{1.571/2}$$

$$= 0.125 \text{ m}$$

It is interesting to note and to remember that in the case of pipes flowing full or half full, the hydraulic radius is exactly equal to one-quarter of the diameter. This follows from the properties of a circle, as is determined below.

$$A = \frac{\pi \times d^2}{4}$$

and

$$P = \pi \times d$$

for the full pipe and

$$A = \frac{\pi \times d^2}{2 \times 4}$$

and

$$P = \frac{\pi \times d}{2}$$

when the pipe is half full.
    Hence

$$R = \frac{(\pi \times d^2)/4}{(\pi \times d)/2}$$

or

$$R_{\text{full}} = \frac{d}{4}$$

Similarly,

$$R_{\text{half full}} = \frac{(\pi \times d)/(2 \times 4)}{(\pi \times d)/2}$$

$$= \frac{d}{4}$$

(c) If the depth of water in the pipe is 150 mm, P is the length of the arc and A is the area of the segment as shown in Fig. 6.3a. Their length and area can be calculated from geometric formulas, and

$$R = \frac{0.098 \text{ m}^2}{0.680 \text{ m}}$$

$$= 0.144 \text{ m}$$

EXAMPLE 6.2

For the channel in Fig. 6.3b, determine the hydraulic radius, if the width of the channel $b = 2.5$ m and the depth of flow $D = 0.85$ m.

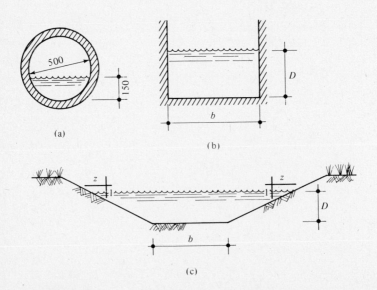

(a)

(b)

(c)

FIGURE 6.3

SOLUTION

The cross-sectional area of flow is

$$A = 2.5 \times 0.85$$
$$= 2.125 \text{ m}^2$$

and the wetted perimeter is

$$P = 2.5 + (2 \times 0.85)$$
$$= 4.2 \text{ m}$$

then

$$R = \frac{2.125 \text{ m}^2}{4.2 \text{ m}}$$
$$= 0.51 \text{ m}$$

EXAMPLE 6.3

Determine the hydraulic radius for the channel in Fig. 6.3c if the base width $b = 1.75$ m, the depth of flow $D = 0.95$ m, and the side slope of the channel sides $z = 2$.

SOLUTION

$$A = (1.75 \times 0.95) + (0.95 \times 0.95 \times 2)$$
$$= 3.47 \text{ m}^2$$
$$P = 1.75 + 2\sqrt{0.95^2 + (2 \times 0.95)^2}$$
$$= 5.60 \text{ m}$$

and

$$R = \frac{3.47 \text{ m}^2}{6.00 \text{ m}}$$
$$= 0.58 \text{ m}$$

It should be noted that in the case of trapezoidal channels, the value of R will approach the value of b as the width of the channel increases. This can be readily visualized by considering the calculations of A and P above. In both formulas the terms related to sides of the channel are independent of b and will remain constant if b is increased.

The Reynolds Number

The distinction between laminar-flow and turbulent-flow conditions has been known by scientists for a very long time. However, it was Osborne Reynolds, a French scientist, who in the late eighteenth century demonstrated and formalized the turbulent- and laminar-flow relationships.

Reynolds' experiment, a simple arrangement that can be easily duplicated in any laboratory and that led to the development of a clear and scientific definition of laminar and turbulent flow, is shown in Fig. 6.4.

In the experiment, water in the tank B is allowed to flow through tube C, by opening valve A. The amount of water flowing is controlled by the opening or closing of valve A. When flow has been established in tube C, opening valve E allows a fine stream of dye to enter the flow in tube C.

Reynolds found that when the flow in tube C was very low and the water velocity was small, the dye would flow in a very defined fine stream within the flowing water in C. In other words, the flow was such that the particles of dye were all following essentially parallel path lines and the flow was laminar.

When the valve A was opened further and the velocity of flow in tube C was increased, a point would be reached, where the fine dye stream would suddenly disperse throughout all of the water in tube C. This indicated that flow conditions were now such that the path lines intersected, distributing dye particles throughout the tube C, and that turbulent flow was established.

Subsequently, when valve A was gradually closed, a point would again be reached when the dye stream returned to its original fine line, indicating that the flow in tube C had returned to the laminar condition.

Numerous experiments with this equipment led Reynolds to the following conclusions:

1. Changeover from laminar flow to turbulent flow—and vice versa, from turbulent to laminar flow—does not happen at the same rate of flow in tube C. The change from laminar to turbulent flow occurs at a lesser velocity than when the flow goes from turbulent to laminar.

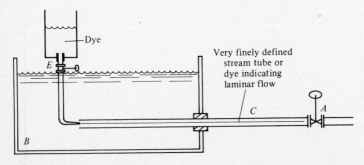

FIGURE 6.4

2. The points at which these changes in flow occur are directly related to the diameter of tube C, the velocity of flow in tube C, and the density of the fluid and inversely to the viscosity of the fluid.

On the basis of these experiments, a relationship defining the distinction between laminar and turbulent flow known as the Reynolds number, R_N, was formulated.

For pressure flow of liquids in circular pipes, R_N is

$$R_N = \frac{dv\rho}{\mu} \tag{6.2}$$

where d = pipe diameter

v = velocity of flow in the pipe

ρ = density of the liquid

μ = dynamic viscosity of the fluid

If the kinematic viscosity $v = \mu/\rho$, is substituted in it, Eq. (6.2) yields

$$R_N = \frac{dv}{v} \tag{6.3}$$

A close examination of the Reynolds number reveals that it is a dimensionless quantity, since the units of the kinematic viscosity are m^2/s and those of d and v are m and m/s, respectively.

For fluids flowing in circular pipes the following limits have been established to determine whether the flow is laminar or turbulent:

Laminar flow occurs when $R_N < 2000$.
Turbulent flow occurs when $R_N > 4000$.

Values of R_N between 2000 and 4000 indicate a transition zone between the two flow conditions and are related to Reynolds' conclusions indicated as item 1, above.

EXAMPLE 6.4

A fluid flows with a velocity of 0.30 m/s in a pipe with a diameter of 25 mm. Determine whether this flow is laminar or turbulent, if (a) the fluid is water, with $v = 1.31 \times 10^{-6}$ m^2/s, (b) the fluid is a heavy fuel oil with $v = 205.0 \times 10^{-6}$ m^2/s.

SOLUTION

(a) If the fluid is water, then, according to Eq. (6.3),

$$R_N = \frac{0.025 \text{ m} \times 0.30 \text{ m/s}}{1.31 \times 10^{-6} \text{ m}^2/\text{s}}$$

$$= 5725 > 4000$$

and consequently the flow is turbulent.

(b) If the fluid is a heavy oil, then, similarly,

$$R_N = \frac{0.025 \text{ m} \times 0.30 \text{ m/s}}{205.0 \times 10^{-6} \text{ m}^2/\text{s}}$$

$$= 37 < 2000$$

and the flow is laminar.

The Reynolds Number for Open Channels

The Reynolds number, as mentioned previously, was developed from experiments related to pipe flow. Therefore, one term of the formula for R_N includes the pipe diameter d.

In order to apply the Reynolds number to open-channel-flow conditions, it is necessary to replace the pipe diameter in the equation with another term, dimensionally correct. A logical choice for this term is the hydraulic radius R. It is dimensionally the same as the diameter and further because of its relationship between cross-sectional area of flow and wetted perimeter, it introduces some consideration of shape into the formula.

For open-channel-flow conditions, therefore, the value of the Reynolds number is generally accepted to be

$$R_N = \frac{vR}{v} \qquad (6.4)$$

where v is the velocity of flow, R is the hydraulic radius, and v is the kinematic viscosity of the fluid.

Since, from discussions in Example (6.1), the hydraulic radius for a circular conduit is

$$R = \frac{d}{4}$$

the relationship between R_N for open channels and pipes is as follows:

$$R_{N(\text{open channel})} = \frac{R_{N(\text{pipe})}}{4}$$

Consequently it can be said that in open channels laminar flow occurs when $R_N < 500$ and turbulent flow occurs when $R_N > 1000$.

Experiments indicate that the lower limit is reasonably correct. The upper limit, however, has been much more difficult to define. On the basis of numerous experiments, this upper limit has been estimated to be near 2000.

EXAMPLE 6.5

The velocity of flow in the trapezoidal channel of Example 6.2 is 0.30 m/s and the liquid is water with a kinematic viscosity of 1.31×10^{-6} m²/s. Determine whether the flow in this channel is laminar or turbulent.

SOLUTION

From Example 6.2, the hydraulic radius $R = 0.51$ m; hence

$$R_N = \frac{0.30 \text{ m/s} \times 0.51 \text{ m}}{1.31 \times 10^{-6} \text{ m}^2/\text{s}}$$

$$= 116\,800 > 2000$$

and therefore the flow is turbulent.

6.3   CAPACITY OF FLOW AND VELOCITY OF FLOW

Capacity of Flow (Q)

The capacity of flow is the amount of fluid passing a particular point in a stream during a certain time interval. Capacity of flow is also referred to as the rate of flow, the discharge, or the capacity. It is generally expressed in units of volume per unit time, such as cubic meters per second (m³/s) or liters per second (L/s).

Velocity of Flow

The velocity of flow in a flowing liquid is generally understood to refer to the average velocity of all the particles of the flowing fluid.

Obviously, when dealing with turbulent-flow conditions and considering a number of particles with path lines similar to the one shown in Fig. 6.2, no two particles would have the same velocity at any particular point along the stream tube. Careful measurements and observations have shown that, even in laminar flow, not all particles in the stream move along at the same speed.

Figure 6.5 shows results of typical velocity measurements in open-channel

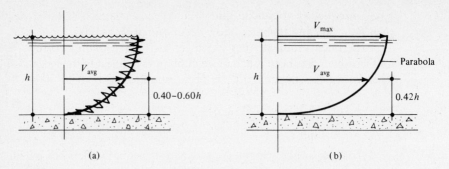

FIGURE 6.5   Typical open-channel velocity profiles. (a) Turbulent flow.
(b) Laminate flow.

types of flow conditions. In Fig. 6.5b is shown a velocity profile of laminar open-channel flow. The velocity distribution forms a parabola, and the relationships between average and surface velocity indicated can be derived from the parabolic equation. When the flow is turbulent, the velocity profile is more like that shown in Fig. 6.5a. Similar observations for pipe-flow conditions are shown schematically in Fig. 6.6. Velocity distribution patterns in natural streams are shown in Fig. 6.7.

FIGURE 6.6   Typical pipe flow velocity profiles. (a) Turbulent flow. (b) Laminar flow.

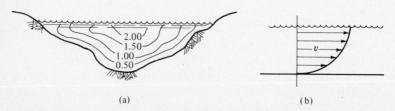

FIGURE 6.7   Typical natural channel velocity patterns. (a) Velocity contours.
(b) Velocity profile.

Capacity-Velocity Relationship

Consider the rectangular channel in Fig. 6.8 and water flowing from point 1 to point 2. The cross section of the flowing mass of water is $KLMN$ and has an area equal to A. If all the water particles in the cross section (at point 1) reach the point 2 in an average time of t seconds, then the volume V of water that has passed point 2 in that time is equal to the prism $KLMNK'L'M'N'$, or

$$V = A \times l \tag{6.5}$$

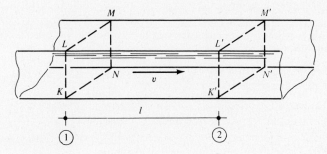

FIGURE 6.8

Hence the volume of water passing point 2 in 1 second is

$$Q = \frac{A \times l}{t} \tag{6.6}$$

where Q is the capacity of flow.

Since l is the distance traveled by the water particles in t seconds,

$$v = \frac{l}{t} \tag{6.7}$$

where v is the average velocity of the flowing water. Substituting the value of l/t from Eq. (6.7) into Eq. (6.6) yields

$$Q = Av \tag{6.8}$$

Equation (6.8) represents a general expression of the relationship between capacity and velocity of flow. Other forms of this equation are

$$v = \frac{Q}{A} \tag{6.9}$$

and

$$A = \frac{Q}{v} \qquad (6.10)$$

EXAMPLE 6.6

In Fig. 6.8, if $KN = 1.25$ m, $KL = 1.00$ m and $l = 1.85$ m, and it takes an average of 3 seconds for the water to travel from point 1 to point 2, determine the velocity and capacity of flow.

SOLUTION

The cross-sectional area A of the flow is

$$A = 1.25 \text{ m} \times 1.00 \text{ m}$$
$$= 1.25 \text{ m}^2$$

The volume of water passing point 2 in 3 seconds is

$$V = 1.25 \text{ m}^2 \times 1.85 \text{ m}$$
$$= 2.32 \text{ m}^3$$

Consequently,

$$Q = \frac{2.32 \text{ m}^3}{3s}$$
$$= 0.77 \text{ m}^3/s$$
$$= 770 \text{ L/s}$$

and, according to Eq. (6.9)

$$v = \frac{0.77 \text{ m}^3/s}{1.25 \text{ m}^2}$$
$$= 0.62 \text{ m/s}$$

EXAMPLE 6.7

A pipe with a diameter of 500 mm has a rate of flow of 500 L/s of water. Determine the velocity of flow and whether the flow is laminar or turbulent Assume the kinematic viscosity of the water to be 1.31×10^{-6} m^2/s.

SOLUTION

The area of flowing water is

$$A = \frac{(0.5 \text{ m})^2 \times \pi}{4}$$

$$= 0.196 \text{ m}^2$$

and according to Eq. (6.9),

$$v = \frac{0.5 \text{ m}^3/\text{s}}{0.196 \text{ m}^2}$$

$$= 2.55 \text{ m/s}$$

The Reynolds number is

$$R_N = \frac{0.5 \text{ m} \times 2.55 \text{ m/s}}{1.31 \times 10^{-6} \text{ m}^2/\text{s}}$$

$$= 973\,300 > 4000$$

and therefore the flow is turbulent.

EXAMPLE 6.8

Determine the pipe diameter required to increase the velocity in the pipe of Example 6.7 to 4.0 m/s, assuming that Q does not change.

SOLUTION

From Eq. (6.10),

$$A = \frac{0.5 \text{ m}^3/\text{s}}{4.0 \text{ m/s}}$$

$$= 0.125 \text{ m}^2$$

Consequently,

$$d = \left(\frac{4 \times 0.125 \text{ m}^2}{\pi} \right)^{1/2}$$

$$= 0.40 \text{ m}$$
$$= 400 \text{ mm}$$

Steady Flow

The flow is said to be steady when the capacity at a specific cross section in the stream does not vary with time. For instance, if the amount of water passing point 1 in Fig. 6.8 was always the same, the flow would be steady. Steady flow occurs almost always in most applications of pressure flow in the civil engineering field. Typical examples of unsteady flow occur in cases of the passing of a flood wave through a reach of a river.

Uniform Flow

Uniform flow relates to a flow condition over a certain length or reach of a stream and can occur only during steady-flow conditions. Uniform flow exists when the average velocity at all the cross sections of a stream is constant.

Considering Eq. (6.8),

$$Q = Av$$

it follows that if Q is constant, steady flow and v is constant, uniform flow, the cross-sectional area A must also be constant. Uniform flow therefore can exist only when the cross section of the conduit carrying the fluid is constant. If a steady flow of liquid occurs in a conduit with varying cross section, the velocity will not be constant and hence the flow is said to be nonuniform or varying.

Obviously, the above definitions can be applied only to incompressible fluids. In other words, uniform flow generally exists only when the fluid is liquid. Since gases are compressible, they will undergo density and volumetric changes as flow conditions and changes in cross section dictate. Truly uniform flow will seldom occur when the fluid is a gas.

Continuous Flow

Continuous flow exists when the mass of fluid passing a given cross section is constant. In Fig. 6.9, if the mass of fluid passing point 1 is equal to that passing point 2, continuity exists and

$$M_1 = M_2 \tag{6.11}$$

where M_1 and M_2 are the mass of fluid passing points 1 and 2, respectively,

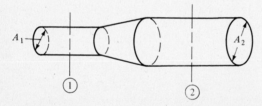

FIGURE 6.9

in a given time. Consequently,

$$M_1 = \rho_1 A_1 v_1 \tag{6.12}$$

and

$$M_2 = \rho_2 A_2 v_2 \tag{6.13}$$

Hence it follows that

$$\rho_1 A_1 v_1 = \rho_2 A_2 v_2 \tag{6.14}$$

In Eq. (6.12), if the fluid is an incompressible liquid, the density will remain constant and the relationship indicating continuity can be expressed as

$$Q = A_1 v_1 = A_2 v_2 = A_3 v_3 = \cdots \tag{6.15}$$

If, as shown in Fig. 6.10, flow is added or taken away, by branches, between points 1 and 2, the flow will no longer be continuous and

$$Q_1 \neq Q_2 \neq Q_3 \tag{6.16}$$

and

$$A_1 v_1 \neq A_2 v_2 \neq A_3 v_3 \neq \cdots \tag{6.17}$$

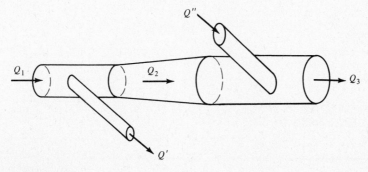

FIGURE 6.10

EXAMPLE 6.9

A pipe varies in diameter from 250 mm at point 1 to 500 mm at point 2. Water flows from point 1 to point 2 at a constant rate of 100 L/s. Determine the velocity at points 1 and 2.

SOLUTION

From Eqs. (6.15) and (6.9),

$$A_1 v_1 = A_2 v_2$$

and

$$v_1 = \frac{Q}{A_1}$$

Consequently

$$v_1 = \frac{0.10 \times 4}{0.25^2 \times \pi}$$

$$= 2.04 \text{ m/s}$$

and, rearranging Eq. (6.15),

$$v_2 = \frac{v_1 A_1}{A_2}$$

$$= v_1 \frac{(\pi d_1^2)/4}{(\pi d_2^2)/4}$$

$$= v_1 \left(\frac{d_1}{d_2}\right)^2$$

$$= 2.04 \left(\frac{0.25}{0.50}\right)^2$$

$$= 0.51 \text{ m/s}$$

EXAMPLE 6.10

Water flows through a pipe that is 30.0 m long and gradually changes in diameter from 750 mm to 150 mm. With the rate of flow at 500 L/s, determine the velocity at various points along the length of this pipe and draw a graph representing the d versus v relationship.

SOLUTION

If this pipe is divided into four sections of equal length, then each section will be 7.5 m long and the diameters will decrease by equal amounts of 150 mm. Referring to the point in the pipe where $d = 750$ mm as point 1, then the diameters at points 2, 3, 4, and 5 will be 600 mm, 450 mm, 300 mm and 150 mm, respectively.

$$v_1 = \frac{0.5 \times 4}{0.75^2 \times \pi}$$

$$= 1.13 \text{ m/s}$$

Using the relationship between the velocities and the square of the diameters, developed in Example 6.5, the velocities can easily be calculated for the other points. A tabular arrangement of these calculations is shown below.

Point	d	$(d_1/d_n)^2$	v
1	0.75		1.13
2	0.60	1.56	1.77
3	0.45	1.78	3.15
4	0.30	2.25	7.09
5	0.15	4.00	28.36

A graphical representation of these values is shown in Fig. 6.11.

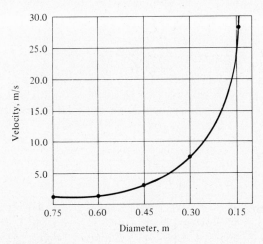

FIGURE 6.11

Continuous Flow in Compressible Fluids

As mentioned in previous discussions of continuous-flow conditions, compressible fluids undergo changes in density. It follows, therefore, from Eq. (6.14) that

$$\rho_1 A_1 v_1 = \rho_2 A_2 v_2 = \rho_3 A_3 v_3 \tag{6.18}$$

Since, as shown in Chapter 2, $\gamma = \rho \times g$, where g is the acceleration due to gravity constant, multiplying Eq. (6.16) by g yields

$$\gamma_1 A_1 v_1 = \gamma_2 A_2 v_2 = \gamma_3 A_3 v_3 \qquad (6.19)$$

EXAMPLE 6.11

A 40-m-high chimney has a diameter of 9.0 m at the bottom and 7.0 m at the top. Gas flows upward in this chimney with a velocity of 4.0 m/s at the bottom. The mass density of the gas is 0.005 kg/m³ when entering and 0.007 kg/m³ when leaving the chimney. The density varies gradually between these two values as the gas rises. Determine the average velocity of the gas at every 10.0 m in the chimney.

SOLUTION

Rearranging of Eq. (6.18) yields

$$v_2 = \frac{\rho_1 A_1}{\rho_2 A_2} \times v_1$$

or

$$v_2 = \frac{\rho_1}{\rho_2} \times \left(\frac{d_1}{d_2}\right)^2 \times v_1$$

As in the previous example, the calculations, in this instance using the above equation can be conveniently arranged in tabular form as follows:

Point	d	$\left(\dfrac{d_1}{d_n}\right)^2$	ρ	$\dfrac{\rho_1}{\rho_2}$	v
Bottom	9.00		0.0050		4.00
10 m	8.50	1.12	0.0055	0.909	4.08
20 m	8.00	1.13	0.0060	0.917	4.22
30 m	7.50	1.14	0.0065	0.923	4.43
Top	7.00	1.15	0.0070	0.929	4.72

6.4 ENERGY, WORK, AND POWER IN FLUID FLOW

Energy, Work, and Power

In order to illustrate the concepts of energy, work, and power, consider the simple form of pile driver in Fig. 6.12.

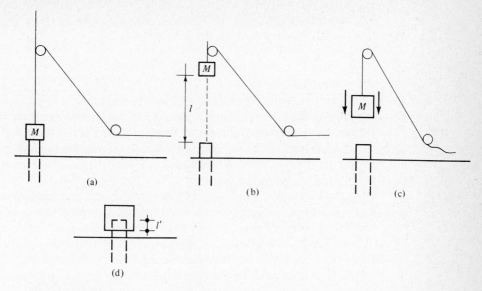

FIGURE 6.12

In Fig. 6.12*a*, the man commences to pull the hammer with a mass *M* upward to the elevation shown in Fig. 6.12*b*. In doing so, the man has expended *energy* or performed *work*. The amount of work done by the man is in direct relation to the weight of the hammer and the distance this weight was lifted. Mathematically, the amount of work performed can be expressed as

$$W = F \times l \qquad (6.20)$$

where *F* is the force of gravity acting on the hammer, or the weight of the hammer.

From the above it follows that the units of work are represented as force × distance, or newtons × meters. In SI units,

$$1 \text{ N·m} = 1 \text{ J}$$

where J is the symbol for joule, the unit of work.

When the hammer is at the position shown in Fig. 6.12*b* and the man releases the rope, the hammer will fall, strike the top of the pile, and drive it into the ground a certain distance.

At the point of release, no work is performed by the man. Yet the hammer, whose weight has not changed, will travel a distance *l* equal to the distance it was raised by the man. Consequently, when the hammer is at the top position it can perform an amount of work equal to the work performed by the man in raising it to that position. In other words, at the top position work is stored in the hammer or the hammer has the potential to perform work. This potential to perform work or expend energy is called *potential energy*.

In the form of an equation, the above can be expressed as

$$PE = F \times l \tag{6.21}$$

where PE = potential energy.

If the man were to lower the hammer slowly and allow it to come to rest gently on the pile, the pile would not be driven into the ground. No work would be performed by the hammer. The energy that was stored in the block was used up by the resistance to falling provided by the man in lowering the hammer. In this instance the work performed by the man will again be equal to the original amount required in raising the hammer.

When the hammer is allowed to fall, however, it will gain speed and will strike the pile with a certain velocity. Even though at impact the hammer is back down to its original position and work equal to raising it was done, the pile will be driven into the ground. In other words, work will still be performed.

What has happened is that the potential energy in the hammer at the raised position has been converted into energy related to the velocity with which the hammer strikes the pile. This form of energy is called *kinetic energy*. Its value is in direct relationship to the mass of the moving body and the square of its velocity, or kinetic energy, KE, is

$$KE = \frac{Mv^2}{2} \tag{6.22}$$

A consideration of the units of the terms of KE, indicates

$$KE = \frac{kg \times m^2/s^2}{2} = \frac{1}{2} \times \frac{kg \cdot m}{s^2} \times m$$

$$= N \cdot m = J$$

As shown in Fig. 6.12*d*, the work done by the hammer when striking the pile will be

$$W = F \times l' \tag{6.23}$$

which appears to be considerably less than the work performed by the man or the potential energy stored in the hammer when at the top of its stroke. However, the resistance that is provided by the soil in preventing the pile to be driven into the ground is responsible for absorbing the energy represented by the difference between Fl and Fl'.

Obviously, a very powerful man would be able to lift the hammer to its top position in a shorter time than a weak man. In performing the work faster the powerful man provides more power to the task. *Power*, then, is the work performed in a unit of time. Its units are joules per second (J/s) or watts (W).

EXAMPLE 6.12

If the hammer in Fig. 6.12 has a mass of 40.0 kg and if $l = 10.0$ m, determine the following.

(a) The work required to lift the hammer from its position in Fig. 6.12*a* to that in Fig. 6.12*b*.

(b) The potential energy stored in the hammer when in the position of Fig. 6.12*b*.

(c) The kinetic energy of the hammer at the point of impact with the top of the pile.

(d) The power required to lift the hammer to its top position in 10 seconds.

SOLUTION

(a) The work required to lift the hammer from the top of the pile to the position of Fig. 6.12*b* is expressed by Eq. (6.20); hence

$$W = 40.0 \text{ kg} \times 9.81 \text{ m/s}^2 \times 10.0 \text{ m}$$

$$= 3924 \text{ N} \cdot \text{m}$$

$$= 3.9 \text{ kN} \cdot \text{m}$$

$$= 3.9 \text{ kJ}$$

(b) The potential energy of the hammer in its raised position will be expressed in relation to the top of the pile. It is expressed by Eq. (6.23). In this instance the potential energy of the hammer in the raised position is equal to the work done in raising it, or

$$PE = 40.0 \text{ kg} \times 9.81 \text{ m/s}^2 \times 10.0 \text{ m}$$

$$= 3.9 \text{ kJ}$$

(c) Assuming that the friction between the hammer and the surrounding air and that between the pulleys and the rope is negligible, it is known from physics that the hammer, when falling, will accelerate at a rate of $g = 9.81$ m/s² and the relationships below result.

$$l = \frac{gt^2}{2}$$

and

$$v = gt$$

Consequently, the time required for the hammer to fall to the top of the pile is

$$t = \left(\frac{2 \times 10.0 \text{ m}}{9.81 \text{ m/s}^2} \right)^{1/2}$$

$$= 1.43 \text{ s}$$

Its velocity at impact will be

$$v = 9.81 \text{ m/s}^2 \times 1.43 \text{ s}$$

$$= 14.01 \text{ m/s}$$

and, from Eq. (6.22),

$$KE = \frac{40.0 \text{ kg} \times (14.01 \text{ m/s})^2}{2}$$

$$= 3924 \text{ kg} \cdot \text{m}^2/\text{s}^2$$

$$= 3.9 \text{ kJ}$$

It is interesting to note that this value for the kinetic energy in the hammer is exactly equal to its potential energy when at the top position. This was to be expected, of course, since in its downward trip no energy was added to the hammer. In fact what happened is that the potential energy stored in the hammer was transformed into an equal amount of kinetic energy.

(d) If the work of lifting the hammer to its top position is done in 10 seconds, then the power exerted per unit time will be

$$\text{Power} = \frac{3.9 \text{ kJ}}{10 \text{ s}}$$

$$= 0.39 \text{ kJ/s}$$

$$= 0.39 \text{ kW}$$

Elevation and Pressure Energy

In the foregoing section two forms of energy—potential and kinetic—were discussed. The potential energy in the hammer of Fig. 6.12b was related to its elevation above the top of the pile. In hydraulics this form of energy is often referred to as *elevation energy*.

It is evident that elevation energy must refer to a point of reference, or datum. Obviously, in the case of the pile driver of Fig. 6.12 the potential energy in the hammer when in the raised position would not be the same if the difference in elevation between the hammer and the top of the pile changed.

There are other forms of potential energy, related to internal conditions of the fluid or its own composition. Among these forms of energy are chemical energy, heat energy, and pressure energy. Heat energy is of importance to the civil engineer only as a method of explaining the phenomena of energy losses in fluid flow, and will be discussed later.

Pressure energy is a form of potential energy that plays an important part

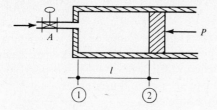

FIGURE 6.13

in the flow of fluids. Pressure in a fluid has the potential to make a fluid flow and perform work, as can be visualized by considering the cylinder in Fig. 6.13.

If valve A is opened, water is allowed to enter this cylinder under pressure and in so doing the piston is moved to the right, against the force P, from point 1 to point 2. In effect, the force P is moved a distance l. Consequently, work has been done and the amount of this work is

$$W = Pl \qquad (6.24)$$

If p is the unit pressure in the water entering the cylinder, then the total pressure P on the piston will be as given earlier in Eq. (3.2); that is,

$$P = pA$$

where A is the area of the piston. Substitution of this value for P in Equation (6.24) yields

$$W = pAl \qquad (6.25)$$

A check of the units in this expression reveals, as expected, that W will be expressed in units of $N \cdot m$ or J. Since W in Eq. (6.24) is the work or energy due to pressure, it follows that the pressure energy PE in this instance can be expressed as

$$PE = pAl \qquad (6.26)$$

Total Energy in Fluid Flow

Any point in a flowing fluid contains all three forms of energy discussed above, and the total sum of these forms of energy is mainly responsible for the fluid to flow or perform work. Therefore the total energy in a flowing fluid is

$$\text{Total energy} = EE + KE + PE \qquad (6.27)$$

where EE, KE, and PE represent the elevation, kinetic, and pressure energy, respectively.

Energy Head

All the terms in Eq. (6.27) are in units of joules or newton-meters. If these terms are divided by newtons, they will all be expressed in terms of meters.

For example, the pressure energy, PE, will now be expressed as a linear quantity. This concept was already discussed and used in Chapter 3. The linear quantity was given the name of pressure head.

Similar linear forms of expressing elevation and kinetic energy will result and are called *elevation head* and *velocity head*, respectively.

ELEVATION HEAD

Consider the small mass of liquid in Fig. 6.14 with a mass M, and at an elevation z, above the datum. According to previous discussions, its elevation energy is

$$EE = Mgz \qquad (6.28)$$

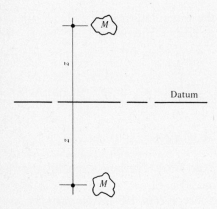

FIGURE 6.14

In this equation Mg represents the weight of the mass of liquid and is expressed in newtons. Dividing Eq. (6.28) by Mg will yield

$$\text{Elevation head} = z \qquad (6.29)$$

In the event that the mass M were located a distance z, below the datum, it is obvious that work would need to be performed to bring that mass to the elevation of the datum. It can therefore be said in such condition, that the mass M has a negative elevation energy and the elevation head would also be negative or equal to $-z$.

PRESSURE HEAD

Pressure energy was discussed earlier, and Eq. (6.26) was developed to read

$$PE = pAl$$

From Eq. (3.3), $p = \gamma h$, where $\gamma = \rho g$, the specific weight of the fluid, and h is the pressure head in the fluid. Substituting this value for p in Eq. (6.26) yields

$$PE = \rho g h Al \qquad (6.30)$$

The right side of this equation can be written as

$$= h \times \rho g Al \qquad (6.31)$$

where $\rho g Al$ represents the weight of the water displaced, in newtons. Therefore dividing both sides of Eq. (6.31) by newtons yields

$$\text{Pressure head} = h \qquad (6.32)$$

or, from Eq. (3.4),

$$\text{Pressure head} = \frac{p}{\gamma} \qquad (6.33)$$

VELOCITY HEAD

According to Eq. (6.24), the kinetic energy of a moving mass M is

$$KE = \tfrac{1}{2}Mv^2$$

However, $M = W/g$. Substituting this value for M in Eq. (6.22) yields

$$KE = \frac{1}{2}\frac{W}{g}v^2 \qquad (6.34)$$

Dividing both sides of this equation by W, or newtons, gives the expression for velocity head

$$\text{Velocity head} = \frac{v^2}{2g} \qquad (6.35)$$

Again, an investigation of $v^2/2g$ reveals that this term is a linear quantity.

TOTAL ENERGY HEAD (E)

Since the total energy is equal to the sum of the individual forms of energy contained in a flowing fluid, the total energy head E will also equal the sum of the energy heads, or

$$E = \frac{v^2}{2g} + \frac{p}{\gamma} + z \qquad (6.36)$$

Equation (6.36) is a very important relationship in the study of flowing fluids. It expresses the total energy inherent in a mass of flowing fluid equal to one unit in weight.

Although the energy head is normally expressed in linear terms, it must be remembered that its units really are N·m/N or J/N.

POWER IN FLUID FLOW

Power is the amount of work performed in a unit of time. If a flow of fluid has a capacity of Q m³/s and the fluid has a specific weight of γ N/m³, then $Q\gamma$ represents the number of newtons of fluid flowing per second. The power inherent in a flowing fluid is therefore

$$\text{Power} = Q\gamma E \qquad (6.37)$$

EXAMPLE 6.13

A horizontal pipe, at an elevation of 135.25 m above sea level, has water flowing in it at a rate of 500.0 L/s and at a pressure of 600.0 kPa. The pipe diameter is 500 mm. Determine
   (a) The energy head and the power of the flowing water.
   (b) The energy head, if the datum is considered to be at the centerline of the pipe.
   (c) The energy head and the power of the flow if the fluid is oil, with $s = 0.85$.

SOLUTION

(a) The velocity of flow is

$$v = \frac{0.50 \times 4}{\pi \times 0.5^2}$$

$$= 2.55 \text{ m/s}$$

and therefore, from Eq. (6.35), the velocity head is

$$\frac{v^2}{2g} = \frac{2.55^2}{2 \times 9.81}$$

$$= 0.33 \text{ m}$$

Similarly, from Eq. (6.33), the pressure head is

$$\frac{p}{\gamma} = \frac{600.0}{9.81}$$

$$= 61.16 \text{ m}$$

and, from Eq. (6.29) the elevation head relative to the datum is

$$z = 135.25 \text{ m}$$

Hence the total energy head, according to Eq. (6.36), is

$$E = 0.33 + 61.16 + 135.25$$

$$= 196.74 \text{ m}$$

Therefore for water, with $\gamma = 9.81$ kN/m³, the weight of the flowing fluid is

$$Q\gamma = 0.50 \times 9.81$$

$$= 4.91 \text{ kN/s}$$

and, from Eq. (6.37),

$$\text{Power} = 4.91 \times 196.74$$

$$= 966.0 \text{ kJ/s}$$

$$= 966.0 \text{ kW}$$

(b) If the datum were at the centerline of the pipe, $z = 0$ and

$$E = 0.33 + 61.16$$

$$= 61.49 \text{ m}$$

Raising the datum has not altered the capacity of flow, only the energy head. Consequently,

$$\text{Power} = 4.91 \times 61.49$$

$$= 302.0 \text{ kW}$$

In order to visualize the difference in the power when the datum is at sea level, or at the centerline of the pipe, consider what is required to pump this much water. Obviously, if the pump and water source are located at sea level, it will take a great deal more work and power to pump the water at this rate of flow through the pipe than if the pump and source are the same elevation as the pipe.

(c) If the fluid is oil, with $s=0.85$,

$$\gamma = 9.81 \times 0.85$$

$$= 8.34 \text{ kN/m}^3$$

and the pressure head will be

$$\frac{p}{\gamma} = \frac{600.0}{8.34}$$

$$= 71.94 \text{ m}$$

The elevation head and the velocity head are independent of the relative density of the fluid and will remain as in (a). Therefore

$$E = 0.33 + 71.94 + 135.25$$

$$= 207.52 \text{ m}$$

The power will be

$$\text{Power} = 0.50 \times 8.34 \times 207.52$$

$$= 865 \text{ kW}$$

EXAMPLE 6.14

If the energy head in the pipe of Example 6.13 is 197.15 m and the pressure and elevation remain the same, determine the capacity of flow.

SOLUTION

Solving Eq. (6.36) for the velocity head yields

$$\frac{v^2}{2g} = E - \frac{p}{\gamma} - z$$

Consequently,

$$\frac{v^2}{2g} = 197.15 - 61.16 - 135.25$$

$$= 0.74 \text{ m}$$

and

$$v = (0.74 \times 2 \times 9.81)^{1/2}$$

$$= 3.81 \text{ m/s}$$

Hence

$$Q = 3.81 \ \text{m/s} \times \frac{0.5^2 \times \pi}{4}$$
$$= 0.748 \ \text{m}^3/\text{s}$$
$$= 748 \ \text{L/s}$$

EXAMPLE 6.15

The vertical tube connected to the pipe in Fig. 6.15 is a piezometer. The pipe diameter is 250 mm and its centerline is 25.0 m above a certain datum. Water flows in the pipe at the rate of 300 L/s, and at the point immediately below the piezometer the energy head is 75.85 m. Determine the pressure in the pipe and the height to which the water will rise in the piezometer.

SOLUTION

Rearranging Eq. (6.36) to solve for the pressure head yields

$$\frac{p}{\gamma} = E - \frac{v^2}{2g} - z$$

The velocity in the pipe is

$$v = \frac{0.30 \times 4}{0.25^2 \times \pi}$$
$$= 6.11 \ \text{m/s}$$

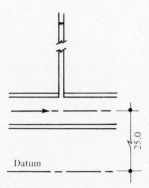

Datum

25.0

FIGURE 6.15

and

$$\frac{v^2}{2g}=\frac{6.11^2}{2\times9.81}$$

$$=1.90 \text{ m}$$

Therefore

$$\frac{p}{\gamma}=75.85-1.90-25.00$$

$$=48.95 \text{ m}$$

and

$$p=48.95\times9.81$$

$$=480 \text{ kPa}$$

Since the pressure head at the piezometer is 48.9 m, that is how high the water would rise in the piezometer.

Graphical Representation of Energy Head

In Fig. 6.16 consider the particle of water A in a flowing stream. If v_A, p_A, and z_A, are the velocity, pressure and elevation head of this particle, then the

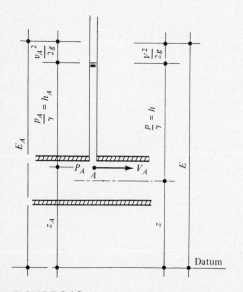

FIGURE 6.16

energy head of the particle, E_A, will

$$E_A = \frac{v_A{}^2}{2g} + \frac{p_A}{\gamma} + z_A \tag{6.38}$$

Since all the values in this equation represent linear quantities, E_A can be expressed graphically, in linear distances from the datum, as shown in Fig. 6.16. A piezometer located at this point would have water rise in it to a distance above the centerline of the pipe equal to

$$h_A = \frac{p_A}{\gamma}$$

In pipe-flow applications it is standard practice to refer to the centerline of the pipe for dimensioning and for indicating pipe elevations. If v is the average velocity of all the particles of the flow in the pipe, then the total energy head will be as expressed in Eq. (6.36) and shown graphically in Fig. 6.16. The graphical presentation of energy head in open-channel-flow conditions is shown in Fig. 6.17.

Open-channel elevations are generally referred to the invert or bottom of the channel, since the depth of flow is variable and there is no constant horizontal centerline. In other words, all the terms of Eq. (6.36) refer to the bottom of the channel.

As can be seen from Fig. 6.17, the pressure head will be equal to the depth of flow D, and therefore, in open-channel flow,

$$E = \frac{v^2}{2g} + D + z \tag{6.39}$$

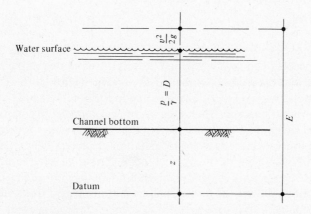

FIGURE 6.17

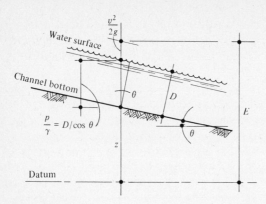

FIGURE 6.18

It must be pointed out that, as can be seen from Fig. 6.18, equating the pressure head in open-channel flow with the depth D introduces an error. Actually the true value for the pressure head is

$$\frac{p}{\gamma} = \frac{D}{\cos \theta} \tag{6.40}$$

However in most civil engineering applications the channel bottom has a gentle slope, in the order of 1 percent or flatter. At a slope of 1 percent, or 0.01, $\theta = 0.57°$ and $\cos \theta = 0.99995$. Consequently in all applications of this type of channel considering the pressure equal to D will not introduce any appreciable error.

6.5   CONSERVATION OF ENERGY IN FLUID FLOW (BERNOULLI'S THEOREM)

At the beginning of Section 6.4 the example of a simple pile driver was used to illustrate the concepts of energy, work, and power. It was shown in the accompanying example that the hammer of the pile driver, when allowed to drop, will at the point of impact have a kinetic energy equal to the potential energy stored in it when at the raised position. This observation was qualified by stating that small amounts of energy lost due to friction of the air and pulleys was ignored.

The above observation leads to the formulation of a fundamental principle underlying numerous scientific theories: "the principle of conservation of energy." In simple terms this principle states that no new energy can be created and no energy can be destroyed. Rather, energy is transformed from one form to another. The hammer of the pile driver transformed elevation energy, stored in it at the top of its stroke, to an equal amount of kinetic energy during its fall to the top of the pile. In many instances this transformation of energy is not as obvious as in the pile-driver example—particularly when mechanical energy, such as pressure, elevation and kinetic energy, is changed to heat or chemical forms of energy, and vice versa.

For instance, in braking a car, the kinetic energy inherent in the moving vehicle due to its velocity is rapidly reduced. Yet, no other form of energy is readily apparent to take its place. As a matter of fact, additional energy will be required to bring the car back to its original speed. The kinetic energy apparently lost in the braking action has, among other things, brought about a substantial increase in heat in the brakes. As yet no practical means exist to reuse this heat in bringing the car back to speed; hence the heat energy in the brakes is dissipated to the surrounding air.

Bernoulli's Theorem for Ideal Fluids

Daniel Bernoulli, in the early part of the eighteenth century, expressed the conservation-of-energy principle in terms of flow in water and hydraulics. Essentially, his theorem states that in a fluid with steady, continuous flow, the mechanical energy at any point along its path of flow is constant. In so doing he ignored that certain energy losses take place due to resistance to flow because of the viscosity of the fluid and other phenomena. If the fluid is an ideal fluid with no viscosity and no other forms of energy loss, this principle can be mathematically expressed in terms of energy head at two points in the stream, as follows:

$$\frac{v_1{}^2}{2g} + \frac{p_1}{\gamma} + z_1 = \frac{v_2{}^2}{2g} + \frac{p_2}{\gamma} + z_2 \qquad (6.41)$$

In order to visualize the veracity of the above statement, consider the total mechanical energy in the small mass of liquid M in a streamline, as in Fig. 6.19. When the small mass is at a distance a above the channel bottom, it is subjected to two vertical forces, F and F_1, where F is the force of gravity and

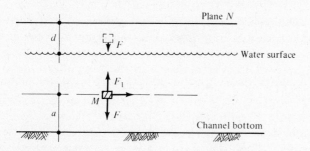

FIGURE 6.19

F_1 is the buoyancy force. The force F, or weight of the small mass M is

$$F = M \times \gamma \tag{6.42}$$

where γ is the specific weight of the particle.

The force F_1 acts upward—in a direction opposite to F—and, from Archimedes principle, discussed earlier, is equal to

$$F_1 = M \times \gamma \tag{6.43}$$

where γ is the specific weight of the fluid.

Since the small mass and the liquid in the channel are the same fluids, the values of their relative densities and specific weights are equal; hence

$$F = F_1 \tag{6.44}$$

Obviously, therefore, no work will be required to raise or lower the small mass within the liquid. Only when the mass M is raised above the liquid surface will work be performed. Consequently, in relation to the reference plane N in Fig. 6.19, the potential energy or elevation energy (EE) of M is

$$EE = M \times \gamma \times (-d) \tag{6.45}$$

and its elevation energy head is $-d$.

If the mass M moves with a velocity v, then from Eq. (6.35) the velocity head is $v^2/2g$. Therefore the total mechanical energy head of the mass M in relation to the reference plane N is

$$E = -d + \frac{v^2}{2g} \tag{6.46}$$

If the mass M moves in the stream from point 1 to point 2, as illustrated in Fig. 6.20 and if the velocities at points 1 and 2 are v_1 and v_2, respectively, then, since no mechanical energy is added or removed, the total mechanical energy at point 1 must be equal to that at point 2. In relation to plane N,

$$-d_1 + \frac{v_1{}^2}{2g} = -d_2 + \frac{v_2{}^2}{2g} = \text{constant} \tag{6.47}$$

In relation to the datum plane, which is a distance n below the plane N, Eq. (6.47) will read

$$-d_1 + \frac{v_1{}^2}{2g} + n = -d_2 + \frac{v_2{}^2}{2g} + n = \text{constant} \tag{6.48}$$

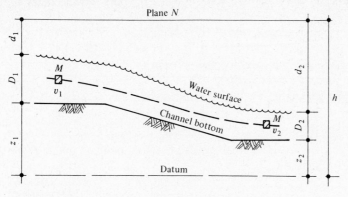

FIGURE 6.20

From Fig. 6.20 it is found that

$$n = z_1 + D_1 + d_1 = z_2 + D_2 + d_2 \tag{6.49}$$

Substituting these values for n in Eq. (6.48) yields

$$\frac{v_1^2}{2g} + D_1 + z_1 = \frac{v_2^2}{2g} + D_2 + z_2 \tag{6.50}$$

According to the section on graphical representation of energy head, D for open-channel flow is equal to the pressure head p/γ. Hence the generalized form of Eq. (6.41) results, or

$$\frac{v_1^2}{2g} + \frac{p_1}{\gamma} + z_1 = \frac{v_2^2}{2g} + \frac{p_2}{\gamma} + z_2 \tag{6.51}$$

Equation (6.51), the general expression for Bernoulli's theorem, provides an invaluable tool in solving fluid flow problems. It is the foundation of most of the theories and principles dealing with pressure and open channel flow. In the form of Eq. (6.51) it is, of course, applicable only to those problems dealing with ideal liquids.

EXAMPLE 6.16

Water flows at the rate of 150 L/s in the pipe of Fig. 6.21 from point 1 to point 2. The pressure at point 1 is 450 kPa. Assuming that there is no energy loss or gain between points 1 and 2, determine the pressure at point 2.

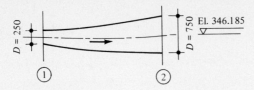

FIGURE 6.21

SOLUTION

Equation (6.51), when written between points 1 and 2 in the pipe, must be satisfied, or

$$\frac{v_1{}^2}{2g}+\frac{p_1}{\gamma}+z_1=\frac{v_2{}^2}{2g}+\frac{p_2}{\gamma}+z_2$$

For point 1 the velocity head, the pressure head, and elevation can be determined as follows:

$$v_1=\frac{0.15\times4}{\pi\times0.25^2}$$

$$=3.06\text{ m/s}$$

and

$$\frac{v_1{}^2}{2g}=\frac{(3.06)^2}{2\times9.81}$$

$$=0.48\text{ m}$$

$$\frac{p_1}{\gamma}=\frac{450}{9.81}$$

$$=45.87\text{ m}$$

$$z_1=346.18\text{ m}$$

The velocity and elevation heads for point 2 are found as for point 1 and are as follows:

$$v_2=\frac{0.15\times4}{\pi\times0.75^2}$$

$$=0.34\text{ m/s}$$

and

$$\frac{v_2{}^2}{2g}=\frac{(0.34)^2}{2\times9.81}$$

$$=0.006\text{ m}$$

$$z_2 = 346.18 \text{ m}$$

Substituting all these known values into Eq. (6.51) yields

$$0.48 + 45.87 + 346.185 = 0.006 + \frac{p_2}{\gamma} + 346.185$$

or

$$\frac{p_2}{\gamma} = 46.34 \text{ m}$$

and

$$p_2 = 46.34 \times 9.81$$
$$= 455 \text{ kPa}$$

In this instance, with points 1 and 2 on the same elevation,

$$z_1 = z_2$$

and these terms could have been omitted from Eq. (6.51). In other words, the datum could have been considered to be at the centerline of the pipe. Equation (6.51) would then have read

$$0.48 + 45.872 = 0.006 + \frac{p_2}{\gamma}$$

and

$$p_2 = 455 \text{ kPa}$$

The choice of datum in solving problems related to energy head is of great importance, in that it can lead to simplification of the calculations involved. It must, of course, be emphasized that all the energy head terms in any one equation must refer to the same datum in order to have valid comparisons.

EXAMPLE 6.17

The fountain in Fig. 6.22 must spray water to a height of 3.0 m above the concrete floor. Determine the amount of water flow required to accomplish this and the minimum diameter d_1, required of the opening in the floor. Neglect friction losses between the water and the air.

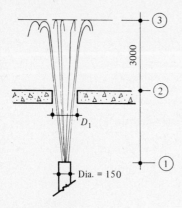

FIGURE 6.22

SOLUTION

Considering Bernoulli's theorem between the top of the pipe and the top of the water spray in Fig. 6.22 leads to the following format for Eq. (6.51).

$$\frac{v_1^2}{2g} + \frac{p_1}{\gamma} + z_1 = \frac{v_3^2}{2g} + \frac{p_3}{\gamma} + z_3$$

At the top of the water spray, the water has just reached the point where its upward movement is changing to downward. Therefore the velocity at this point is zero. Also, the pressure at the top of the water spray is atmospheric. Consequently,

$$\frac{v_3^2}{2g} = 0$$

$$\frac{p_3}{\gamma} = 0$$

$$z_3 = 6.0 \text{ m}$$

At the point where the water leaves the pipe the velocity is unknown, but, again, the pressure is atmospheric. Hence

$$\frac{p_1}{\gamma} = 0$$

and, if the datum is chosen to be at this point,

$$z_1 = 0$$

Substituting these values in Eq. (6.51) as revised above leads to

$$\frac{v_1^2}{2g} + 0 + 0 = 0 + 0 + 6.0$$

or

$$v_1 = (6.0 \times 2 \times 9.81)^{1/2}$$
$$= 10.85 \text{ m/s}$$

Hence, from Eq. (6.8),

$$Q = 10.85 \times \frac{0.15^2 \times \pi}{4}$$
$$= 0.192 \text{ m}^3/\text{s}$$
$$= 192 \text{ L/s}$$

In order to determine the opening required in the floor, Bernoulli's equation can be written between points 1 and 2. Then

$$\frac{v_1^2}{2g} + \frac{p_1}{\gamma} + z_1 = \frac{v_2^2}{2g} + \frac{p_2}{\gamma} + z_2$$

and

$$6.0 + 0 + 0 = \frac{v_2^2}{2g} + 0 + 3.0$$

Hence

$$v_2 = (3.0 \times 2 \times 9.81)^{1/2}$$
$$= 7.67 \text{ m/s}$$

Therefore, from Eq. (6.10),

$$A = \frac{Q}{v} = \frac{0.192}{7.67}$$
$$= 0.025 \text{ m}^2$$

and

$$d_1 = 0.178 \text{ m}$$
$$= 178 \text{ mm}$$

Bernoulli's Theorem for Real Fluids

In the above discussions it was assumed that no energy was added or taken away from the stream of flowing fluid between points 1 and 2. In reality, however, there is always a certain amount of energy loss when a fluid flows. This loss of energy can be readily observed in instances where flowing water is used to operate turbines, for instance.

However even when flowing in a water main or a sewer pipe, the water's mechanical energy is diminished. As will be discussed in some more detail in Chapter 7, this lost mechanical energy is transformed to other forms of energy, such as heat energy, and is not recoverable by the flowing fluid.

Such energy loss must, of course be taken into account when dealing with applications of Bernoulli's theorem to real fluids. If in Eq. (6.51) the direction of flow is taken to be from point 1 to point 2 and energy losses take place during this flow, then the total energy head at point 2 must be less than that at point 1. Consequently, the equal sign in Eq. (6.51) is no longer applicable. Some quantity, obviously the amount of energy head lost, must be added to the quantities at point 2, or

$$\frac{v_1^2}{2g} + \frac{p_1}{\gamma} + z_1 = \frac{v_2^2}{2g} + \frac{p_2}{\gamma} + z_2 + H_l \tag{6.52}$$

where H_l represents the energy head lost between points 1 and 2.

Similarly, if energy is added to the stream—by means of a pump, for instance—this added energy must be accounted for in Eqs. (6.51) and (6.52). In this case the added energy head term H_u is added to the values of the energy heads at point 1. Hence

$$\frac{v_1^2}{2g} + \frac{p_1}{\gamma} + z_1 + H_u = \frac{v_2^2}{2g} + \frac{p_2}{\gamma} + z_2 + H_l \tag{6.53}$$

Equation (6.53) is the general form of mathematical statement expressing Bernoulli's theorem as applied to real fluids.

EXAMPLE 6.18

Water flows out of the reservoir in Fig. 6.23 through a 250-mm pipe. Energy-head losses have been observed as follows: between points 2 and 3, 1.45 m; between points 3 and 4, 1.65 m; and between points 4 and 5, 1.90 m. Determine (a) the amount of water flowing out of the reservoir and (b) the pressure at points 3 and 4.

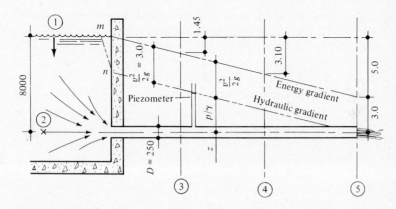

FIGURE 6.23

SOLUTION

In order to solve the first part of this problem, it is necessary to write Bernoulli's equation between two points in the fluid. At points 2, 3, and 4 information other than the values of z is not readily available. However at points 1 and 5 considerably more values of the energy head are known or readily determined.

First, it is necessary to determine the datum to which all energy heads will be referred. In this instance an obviously convenient choice for this datum is the centerline of the pipe. Then, for point 1 at the surface of the reservoir,

$$z_1 = 8.0 \text{ m}$$

$$\frac{p_1}{\gamma} = 0$$

Since the area of the reservoir surface is large, the water level in the reservoir will drop slowly. In other words, the velocity at point 1 will be small, and therefore in most instances the velocity head at point 1 will have a negligible value, or

$$\frac{v_1^2}{2g} = 0$$

At point 5, where the water has just left the pipe,

$$z_5 = 0$$

and

$$\frac{p_5}{\gamma} = 0$$

Writing the Bernoulli equation, Eq. (6.53), between points 1 and 5 yields

$$\frac{v_1^2}{2g} + \frac{p_1}{\gamma} + z_1 + H_u = \frac{v_5^2}{2g} + \frac{p_5}{\gamma} + z_5 + H_l$$

and substituting the known values for energy heads at points 1 and 5 as determined above, this becomes

$$0 + 0 + 8.0 + H_u = \frac{v_5^2}{2g} + 0 + 0 + H_l$$

No energy is added between points 1 and 5; consequently,

$$H_u = 0$$

The energy losses between points 1 and 5 are made up of the sum of the energy losses between points 2 and 3, 3 and 4, and 4 and 5, or

$$H_{l(1-5)} = H_{l(2-3)} + H_{l(3-4)} + H_{l(4-5)}$$

$$= 1.45 + 1.65 + 1.90$$

$$= 5.0 \text{ m}$$

Hence, between points 1 and 5 Bernoulli's equation is reduced to

$$8.0 = \frac{v_5^2}{2g} + 5.0$$

and

$$v_5 = 7.67 \text{ m/s}$$

Remembering that $Q = Av$, it follows that

$$Q = 0.376 \text{ m}^3/\text{s}$$

$$= 376 \text{ L/s}$$

In order to find the pressure at other points along the pipe, it is only necessary to apply Eq. (6.53) to two successive points, keeping in mind that, since the pipe diameter and Q are constant, the velocity head is also constant.

For instance, the pressure at point 3 could be found as follows

$$\frac{v_1^2}{2g} + \frac{p_1}{\gamma} + z_1 + H_u = \frac{v_3^2}{2g} + \frac{p_3}{\gamma} + z_3 + H_l$$

$$0 + 0 + 8.0 + 0 = 3.0 + \frac{p_3}{\gamma} + 0 + 1.45$$

and

$$p_3 = 34.8 \text{ kPa}$$

To find p_3, Eq. (6.53) could also have been written between points 3 and 5, as follows:

$$\frac{v_2^3}{2g} + \frac{p_3}{\gamma} + z_3 + H_u = \frac{v_5^2}{2g} + \frac{p_5}{\gamma} + z_5 + H_l$$

or

$$3.0 + \frac{p_3}{\gamma} + 0 + 0 = 3.0 + 0 + 0 + 3.55$$

and

$$p_3 = 34.8 \text{ kPa}$$

Similar reasoning and operations will find the values of p_2 and p_4.

In solving problems of this nature it is often advantageous to express the equations or other relationships in table form. This will help in visualizing the problem, reduce the work, and allow for easy checking when required. The present problem could have been solved in the following table format.

Point	$\dfrac{v^2}{2g}$	$\dfrac{p}{\gamma}$	z	H_u	H_l	E
1	0.0	0.0	8.0	—	—	8.0
2	0.0	8.0	0.0	—	0.0	8.0
3	3.0	3.55	0.0	—	1.45	6.55
4	3.0	1.90	0.0	—	3.10	4.90
5	3.0	0.0	0.0	—	5.0	3.0

The lines for points 1 and 5 can be readily filled in, remembering that the sum of all the energy head terms must equal the total energy head E. For instance, it is given that the total head loss from 1 to 5 is equal to 5.0 m. The total head at 1 is made up only of elevation head and must equal 8.0 m. Therefore the total head at point 5 has to be 3.0 m. Since the pressure head and the elevation head are both equal to zero at 5, it follows that the velocity head has to be equal to 3.0 m.

Lines for all the other points can now be filled in without the need of rewriting Eq. (6.53) for each case.

In the case of the line of point 2, to state that the pressure head is equal to 8.0 m is not exactly correct. As can be seen from the streamlines sketched in Fig. 6.23, some energy at this point must be due to velocity head. The amount of this head cannot be determined accurately and in most instances has no significance. Suffice it to say that at point 2 the elevation head of point 1 is in the process of changing to the velocity head at point 3 and indeed the velocity head immediately inside the pipe.

EXAMPLE 6.19

The pump in Fig. 6.24 delivers 65 L/s of water to the elevated storage tank. The energy losses in the suction pipe and the discharge pipe are 3 and 6 times their respective velocity heads. The pump efficiency is 72 percent and that of the motor is 76 percent. If the power cost is $0.15/kWh, determine the daily pumping cost.

SOLUTION

Writing Eq. (6.53) between point 1 on the surface of the reservoir and point 2 on the surface of the storage tank will yield

$$\frac{v_1^2}{2g} + \frac{p_1}{\gamma} + z_1 + H_u = \frac{v_2^2}{2g} + \frac{p_2}{\gamma} + z_2 + H_l$$

Again, velocity heads at points 1 and 2 can be ignored, and hence

$$0 + 0 + 146.40 + H_u = 0 + 0 + 163.50 + H_l$$

The value of H_l can be calculated from the given data as follows

$$v_{200} = \frac{0.065 \times 4}{0.20^2 \times \pi}$$

$$= 2.07 \text{ m/s}$$

$$H_{l(200)} = 3 \times \frac{2.07^2}{2 \times 9.81}$$

$$= 0.66 \text{ m}$$

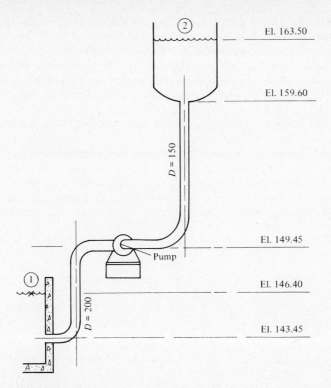

FIGURE 6.24

and

$$V_{150} = \frac{0.065 \times 4}{0.15^2 \times \pi}$$

$$= 3.68 \text{ m/s}$$

$$H_{l(150)} = 6 \times \frac{3.68^2}{2 \times 9.81}$$

$$= 4.14 \text{ m}$$

and

$$H_l = 0.66 + 4.14$$

$$= 4.80 \text{ m}$$

Hence, after substituting in the above Bernoulli equation and solving for H_u, the energy added by the pump is as follows:

$$H_u = 163.5 + 4.80 - 146.40$$

$$= 21.90 \text{ m}$$

Consequently, from Eq. (6.37),

$$Power = 0.065 \times 9.81 \times 21.90$$

$$= 14.0 \text{ kJ/s}$$

$$= 14.0 \text{ kW}$$

The pump has to put out 14 kW, but because of efficiency losses, it will be necessary to supply more power to the pump. Similar conditions exist for the motor. Consequently, the power input to the motor will be

$$Power \ input = \frac{14.0}{0.76 \times 0.72}$$

$$= 26 \text{ kW}$$

The pumping cost will therefore be as follows:

$$Cost = 26.0 \text{ kW} \times 24 \text{ h/day} \times \$0.15/\text{kWh}$$

$$= \$93.60/\text{day}$$

EXAMPLE 6.20

Determine the pumping costs for the pump in Example 6.19 if 100 L/s of oil with $s = 0.80$, is being pumped from the reservoir to the storage tank.

SOLUTION

Equation (6.53) can be applied as before and values of the velocity heads and energy losses in the system can be calculated as for Example 6.17. Then

$$H_l = 11.36$$

and

$$H_u = 163.50 + 11.36 - 146.40$$

$$= 28.46 \text{ m}$$

Consequently, taking into account the density of the oil and the pump and motor efficiencies, the power required will be

$$Power = \frac{0.100 \times 9.81 \times 0.80 \times 28.46}{0.76 \times 0.72}$$

$$= 40.8 \text{ kW}$$

and

$$Cost = 40.8 \times 24 \times 0.15$$
$$= \$146.88 \text{ per day}$$

6.6 HYDRAULIC AND ENERGY GRADIENTS

Energy Gradient

At point 3 in the pipe of Fig. 6.23 and Example 6.16 the total energy head in relation to the centerline of the pipe is made up of the sum of the pressure head and the velocity head, or 3.55 m and 3.0 m, respectively. As discussed earlier, this energy head can be represented graphically as a point a linear distance above the pipe at point 3. Similar graphical representations of the energy heads at points 2, 4, and 5 can be made.

It is generally assumed that energy losses in a flowing fluid take place uniformly along the length of the flow path, or pipe. Consequently, drawing a straight line between the total energy at point 3 and point 4, for instance, will produce a line representing the total energy head at any point between 3 and 4. Similar lines can be drawn for sections 2–3 and 4–5 of the pipe. This line is called the *energy gradient*, and the distance between it and the datum represents the energy head.

In the case of Fig. 6.23, where the datum is the centerline of the pipe, $z = 0$ and the total energy head

$$E = \frac{v^2}{2g} + \frac{p}{\gamma} \tag{6.54}$$

Hydraulic Gradient

Subtracting the velocity head from E in Eq. (6.54) will leave the pressure head p/γ. This can also be represented graphically as is shown in Fig. 6.23. If the velocity is constant along a section of the pipe or flow path, then the velocity head will also be constant. Therefore a line drawn a distance equal to the velocity head below the *energy gradient* will be parallel to it. This line is called the *hydraulic gradient*.

The vertical distance between the hydraulic gradient and the centerline of the pipe represents graphically the pressure head at any point along the pipe. As discussed earlier, at the entrance to the pipe there is a short transition zone where full pipe velocity is developing, and this is represented in Fig. 6.23 by the line *mn*.

The hydraulic and energy gradients are important tools in hydraulic engineering, and their use will be illustrated in future applicable chapters.

Slope of the Energy Gradient (s)

The energy gradient between points 3 and 4 in Fig. 6.23 has a slope s equal to

$$s = \frac{H_{l(3-4)}}{l_{(3-4)}} \qquad (6.55)$$

or, in general, the slope of the energy gradient

$$s = \frac{H_l}{l} \qquad (6.56)$$

If the flow is uniform when v is constant, the velocity head will be constant and the hydraulic gradient will be parallel to the energy gradient. In such instances the slope of the energy gradient will also be the slope of the hydraulic gradient.

6.7   SUMMARY OF THE CHAPTER IN BRITISH UNITS

Definitions

The definitions discussed in Section 6.2 apply to applications of fluid-flow principles in British Units.

Hydraulic Radius

The hydraulic radius R in a conduit is given by Eq. (6.1); that is,

$$R = \frac{A}{P}$$

where A is the cross-sectional area in square feet and P is the wetted perimeter in feet. Hence the units of R will be those of length, or feet.

EXAMPLE 6.21

Determine the hydraulic radius for a circular pipe with a diameter of 12 inches when the pipe is flowing full and when the depth of flow is 4 inches.

SOLUTION

When the pipe is flowing full,

$$A = \frac{1.0^2 \times \pi}{4}$$

$$= 0.79 \text{ ft}^2$$

$$P = 1.0 \times \pi$$

$$= 3.14 \text{ ft}$$

and

$$R = \frac{0.79 \text{ ft}^2}{3.14 \text{ ft}}$$

$$= 0.25 \text{ ft}$$

When the depth of flow is 4 inches, the area A is the area of the segment and the wetted perimeter is the length of the arc, as shown in Fig. 6.25. Their values can be calculated from geometric relationships:

$$A = 0.23 \text{ ft}^2$$

$$P = 1.23 \text{ ft}$$

and therefore

$$R = 0.19 \text{ ft}$$

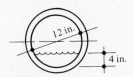

FIGURE 6.25

EXAMPLE 6.22

Determine the hydraulic radius for the channel in Fig. 6.26.

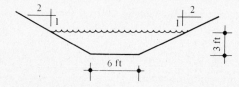

FIGURE 6.26

SOLUTION

$$A = (3.0 \times 6.0) + (3.0 \times 6.0)$$
$$= 36.0 \text{ ft}^2$$
$$P = 6.0 + ((3.0 \times 2.0)^2 + 3.0^2)^{1/2} \times 2$$
$$= 19.4 \text{ ft}$$

and

$$R = \frac{36 \text{ ft}^2}{19.4 \text{ ft}}$$
$$= 1.86 \text{ ft}$$

The Reynolds Number

The Reynolds number R_N is dimensionless, as pointed out earlier in this chapter and as can be determined from Eq. (6.3) or Eq. (6.4). Hence in the British system of units, for circular pipes flowing full,

$$R_N = \frac{d \times v}{v}$$

and, for open-channel flow,

$$R_N = \frac{R \times v}{v}$$

where d = pipe diameter in feet
$\quad v$ = velocity of flow in feet per second
$\quad R$ = hydraulic radius in feet
$\quad v$ = the kinematic viscosity in square feet per second

Therefore, the ranges of values of Reynolds number, distinguishing between laminar or turbulent flow are the same as those in the SI system of units. For pipe flow,

Laminar flow occurs when $R_N < 2000$.
Turbulent flow occurs when $R_N > 4000$.

and for open-channel flow,

Laminar flow occurs when $R_N < 500$.
Turbulent flow occurs when $R_N > 1000$.

EXAMPLE 6.23

Oil flows in a 6-feet diameter pipe, with a velocity of 0.56 ft/s. The oil has a kinematic viscosity of 3.8×10^{-5} ft²/s. Is the flow laminar or turbulent?

SOLUTION

From Eq. (6.3),

$$R_N = \frac{0.5 \times 0.56}{3.8 \times 10^{-5}}$$

$$= 7370 > 4000$$

Hence the flow is turbulent.

EXAMPLE 6.24

If the water in the channel of Example 6.22 flows with a velocity of 2.5 ft/s, is the flow laminar or turbulent? ($\nu_{water} = 1.41 \times 10^{-6}$ ft²/s.)

SOLUTION

From Example 6.22 the hydraulic radius for this channel is

$$R = 1.86 \text{ ft}$$

Hence, from Eq. (6.4),

$$R_N = \frac{1.86 \times 2.5}{1.41 \times 10^{-6}}$$

$$= 3\,300\,000 > 1000$$

and the flow is turbulent.

Capacity of Flow and Velocity

The rate of flow, or capacity, in British units is generally expressed in terms of gallons or cubic feet per unit of time, such as

GPM = gallons per minute
GPD = gallons per day
MGD = million gallons per day
cfs = cubic feet per second

When converting from gallons to cubic feet, the following factors are generally accepted in the civil engineering field:

$$1.0 \text{ cubic foot} = 7.48 \text{ U.S. gallons}$$
$$= 6.42 \text{ imperial gallons}$$

Keeping the above in mind, the capacity velocity relationship expressed in Eqs. (6.8), (6.9), and (6.10) apply in the British System of units as well.

EXAMPLE 6.25

In Fig. 6.8, if $KL = 3.0$ ft, $KN = 2.4$ ft, $l = 4.44$ ft, and it takes an average of 3 seconds for the water particle to travel from point 1 to point 2, determine the velocity and capacity of flow.

SOLUTION

$$A = 3.0 \text{ ft} \times 2.4 \text{ ft}$$
$$= 7.2 \text{ ft}^2$$
$$v = \frac{4.44 \text{ ft}}{}$$
$$= 1.48 \text{ ft/s}$$

Hence, from Eq. (6.8),

$$Q = 7.2 \text{ ft}^2 \times 1.48 \text{ ft/s}$$
$$= 10.6 \text{ cfs}$$
$$= 10.6 \times 7.48 \times 60$$
$$= 4760 \text{ U.S. GPM}$$

EXAMPLE 6.26

A pipe, with a diameter of 15 inches, has a rate of flow of 350 U.S. GPM of water. Determine whether the flow is laminar or turbulent.

SOLUTION

$$A = \frac{1.25^2 \times \pi}{4}$$
$$= 1.23 \text{ ft}^2$$

$$Q = \frac{350}{7.48 \times 60}$$

$$= 0.78 \text{ cfs}$$

Hence

$$v = \frac{0.78}{1.23}$$

$$= 0.63 \text{ ft/s}$$

and

$$R_N = \frac{1.25 \times 0.63}{1.41 \times 10^{-6}}$$

$$= 558\,500 > 4000$$

The flow is turbulent.

Energy, Work, and Power

In the British System of units the units for work and power commonly used are

Energy and work: feet × pounds or ft·lb

Power: feet × pounds per unit of time or ft·lb/s

The usually accepted conversion for the unit of horsepower, hp, in civil engineering applications is

$$1.0 \text{ hp} = 550 \text{ ft·lb/s} \qquad (6.57)$$

Hence the energy in fluid flow is expressed in foot pounds, from which follows that the energy head E which is the energy per unit weight of fluid will have units of feet per pound of fluid. Therefore, the total energy head in a system of flowing fluid as expressed in Eq. (6.36) applies here as well, or

$$E = \frac{v^2}{2g} + \frac{p}{\gamma} + z$$

but in this instance

$$\frac{v^2}{2g} = \text{kinetic energy head in feet}$$

$$\frac{p}{\gamma} = \text{pressure head in feet}$$

$$z = \text{elevation head in feet}$$

The relationship for power expressed in Eq. (6.37) will also apply here and

$$\text{Power} = Q\gamma E$$

and when the power is expressed in terms of horsepower,

$$\text{Horsepower} = \frac{Q\gamma E}{550} \qquad (6.58)$$

EXAMPLE 6.27

A horizontal pipe, 346.8 ft above sea level, has water flowing in it at a rate of 8.0 cfs, and the pressure at a point 1 is 75 psi. The pipe diameter is 18 inches. For the point 1, determine (a) the energy and the power in the flow; (b) the energy and the power if the pipe were at sea level; (c) the energy head and power if the liquid were oil with $s = 0.85$.

SOLUTION

For this pipe,

$$v = \frac{8.0 \times 4}{1.5^2 \times \pi}$$

$$= 4.53 \text{ ft/s}$$

and

$$\frac{v^2}{2g} = 1.05 \text{ ft}$$

also, at point 1,

$$\frac{p}{\gamma} = 75.0 \times 2.31$$

$$= 173.25 \text{ ft}$$

when the liquid is water.

    (a) The energy head is

$$E = 1.05 + 173.25 + 246.8$$

$$= 421.1 \text{ ft}$$

Therefore, from Eq. (6.37) and (6.58), the power and horsepower are

$$\text{Power} = 8.0 \text{ ft}^3/\text{s} \times 62.4 \text{ lb/ft}^3 \times 421.1 \text{ ft}$$

$$= 210\,200 \text{ ft} \cdot \text{lb/s}$$

or

$$Horsepower = \frac{210\,200}{550}$$

$$= 382 \text{ hp}$$

(b) If the pipe is at sea level, the elevation head $z=0$ and

$$E = 1.05 + 173.25$$

$$= 174.3 \text{ ft}$$

Hence

$$Horsepower = \frac{8.0 \times 62.4 \times 174.3}{550}$$

$$= 158 \text{ hp}$$

(c) When the fluid is oil, with $s = 0.85$,

$$\frac{p}{\gamma} = 75.0 \times 2.31 \times 0.85$$

$$= 147.26 \text{ ft}$$

and

$$E = 1.05 + 147.26 + 346.8$$

$$= 495.1 \text{ ft}$$

Hence

$$Horsepower = \frac{8.0 \times 62.4 \times 0.85 \times 495.1}{550}$$

$$= 382 \text{ hp}$$

SELECTED PROBLEMS IN SI UNITS

6.1   Compute the cross-sectional area A the wetted perimeter P, and the hydraulic radius for the conduits in Fig. P6.1.

6.2   A 50-mm diameter pipe has 8.0 L/s of oil ($v = 205 \times 10^{-6} \text{ m}^2/\text{s}$) flowing in it. Is the flow laminar or turbulent?

6.3   A 500-mm sewer is flowing half full, when the rate of flow is 150 L/s. If $v = 1.31 \times 10^{-6}$ m^2/s, determine whether the flow is laminar or turbulent.

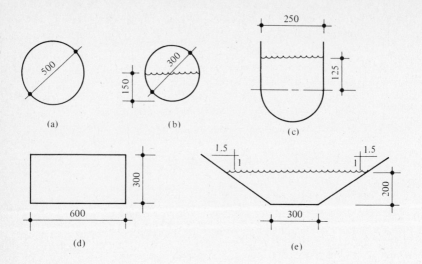

FIGURE P6.1

6.4   A rectangular channel, with a base width of 1000 mm, has water flowing in it at a velocity of 0.65 m/s when the depth of flow is 350 mm. Is the flow laminar or turbulent?

6.5   Determine the velocity of flow for the conduits in Fig. P6.1 if $Q_a = 600$ L/s, $Q_b = 60$ L/s, $Q_c = 150$ L/s, $Q_d = 225$ L/s, and $Q_e = 200$ L/s.

6.6   A pipeline must supply a municipality with 45 million liters of water per day. Determine the velocity of flow if the pipe diameters are (a) 250 mm and (b) 350 mm.

6.7   A municipality wishes to supply 750 000 liters of water per day to a certain area. The velocity in the pipeline must not exceed 2.0 m/s. Determine the diameter required.

6.8   A horizontal pipe, 50 mm in diameter, has its centerline at elevation 3.0 m above the datum. The flow in this pipe is at a rate of 6.0 L/s and the fluid has a relative density $s = 2.0$. At point A in the pipe the pressure is 65 kPa. Determine the power exerted by this flow.

6.9   A horizontal pipe, 150 mm in diameter, is 350 m above the datum and has water flowing in it. The total energy head at point A in this pipe is 45.0 m and $p_A = 35$ kPa. Determine the power exerted by this flow.

6.10   Points A and B on a pipeline are 3.0 km apart. B is 25.0 m higher than A. Water flows from A to B at a rate of 100 L/s. The energy head losses from A to B are 10.0 m per 1000 m of pipe. Determine the pressure at point B if $p_A = 750$ kPa.

6.11   Points A and B are on a pipe that changes in diameter from 150 mm at A, to 450 mm at B. A is 5.0 m lower than B, $p_A = 375$ kPa and $p_B = 300$ kPa and the energy head loss from A to B is 0.5 m. Determine the rate of flow.

6.12 A pipe changes in diameter from 100 mm at point A to 300 mm at point B. A is 25 m lower than B, $p_A = 75$ kPa and $p_B = 49$ kPa. The flow of water in this pipe is at a rate of 8500 L/min. Determine the energy head loss and the direction of flow.

6.13 A 75 mm pipe directs a jet of water vertically upward into the air. If the rate of flow is 50 L/s, determine the diameter of the jet at a point 3.0 m above the pipe end. Determine also the maximum height the water in the jet will reach (neglect air friction).

6.14 If the energy head loss from point A to point B in Fig. P6.14 is 12.0 m, determine the rate of flow.

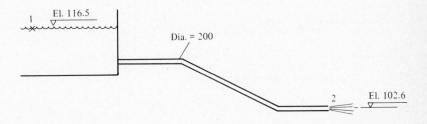

El. 116.5

Dia. = 200

1

2   El. 102.6

FIGURE P6.14

6.15 The pump in Fig. P6.15 delivers 500 L/s of water from reservoir A to reservoir B. The energy head losses in the 200-mm and the 150-mm pipes are 3.5 m and 6.8 m, respectively. The pump and motor efficiencies are 70 and 73 percent. Determine the power required at the motor terminals.

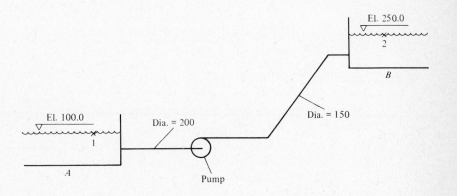

El. 250.0

2

B

El. 100.0

Dia. = 200

Dia. = 150

1

A

Pump

FIGURE P6.15

6.16 The pump in Fig. P6.16 delivers 60 kW to the flowing stream of water. The head losses from A to B are 3.5 m and the velocity in the 300-mm pipe is 4.0 m/s. Determine the maximum height h to which water can be pumped.

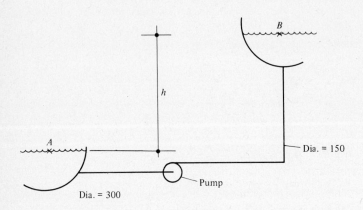

FIGURE P6.16

6.17 For the pipeline in Fig. P6.17 draw the energy and hydraulic gradients and graphically determine the pressure at point 2. The rate of flow is 250 L/s and the energy head losses from 1 to 2 and from 2 to 3 are 65.0 m and 35.0 m, respectively.

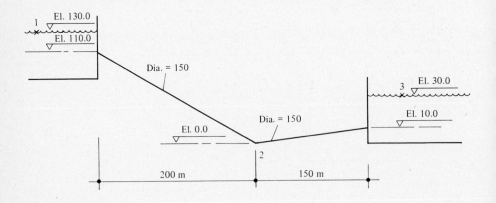

FIGURE P6.17

6.18 For the pipe system in Fig. P6.18 draw the hydraulic gradient, determine the elevation of the water surface at *B*, and graphically find the pressures at points *M* and *N*.
The rate of flow from *A* to *B* is 80 L/s and the energy head losses are 3.7 m from *A*
to *M*, 4.6 m from *M* to *N*, and 3.7 m from *N* to *B*.

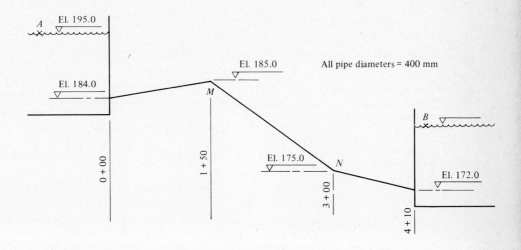

FIGURE P6.18

Fluid-Flow Formulas

7.1 ENERGY LOSSES IN FLUID FLOW

Types of Energy Losses

Energy losses were discussed in Chapter 6, and they must be taken into account when applying Bernoulli's theorem. In the consideration of the energy losses in fluid flow, they are generally broken down into two types: major losses and minor losses.

As was discussed in Chapter 6, it is convenient to express energy considerations in fluid flow in terms of energy head. Energy losses therefore are also expressed in energy head losses or head losses.

Major Head Losses

Major head losses are a form of energy considered to take place continuously along the path of flow. Their nature is discussed in more detail in Section 7.2. Since they appear to be caused by the friction of the liquid with the wall of the conduit, these losses are generally called friction head losses.

Minor Head Losses

Minor head losses are those losses generally created by increased turbulence and resistance to flow at points in the stream where the direction of flow is changed or where other obstructions take place. The most commonly encountered forms of minor head losses are as follows:

h_e, head losses due to a sudden or gradual enlargement of the cross section of flow

h_c, head losses due to a sudden or gradual contraction of the cross section of flow

h_g, head losses due to obstructions in the path of flow, such as gates, valves, metering devices, and so on

h_b, head losses occurring at bends and changes in direction of the flowpath

Total Head Loss (H_l)

From the above, it is evident that the total head loss in a stream of flowing fluid is equal to the sum of all the head losses along its path; that is,

$$H_l = h_f + h_e + h_c + h_g + h_b \qquad (7.1)$$

7.2 ANALYSIS OF MAJOR ENERGY LOSSES

Major Energy Losses or Friction Losses

As mentioned earlier, *friction loss* is not an accurate term to use when describing major head losses in fluid flow. It has, however, gained general acceptance due the similarity between the result of this type of energy loss in fluid flow and that of friction losses in other forms of energy of bodies in motion.

Major head losses are not caused by friction between the wall of the conduit and the fluid, since a fluid wets the conduit wall and fluid particles in contact with the wall have no velocity. Rather, the energy losses take place within the stream itself and can be thought of as energy losses due the changes in kinetic energy of the fluid particles as they bounce against each other. This kinetic energy lost by the particles is transferred in part into other, not readily recoverable, forms of energy such as heat and noise energy.

The phenomena of energy loss in fluid flow are extremely complex and have been the subject of large amounts of research and experimentation. Even today, no means have been developed to express this energy loss in rational mathematical or analytical forms, and much reliance must be placed on experimental investigation and deductions and interpolations based upon these investigations.

The work done in the last 200 years in this area of scientific study provides interesting reading. A detailed discussion of this work and the complex formulas and their development is beyond the scope of this text. However, to provide a thorough comprehension of the formulas and their limitations, the major ones in use today will be discussed from a scientific, as well as historical, point of view.

The Chezy Analysis and Formula

In 1775 Antoine Chezy, a French civil engineer, in attempting to determine the required size of a water supply channel decided to base his design on a comparison of the flow of other similar channels with known velocities and flows. In so doing he developed the formula that bears his name. This formula is generally expressed as follows:

$$v = C\sqrt{Rs} \tag{7.2}$$

where v = average velocity of flow
  R = hydraulic radius
  s = slope of the channel
  C = a coefficient dependent upon the various characteristics of the channel
    and their comparison with those of another similar channel

Actually, in Chezy's study, the value of C was directly proportional to the square of the comparison channel velocity and indirectly to its hydraulic radius and slope. Although it was published, the formula did not receive much attention until shortly before Chezy's death in early nineteenth century.

The Darcy-Weisbach Formula

In about the middle of the nineteenth century Julius Weisbach developed a formula for pipe flow, based to some extent on the Chezy formula. For a pipe with diameter d and length l the Weisbach formula read

$$h_f = f \times \frac{l}{d} \times \frac{v^2}{2g} \tag{7.3}$$

where h_f represents the head losses over a pipe length l and f is a friction factor. The formula appeared simple and easy, since the head losses were expressed in terms of the velocity head and the fact that f seemed dimensionless.

In 1857 the French engineer Henry Darcy, after extensive experimentation, showed that f is not dimensionless but dependent upon numerous parameters, among them the roughness of the conduit wall, the velocity of flow, the viscosity and density of the fluid, and the diameter of the pipe. Since then, much research has been done to determine values for f or to develop a rational formula to enable calculation of its value. No such formulas have been forthcoming, and determination of f is still dependent upon tables and information from experi-

TABLE 7.1   Approximate Values of f in the Darcy-Weisbach Equation for Cold Water Flowing in New Iron, Concrete, or Asbestos Cement Pipe ($k \approx 0.3$ mm)

Diameter, mm	Velocity, m/s						
	0.10	0.25	0.50	1.00	1.50	3.00	5.00
150	0.032	0.029	0.027	0.026	0.024	0.024	0.023
200	0.030	0.027	0.025	0.023	0.022	0.022	0.022
300	0.028	0.024	0.022	0.021	0.020	0.020	0.020
450	0.025	0.021	0.020	0.019	0.018	0.018	0.018
600	0.024	0.020	0.018	0.018	0.017	0.017	0.017
900	0.020	0.018	0.017	0.016	0.016	0.015	0.015

mental investigation. Values for f for water flowing in straight smooth pipe are shown in Table 7.1.

The Hagen-Poiseuille Formula for Laminar Flow

Neither Ludwig Hagen, a German scientist and engineer, nor Jean Louis Poiseuille, a French physician, developed the formula now named after them. It is, however, due to their experiments—carried out independently of each other—that the Hagen-Poiseuille relationship was proved. The formula is a mathematical or analytical derivation of resistance to flow in laminar-flow conditions and is best known in the following form.

$$h_f = \frac{32 l v v}{g d^2} \tag{7.4}$$

Numerous experiments indicate that this relationship holds true for all conditions, as long as the flow is laminar.

If both the numerator and denominator of Eq. (7.4) are multiplied by $2v$ and the result is combined with the relationship

$$R_N = \frac{dv}{v}$$

from Eq. (6.3), then Eq. (7.4) can be rearranged to read

$$h_f = \frac{64}{R_N} \frac{l}{d} \frac{v^2}{2g} \tag{7.5}$$

which is the format of the Darcy-Weisbach formula, with

$$f = \frac{64}{R_N}$$
(7.6)

for laminar flow.

Boundary Layer

Darcy and other earlier scientists already made observations that at the conduit wall the velocity is actually zero. The velocity profiles were, in effect, similar to those shown in Figs. 6.5 and 6.6.

If this is true, however, then it must follow that the friction coefficient f has to be independent of the friction between the conduit wall and the fluid. That this is so in cases of laminar flow is demonstrated by Eq. (7.6), where f is a function only of the Reynolds number.

On the other hand, in turbulent flow there is a definite relationship between the energy losses and the type of conduit or the roughness of conduit wall. Nikuradse and other scientists in the early years of the twentieth century devoted much time to experimentation on this phenomena. Using different gradations of fine sand glued to pipe walls, Nikuradse artificially demonstrated that f depends on the relative roughness of the conduit wall during turbulent flow. Even then, it was observed that the flow in a layer near the wall is laminar. This layer was named the *boundary layer*. The thickness of this layer was found to be dependent on the velocity of flow and on the Reynolds number. The work related to this investigation of the boundary layer led to definitions of two new types of flow conditions: *smooth-pipe flow* and *rough-pipe flow*.

In Fig. 7.1 is shown an enlarged section of conduit wall in contact with flowing fluid. If the value of k, the height of the protuberances created by the wall roughness exceeds the thickness of the boundary layer, the turbulence increases and f is affected by the wall roughness. Flow conditions in this instance are termed rough-pipe flows. Smooth-pipe flow occurs when k is less than the boundary layer thickness. Experiments verify that in smooth-pipe flow f is again independent of the relative roughness of conduit wall.

On the basis of many experiments, Prandtl and Von Karman arrived at empirical formulas for values for f under smooth- or rough-flow conditions.

For smooth-pipe flow, when f is independent of the wall roughness and a

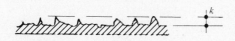

FIGURE 7.1

function only of the Reynolds number,

$$\frac{1}{\sqrt{f}} = 2 \log \frac{N_R \sqrt{f}}{2.51} \tag{7.7}$$

and for rough-pipe flow, when f is independent of Reynolds number but a function of the conduit wall roughness,

$$\frac{1}{\sqrt{f}} = 2 \log (1) \frac{d}{k} \tag{7.8}$$

where d is the pipe diameter.

Table 7.2 lists usually accepted values of k for the more common conduit materials.

TABLE 7.2   Values of k, the Effective Roughness, of Conduit Walls for Determining f in the Darcy-Weisbach Equation

Conduit Material	k feet	millimeters
Asbestos cement	0.001–0.01	0.3–3.0
Cast iron		
Plain	0.00085	0.26
Cement-lined	0.001–0.01	0.3– 3.0
Concrete pipe		
Precast	0.001–0.01	0.3– 3.0
Cast in place		
New	0.001–0.0025	0.3– 0.8
Old	0.025–0.01	0.8– 3.0
Corrugated Metal Pipe		
Plain	0.1 –0.2	30.0–60.0
Paved	0.03 –0.1	9.0–30.0

The White-Colebrook Equation

The values for f established by Prandtl and Von Karman in Eqs. (7.7) and (7.8) were combined by White and Colebrook in the equation that bears their name:

$$\frac{1}{\sqrt{f}} = -2 \log \left(\frac{k}{3.7d} + \frac{2.51}{N_R \sqrt{f}} \right) \tag{7.9}$$

Equation (7.9) is also known as the transition equation because it represents flows between smooth-flow and turbulent-flow conditions.

Obviously, when the value of k/d approaches zero—in other words, when smooth-pipe-flow conditions prevail—the first of the terms in parentheses in Eq. (7.9) becomes zero, and the equation takes the form of Eq. (7.7).

On the other hand, when the flow is very turbulent and the Reynolds number becomes large, the second of the terms in parentheses in Eq. (7.9) approaches zero and the equation takes on the form of that for rough-pipe flow, or Eq. (7.8).

The Moody Diagram

Equations (7.7), (7.8), and, particularly, (7.9) are difficult to work with and can be solved only by time-consuming trial-and-error methods. L. F. Moody developed the diagram shown in Fig. 7.2, which expresses the relationships between Reynolds number and f for various ranges of values of k/d.

A careful study of Fig. 7.2 shows that all curves are concave upward and tend to flatten out as values of N_R increase. In all curves, toward the right-hand side a point is reached where the curves are essentially horizontal and f is independent of the Reynolds number. In other words, full turbulent flow has been reached and f is dependent only on values of k/d, or the relative roughness of the pipe, as in Eq. (7.8). At the left-hand side of the diagram the curves converge into that represented by the smooth-pipe formula, Eq. (7.7).

For values of Reynolds number in the 2000 range, the flow is laminar and f is a function of N_R only and the curves plot as a straight line, as shown. According to Rouse, the point where the curves become essentially horizontal is reached when

$$N_R = 400 \frac{d}{k} \log \left(3.7 \frac{d}{k} \right) \tag{7.10}$$

and this point is shown by the broken line on the Moody diagram.

EXAMPLE 7.1

A municipality wants to supply its water treatment plant, with 10000 m³/day of fresh water through an asbestos cement pipeline with a diameter of 500 mm. The average summer water temperature is 20° C and the average winter temperature is 4°C. Determine the average pressure loss per 1000 m of pipe in summer and in winter.

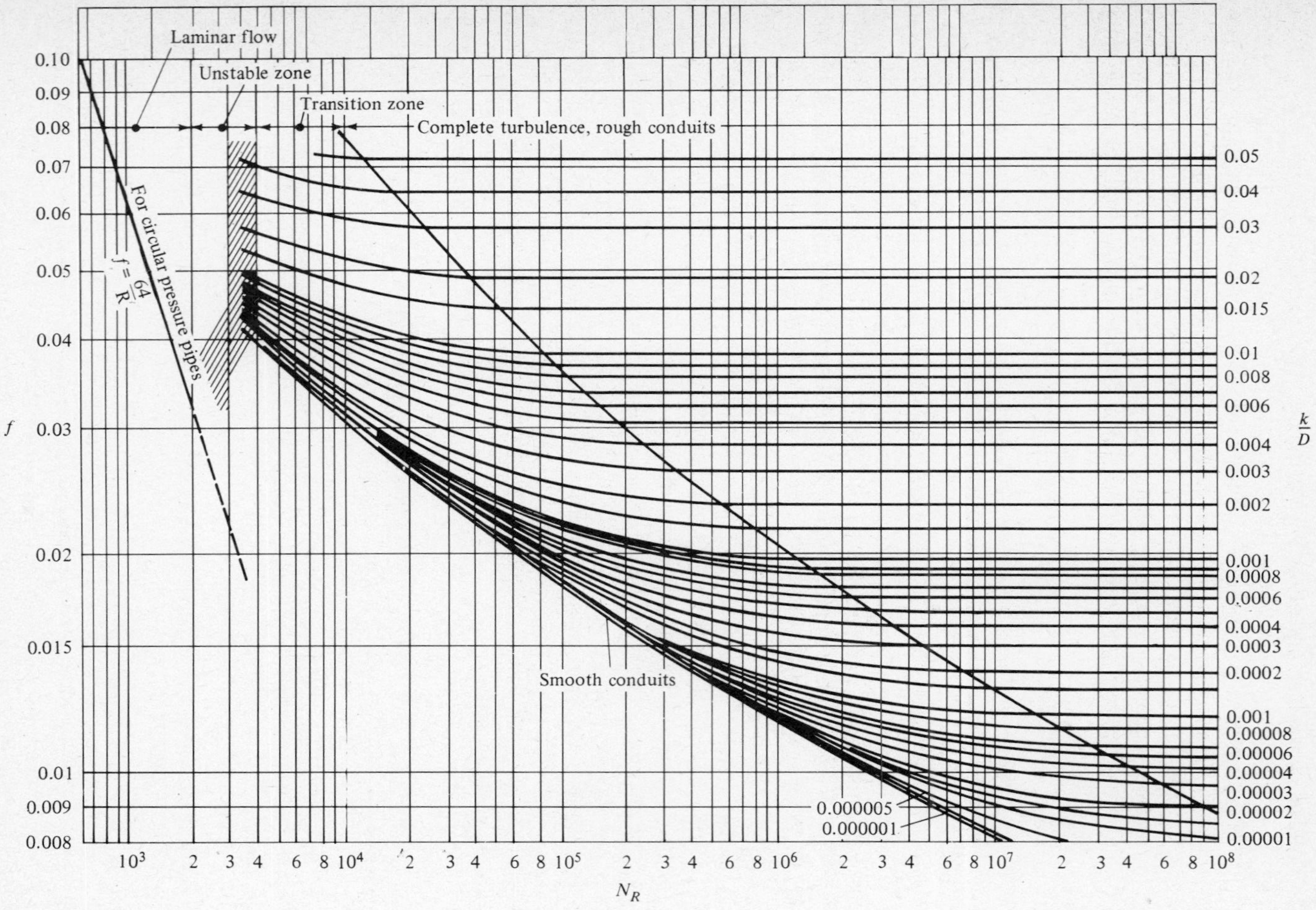
Laminar flow
Unstable zone
Transition zone
Complete turbulence, rough conduits
For circular pressure pipes
$f = \frac{64}{R}$
Smooth conduits
0.000005
0.000001
f
0.10
0.09
0.08
0.07
0.06
0.05
0.04
0.03
0.02
0.015
0.01
0.009
0.008
$\frac{k}{D}$
0.05
0.04
0.03
0.02
0.015
0.01
0.008
0.006
0.004
0.003
0.002
0.001
0.0008
0.0006
0.0004
0.0003
0.0002
0.001
0.00008
0.00006
0.00004
0.00003
0.00002
0.00001
10^3 2 3 4 6 8 10^4 2 3 4 6 8 10^5 2 3 4 6 8 10^6 2 3 4 6 8 10^7 2 3 4 6 8 10^8
N_R

FIGURE 7.2

SOLUTION

In order to use the Darcy-Weisbach formula

$$h_f = f \times \frac{l}{d} \times \frac{v^2}{2g}$$

it is necessary to determine values of v and f. Further, to find f from the Moody diagram of Fig. 7.2, values of the Reynolds number and k/d must be found.

Therefore

$$Q = \frac{10000 \text{ m}^3/\text{day}}{(24 \times 3600) \text{ s/day}}$$

$$= 0.116 \text{ m}^3/\text{s}$$

and

$$v = \frac{0.116 \text{ m}^3/\text{s}}{(0.5^2 \times \pi/4) \text{ m}^2}$$

$$= 0.59 \text{ m/s}$$

Thence, from Eq. (6.3) and Table 2.1,

$$R_{N20} = \frac{0.5 \text{ m} \times 0.59 \text{ m/s}}{1.01 \times 10^{-6} \text{ m}^2/\text{s}}$$

$$= 292\,000$$

and, similarly,

$$R_{N40} = \frac{0.5 \times 0.59}{1.57 \times 10^{-6}}$$

$$= 187\,900$$

From Table 7.2,

$$\frac{k}{d} = \frac{0.03 \text{ mm}}{500 \text{ mm}}$$

$$= 0.00006$$

Consequently, from the Moody diagram, for a temperature of 20°C, $N_R = 292\,100$ and $k/d = 0.00006$,

$$f = 0.0155$$

Thus

$$h_f = 0.0155 \times \frac{1000 \text{ m}}{0.5 \text{ m}} \times \frac{(0.59 \text{ m/s})^2}{9.81 \text{ m/s}^2 \times 2}$$

$$= 0.55 \text{ per } 1000 \text{ m}$$

or

$$\Delta p = 0.55 \times 9.81$$

$$= 5.4 \text{ kPa per } 1000 \text{ m}$$

Similarly for winter conditions, when the temperature averages 4°C and $N_R = 187\,900$, from the Moody diagram,

$$f = 0.0165$$

and

$$h_f = 0.0165 \times \frac{1000}{0.5} \times \frac{(0.59)^2}{9.81 \times 2}$$

$$= 0.59 \text{ m per } 1000 \text{ m}$$

or

$$\Delta p = 0.59 \times 9.81 \ldots$$

$$= 5.7 \text{ kPa per } 1000 \text{ m}$$

EXAMPLE 7.2

A pipeline must deliver 125 L/s of water at 15°C average temperature, through a cement-lined cast-iron pipe. The maximum allowable head loss in the pipe is not to exceed 1.0 m per 1000 m. Determine the diameter required for this pipeline.

SOLUTION

In order to be able to use the Darcy-Weisbach equation it will be necessary to find values for f, d, and v. However, d and v are inter-dependent and a solution by trial and error, assuming values for D, for instance, will be necessary.

From Table 2.1 and for a temperature of 15°C, it is found that $v = 1.15 \times 10^{-6}$ m²/s. Also, from Table 7.2, for cement-lined cast-iron pipe, $k = 0.003$ m. Solving problems of this nature, where trial-and-error solutions are required, it is often advantageous to perform the work in table form, as shown below. The procedure in the solution presented in the table is to assume a diameter and with it compute the corresponding head loss, which can then be compared with the allowable head loss given in the problem.

Often problems of this nature can be solved readily after three trials by constructing a graphical representation of the trials and the resulting head losses, as shown in Fig. 7.3.

d	v	$\dfrac{v^2}{2g}$	N_R	$\dfrac{k}{d}$	f	$\dfrac{h_f}{1000\ m}$
500	0.637	0.021	277 000	0.0006	0.019	0.80
300	1.77	0.159	461 700	0.0001	0.020	10.60
400	0.995	0.050	346 100	0.0007	0.019	2.80
475	0.705	0.025	291 000	0.00063	0.019	1.00

Plotting the first three assumed diameters and the resulting head-losses, as in Fig. 7.3, indicates that the closest diameter to satisfy the design criteria will be 475 mm. This is sustantiated by the checking calculations of the last line in the above table.

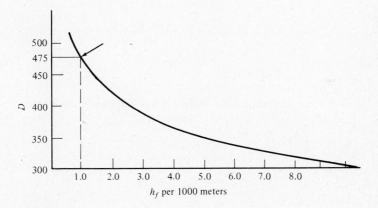

FIGURE 7.3

Other Flow Formulas

Two other flow formulas have received a certain amount of attention from hydraulic engineers. Both formulas are, in essence, attempts to define more accurately the value for the coefficient C in the Chezy formula.

The first of these was developed on the basis of extensive studies of flows in Swiss rivers by E. O. Ganguillet and W. R. Kutter. Known as the Kutter formula, it reads as follows:

$$C = \frac{a + \dfrac{b}{n} + \dfrac{m}{s}}{1 + \left(a + \dfrac{m}{s}\right) \times \dfrac{n}{\sqrt{R}}}$$ (7.11)

where a, b, and m are constants and n is dependent on the roughness of the conduit walls. Values for a, b, and m are, respectively, 23.0, 1.0, and 0.00155 when the liquid is water.

Ganguillet and Kutter were influenced in the development of their formula by very extensive flow measurements performed on the Mississippi River by Lieutenant H. L. Abbott and Captain A. A. Humphreys of the U.S. Army. It was the information supplied by these measurements that led to the inclusion of s, the slope of the energy gradient, in the formula. Subsequent investigations determined that some of the Mississippi data were in error and that the term s is not essential to the formula.

The second formula, known as the Bazin formula, represents another attempt to define C. It reads as follows:

$$C = \frac{A}{1 + \dfrac{m}{\sqrt{R}}}$$ (7.12)

TABLE 7.3   Values of m to Be Used in the Bazin Formula for Determining C in the Chezy Formula

Type of Conduit	m
Concrete	
Smooth	0.06
Rough	0.30
Wood	
Planed	0.06
Planks	0.16
Masonry	
Bricks	0.16
Rubble	0.46
Earth	
Regular surface	0.85
Ordinary surface	1.30
Uneven and overgrown	1.75

where A is a constant and m is a coefficient that is dependent on the roughness of the conduit wall. When the liquid is water, A has a value of 86.7. Values of m in the Bazin formula are shown in Table 7.3 for various pipe materials.

7.3 CIVIL ENGINEERING PIPE-FLOW FORMULAS

Special Considerations in Civil Engineering Hydraulics

The previous paragraphs indicate that a great amount of research and studies have been expended in order to attempt to formulate mathematical or empirical expressions that will accurately describe or predict the energy losses related to flowing fluids. Probably the Moody diagram and the transition-flow formula are the most accurate and most generally applicable to all flow conditions and types of fluids.

The civil engineer deals almost exclusively with water or with other fluids, such as sewage, that for all intents and purposes are water. Furthermore, most civil engineering fluid problems relate to water in its natural state, that is to say at an average temperature range of 8°C to 15°C. Consequently the variations in density and viscosity are so small as to be able to be ignored without losing significant accuracy.

Also, quantities of flow of interest to civil engineers are generally large and are based on estimates that at best can be said to be rough. Consequently, calculations of flow and energy losses to a great degree of accuracy are unwarranted and could well be considered to be misleading.

Under these conditions, therefore, it is hardly justified to work with the cumbersome formulas discussed above, and some more easily manageable ones would be helpful. Two such formulas were developed for specific use in civil engineering applications. These formulas are mentioned below and discussed in greater detail in subsequent chapters.

The Hazen-Williams Formula

The Hazen-Williams formula has been developed specifically for use with water and has been generally accepted as the formula used for pipe-flow problems by the profession in North America. It reads

$$v = 0.849C\,R^{0.63}\,s^{0.54} \tag{7.13}$$

where v = velocity of flow, m/s

R = hydraulic radius, m

s = slope of the energy gradient

C = a roughness coefficient

TABLE 7.4   Values of *C* in the Hazen-Williams Formula and of *n* in the Manning Formula

Pipe Material	C	n
Asbestos Cement	130–140	0.011–0.015
Cast Iron		
Plain	90–100	0.014–0.016
Cement-lined	100–130	0.012–0.015
Concrete		
New	120–130	0.011–0.015
Old	100–120	0.012–0.014
Plastic	140–150	0.009–0.010

Note: For more detailed values of *n*, see Chapter 10.

Values of the Hazen-Williams *C* coefficient are shown in Table 7.4.

The Manning Formula

Developed also for use with water and, specifically in North America, used primarily for open-channel-flow conditions, the Manning formula reads

$$v = \frac{1}{n} R^{2/3} s^{1/2}$$

(7.14)

where v = velocity of flow, m/s

R = hydraulic radius, m

s = slope of the energy gradient

n = a roughness coefficient

Values of Manning's *n* as related to pipe flow are also shown in Table 7.4.

7.4   MINOR LOSSES

The equations in the previous sections are used to determine the energy losses in straight pipe sections. At several points in a pipeline, circumstances dictate a change in pipe diameter, a change in the direction of the pipe, or the need for an obstruction such as a valve or a metering device.

At such points the already turbulent flow lines are further disturbed and additional turbulence occurs, as is indicated in Fig. 7.4. This additional turbulence is responsible for extra energy losses and must be accounted for in some instances. As will be seen later, in large pipelines such as those encountered in a municipal water supply system, such fittings as bends, junctions, or valves occur only at infrequent intervals. Consequently, in this type of application the energy losses due to fittings are only a very small percentage of the overall energy loss in the pipe and hence are considered negligible and are not taken into account in the design.

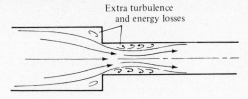

Extra turbulence
and energy losses

FIGURE 7.4

On the other hand, piping systems in a water treatment plant or industrial installations have numerous such fittings, making connections between many small sections of straight pipe. In these installations, the energy losses due to the fittings are often larger than those in straight pipe sections.

Obviously, the faster the flow of the fluid through a fitting, the more extra turbulence is created and the higher the energy loss. Therefore as in the Darcy-Weisbach formula, energy losses in fittings is related to the velocity head. In all cases, the minor losses can be expressed as follows:

$$h_m = k_m \times \frac{v^2}{2g} \tag{7.15}$$

where h_m is the energy head loss due to a fitting, k_m is a loss coefficient, and $v^2/2g$ is the velocity head.

Values for k_m have been developed for numerous types and sizes of fittings. The most common of these are represented in Table 7.5. Applications of the values shown for k_m in Table 7.5 are for the most part self-explanatory. Where v varies in a fitting, the appropriate v to be used in Eq. (7.15) is indicated clearly in drawings accompanying the k_m values. In section 5 of the table certain values of k_m are negative, indicating that in these instances flow in a branch is helped by flow in the other branch. In other words, flow in one branch produces a vacuum or negative pressure loss in the other branch.

In a future chapter, a further discussion and method of determining head losses in fittings is included. At that time a graphical method is presented to find energy losses in fittings. The following examples illustrate the use of Table 7.5 and provide as well a further application of Bernoulli's theorem.

EXAMPLE 7.3

(a) Determine the total head loss in the cement-lined pipeline ($k = 0.9$ mm) of Fig. 7.5 if the flow is 175 L/s of water at 4°C, with $v = 1.13 \times 10^{-6}$ m²/s. Determine also the percentages of major and minor losses in relation to the total head loss.

(b) Perform the exercise described in part (a), but assume that the straight lengths of pipes are 100 times those shown in Fig. 7.5.

TABLE 7.5 Values of k_m for Use with $h_m = k_m \times \dfrac{v^2}{2g}$

1. Enlargements and Contractions

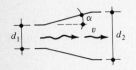

$\dfrac{d_2}{d_1}$	1.0	1.25	1.50	1.75	2.0	2.25	2.50	2.75	3.0
k_m	0.0	0.32	1.56	4.25	9.0	16.5	27.6	43.1	64.0

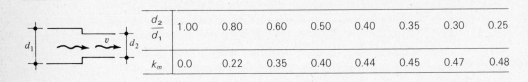

				α			
	3°	5°	7.5°	10°	15°	20°
				n		
$\dfrac{d_2}{d_1}$	0.14	0.20	0.30	0.40	0.70	0.90
1.25	0.04	0.06	0.10	0.13	0.22	0.29
1.50	0.22	0.31	0.47	0.62	1.09	1.40
1.75	0.60	0.85	1.28	1.70	2.98	3.82
2.00	1.26	1.80	2.70	3.60	6.30	8.10
2.25	2.31	3.30	4.95	6.60	11.6	14.8
2.50	3.86	5.52	8.28	11.0	19.3	24.8
2.75	6.03	8.62	12.9	17.2	30.2	38.8
3.00	8.96	12.8	19.2	25.6	44.8	57.6

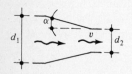

$\dfrac{d_2}{d_1}$	1.00	0.80	0.60	0.50	0.40	0.35	0.30	0.25
k_m	0.0	0.22	0.35	0.40	0.44	0.45	0.47	0.48

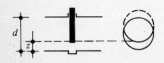

$\dfrac{\alpha}{d_1}$	2.5°	5°	7.5°	10°	15°	20°	30°
k_m	0.06	0.16	0.18	0.20	0.24	0.28	0.32

2. Obstructions

$\dfrac{z}{d}$	1.00	0.80	0.60	0.50	0.40	0.20	0.10
k_m	0.00	0.19	0.90	2.10	5.00	25.0	100

TABLE 7.5 *(Continued)*

3. *Entrances*

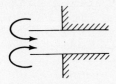

$k_m = 1.0$

$k_m = 0.50$

$k_m = 0.25$

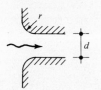

$\dfrac{r}{d}$	0.08	0.16	0.20	0.25 and more
k_m	0.15	0.06	0.03	0.0

α	85°	75°	60°	45°
k_m	0.53	0.59	0.70	0.74

4. *Bends*

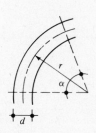

$\dfrac{r}{d}$	α			
	90°	60°	45°	22.5°
1.0	0.30	0.25	0.21	0.13
2.0	0.16	0.13	0.11	0.05
3.0	0.12	0.10	0.08	0.04
4.0	0.11	0.09	0.08	0.04
5.0	0.09	0.07	0.06	0.03
6.0 and more	0.08	0.07	0.06	0.03

TABLE 7.5   (*Continued*)

5. *Branches*

		Q_2/Q	0.00	0.20	0.50	1.00
$\alpha = 90°$		Q_1/Q	1.00	0.80	0.50	0.00
$d_1 = d_2$	k_m		− 0.98	− 0.25	0.45	1.12
			0.00	*0.30*	*0.52*	*0.56*
$d_1 = 2d_2$	k_m		− 0.75	0.55	4.00	16.0
			0.25	*0.50*	*0.95*	*1.40*
$\alpha = 45°$						
$d_1 = d_2$	k_m		− 0.91	− 0.37	0.12	0.38
			0.05	*0.18*	*0.13*	*− 0.54*
$d_1 = 2d_2$	k_m		− 1.00	0.15	3.50	11.0
			0.00	*0.06*	*− 0.60*	*− 4.10*

		Q_2/Q	0.00	0.20	0.50	1.00
$\alpha = 90°$		Q_1/Q	1.00	0.80	0.50	0.00
$d_1 = d_2$	k_m		− 0.98	0.87	0.87	1.27
			0.03	*− 0.03*	*0.02*	*0.35*
$d_1 = 2d_2$	k_m		1.30	1.90	5.00	21.0
			0.15	*− 0.11*	*0.15*	*0.28*
$\alpha = 45°$						
$d_1 = d_2$	k_m		0.90	0.68	0.42	0.47
			0.04	*0.06*	*0.01*	*0.33*
$d_1 = 2d_2$	k_m		0.90	0.50	0.86	10.0
			0.00	*− 0.04*	*0.01*	*0.33*

Note: Values of k_m in italics are Q_1/Q ratios.

SOLUTION

(a) For the 150-mm pipe the Darcy-Weisbach equation will read

$$h_f = f \times \frac{1.6}{0.15} \times \frac{v^2}{2g}$$

For the 300-mm pipe this will read

$$h_f = f \times \frac{2.8}{0.30} \times \frac{v^2}{2g}$$

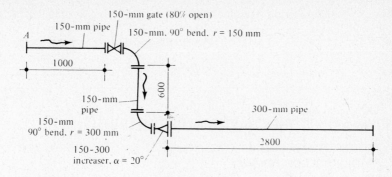

FIGURE 7.5

whereas, for the fittings,

$$h_m = k_m \times \frac{v^2}{2g}$$

Therefore the work can be arranged as in the following table.

Item	Diameter	$\dfrac{v^2}{2g}$	Other Particulars	f or k_m	h_f or h_m
Pipe	150	5.00	$R_N = 1\,315\,000$, $k/d = 0.006$	0.033	1.77
	300	0.31	$R_N = 330\,000$, $k/d = 0.003$	0.027	0.08
Bend	150	5.00	$\alpha = 90°$, $r/d = \frac{150}{150} = 1.0$	0.30	1.50
Bend	150	5.00	$\alpha = 90°$, $r/d = \frac{300}{150} = 2.0$	0.16	0.80
Enlarger	150/300	0.31	$\alpha = 20°$, $d_2/d_1 = \frac{300}{150} = 2.0$	8.10	2.51
Valve	150	5.00	$z/d = 0.8$	0.19	0.95
Total head loss					7.60 m

Hence the percentage of the head loss for the pipes is

$$h_f = 1.85 \text{ m, or } 24\%$$

and for the fittings

$$h_m = 5.75 \text{ m, or } 76\%$$

(b) If the straight sections of the pipes are 100 times those shown, the corresponding values of h_f will be increased 100-fold, whereas the h_m value due to fittings losses will remain constant. Therefore,

$$\text{Total head loss} = 1.85 \times 100 + 5.75$$

$$= 190.75 \text{ m}$$

The corresponding percentages of the head losses are as follows:

$$h_f = 97\% \text{ for pipes}$$

$$h_m = 3\% \text{ for fittings}$$

7.5   SUMMARY OF THE CHAPTER IN BRITISH UNITS

Major Friction Losses

The theoretical discussions earlier in this chapter, concerning major friction losses, are equally applicable to the British System of units. With the exception of those formulas singled out below, all the relationships, tables and formulas discussed earlier are applicable without conversion.

The Darcy-Weisbach Formula

In the British System of units, the Darcy-Weisbach formula retains the form of Eq. (7.3), or

$$h_f = f \times \frac{L}{D} \times \frac{v^2}{2g} \tag{7.3}$$

where h_f, L, D, and $v^2/2g$ are expressed in units of feet.

The Moody Diagram

The Moody diagram is not affected by the system of units used since f, R_N, and the ratio k/d are dimensionless. Therefore values of f can be found from this diagram without any conversion to the British System of units.

EXAMPLE 7.4

Determine the pressure loss through an asbestos cement pipe, with a diameter of 12 in. $Q = 4.5$ cfs, and a length of 500 ft. Assume the water temperature to be 50°F.

SOLUTION

For this pipe, the velocity of flow is

$$v = \frac{4.5 \times 4}{1.0^2 \times \pi}$$

$$= 5.73 \text{ ft/s}$$

Hence, from Eq. (7.3),

$$h_f = f \times \frac{500}{1.0} \times \frac{5.73^2}{2 \times 32.2}$$

In order to determine f, the Reynolds number and the k/d ratio must be determined to enter Moody's diagram.

The Reynolds number R_N (for $v = 1.41 \times 10^{-5}$) is as follows:

$$R_N = \frac{5.73 \times 1.0}{1.41 \times 10^{-5}}$$

$$= 406\,000$$

For asbestos cement pipe, from Table 7.2,

$$k = 0.001$$

and therefore

$$\frac{k}{D} = \frac{0.001}{1.0}$$

$$= 0.001$$

Hence, from the Moody diagram,

$$f = 0.020$$

and

$$h_f = 0.020 \times \frac{500}{1.0} \times \frac{5.73^2}{2 \times 32.2}$$

$$= 16.73 \text{ ft}$$

The Hazen-Williams Formula

The Hazen-Williams formula for the flow of water in conduits, must be converted to be used in the British System of units, as follows:

$$v = 1.318 C R^{0.63} s^{0.54} \qquad (7.16)$$

where v = velocity in feet per second
$\quad R$ = hydraulic radius in feet
$\quad s$ = slope of the energy gradient
$\quad C$ = pipe roughness coefficient, from Table 7.4

The Manning Formula

The Manning formula must also be converted for use in the British System of units. Its converted form is

$$v = \frac{1.486}{n} R^{2/3} s^{1/2} \tag{7.17}$$

where v = velocity in feet per second
   R = hydraulic radius in feet
   s = slope of the energy gradient
   n = pipe roughness coefficient, from Table 7.4

Minor Losses

The minor losses, those due bends, enlargements or contractions, are related to the velocity head as in Eq. (7.15), or

$$h_m = k_m \times \frac{v^2}{2g} \tag{7.15}$$

where h_m = energy head loss in the fitting in feet
   k_m = a loss coefficient, from Table 7.5

Since the values of k_m are related to dimensionless ratios, they do not require conversion into British units.

EXAMPLE 7.5

For the pipe system in Fig. 7.6, determine the total head loss from point A to point B if the flow is 2.5 cfs.

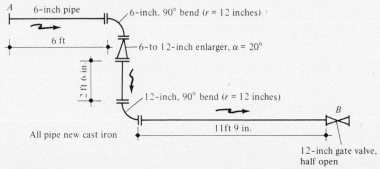

FIGURE 7.6

SOLUTION

For the 6-inch-diameter pipe, the Darcy-Weisbach formula reads

$$h_f = f \times \frac{6.0 \text{ ft}}{0.5 \text{ ft}} \times \frac{v^2}{2g}$$

and for the 12-inch-diameter pipe

$$h_f = f \times \frac{14.25 \text{ ft}}{1.0 \text{ ft}} \times \frac{v^2}{2g}$$

while for the fittings

$$h_m = k_m \times \frac{v^2}{2g}$$

Arranging the work in tabular form yields

Item	Diameter, inches	v	$\dfrac{v^2}{2g}$	Other Particulars	f or k_m	h_f or h_m, feet
Pipe	6	12.7	2.52	R_M=450 000, k/d=0.0017	0.023	0.70
Pipe	12	3.18	0.16	R_M=255 000, k/d=0.00085	0.021	0.05
Bend	6		2.52	α=90°, r/d=2.0	0.16	0.40
Enlarger	12		0.16	α= 20°, d_2/d_1 = 2.0	8.10	1.27
Bend	12		0.16	α=90°, r/d=1.0	0.30	0.05
Valve	12		0.16	z/d=0.5	2.10	0.33
Total H_i						19.1

SELECTED PROBLEMS IN SI UNITS

7.1 A 1000-mm pipeline, 3.0 km long has 600 L/s of fuel oil flowing in it. The pipeline is constructed of steel with an effective roughness coefficient $k = 0.03$ mm. If the kinematic viscosity of the oil is 3.52×10^{-6} m²/s, determine the energy head loss in this pipe.

7.2 A municipality delivers 500 L/s of water from a lake at point A to a reservoir at point B through an asbestos cement (A.C.) pipeline, the profile of which is shown in Fig. P7.2. If the winter and summer water temperatures are 10°C and 20°C respectively, determine the winter and summer pressures at points B and C.

7.3 The pump in Fig. P7.3 pumps fluid at a rate of 150 L/s. The pump and motor efficiencies are 63% and 71%, respectively. The power supplied to the motor terminals is 250 kW. Determine the elevation of the upper liquid surface, if the fluid is (a) water at 10°C, (b) heavy fuel oil at 25°C, and (c) gasoline at 25°C.

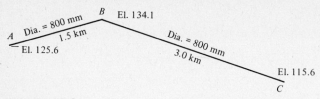

FIGURE P7.2

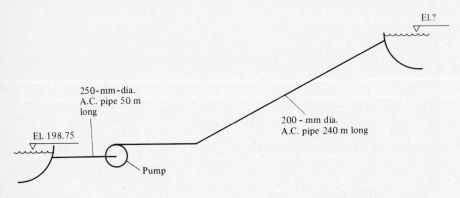

FIGURE P7.3

7.4 A concrete pipeline, 6 km long, must deliver 750 L/s of water (10°C) from point A, elevation 146.35 m, to point B, elevation 153.6 m. The pressures at points A and B are $p_a = 750$ kPa and $p_b = 500$ kPa. Determine the pipe diameter required.

7.5 For the plain cast-iron pipeline in Fig. P7.5, determine the energy loss from A to B, if the flow is (a) 100 L/s of water at 10°C and (b) 150 L/s of gasoline at 25°C.

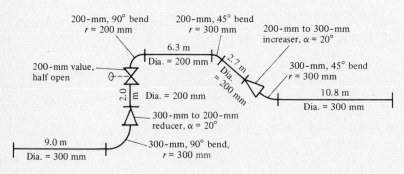

FIGURE P7.5

Orifices, Weirs, and Flow Nets

8.1 FLOW THROUGH ORIFICES

Orifices

Any opening allowing fluid to flow through a wall or plate that holds back the fluid is an *orifice*. A device inserted in a stream of flowing liquid or creating a decrease in the diameter of a pipe, can also be considered an orifice.

The theory related to flow through an orifice has practical applications in problems related to the measurement of flow, water intakes for pumping-stations or treatment plants, flows through nozzles and the flow through control gates on dams.

The orifice flow theory is basically related to the theory of the conservation of energy, or Bernoulli equation, as shown below.

Orifices and the Bernoulli Theorem

A very idealized representation of the flow through an orifice is shown in Fig. 8.1. Application of Eq. (6.41), the Bernouli equation to the flow of a particle of water at the surface at point 1 to point 2 at the discharge side of the orifice will lead to

$$\frac{v_1^2}{2g} + \frac{p_1}{\gamma} + z_1 = \frac{v_2^2}{2g} + \frac{p_2}{\gamma} + z_2$$

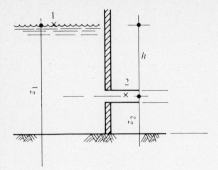

FIGURE 8.1

However, if the reservoir is sufficiently large, then, as already stated on several earlier occasions, the velocity at point 1, and hence the kinetic energy or velocity head at this point, can be considered to be negligible, or

$$\frac{v_1{}^2}{2g} = 0 \qquad (8.1)$$

Also, from earlier discussions,

$$\frac{p_1}{\gamma} = \frac{p_2}{\gamma} = 0 \qquad (8.2)$$

Therefore Eq. (6.42) reduces to

$$0 + 0 + z_1 = \frac{v_2{}^2}{2g} + 0 + z_2 \qquad (8.3)$$

From Fig. 8.1, it can be readily observed that

$$z_1 - z_2 = h \qquad (8.4)$$

and consequently

$$h = \frac{v_2{}^2}{2g} \qquad (8.5)$$

If Eq. (8.5) is expressed in general terms for any point in the discharge through the orifice, it yields

$$v = \sqrt{2gh} \qquad (8.6)$$

Equation (8.6), known as the Torricelli equation, provides a theoretical expression for the conditions of flow through an orifice. This equation, is applicable only to ideal fluids, however, and adjustment is required to provide for friction losses experienced by real fluids when flowing through an orifice. The actual velocity of a real fluid will be somewhat less than the theoretical value obtained from Eq. (8.6).

If A is the cross-sectional area of the orifice and v is the velocity of flow of a liquid through such an orifice, it would appear logical to assume that the capacity of flow Q can be expressed as

$$Q = A\sqrt{2gh} \qquad (8.7)$$

However, Eq. (8.7) does not take into account the actual behavior of streamlines when issuing from an orifice. Figure 8.2 shows such streamlines, and when considering the velocity vectors of the fluid particles, it becomes obvious that the path of the particle a, for instance, will not make an abrupt turn, but will more or less follow the path indicated in the figure. Consequently, at a point a certain distance outside the orifice, the cross-sectional area A of the jet of liquid will be smaller than the area of the orifice itself. The point of minimum area of the jet's cross section is called the *vena contracta*, and in Fig. 8.2 it takes place at the section $n-n$.

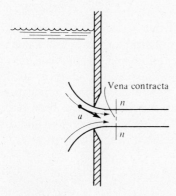

FIGURE 8.2

Orifice Coefficients

To allow for the effects described above—namely, the energy losses due to friction and the fact that the orifice area is larger than the actual area of the stream of fluid leaving the orifice—Eq. (8.7) must be modified.

This modification is accomplished by the introduction of an orifice coefficient C, which in reality is a coefficient of friction combined with a coefficient of contraction. The flow of a liquid through an orifice is therefore generally expressed as

$$Q = CA\sqrt{2gh} \tag{8.8}$$

where Q = flow, in m^3/s
A = cross-sectional area of the orifice, in m^2
h = head on the orifice, in m
C = the orifice coefficient

The actual value of the coefficient C depends on many factors, the most important of which are the head on the orifice, the shape of the orifice, and the type and configuration of the orifice wall. Typical values of C for square, sharp-edged orifices are shown in Table 8.1.

TABLE 8.1   Values of C for Square, Sharp-Edged Orifices

Size of Opening, m	h, m	C
0.30	1.00	0.603
0.30	30.00	0.598
0.03	1.00	0.607
0.03	30.00	0.598

It is evident from the figures in Table 8.1 that the values of C, though different, vary only slightly. For many practical applications a value of $C = 0.6$ for these particular orifices will provide results well within the usually accepted limits of accuracy for civil engineering applications. Only in such areas as laboratory studies and very accurate flow measurement applications would greater accuracy be required. Numerous experiments have been performed in determining the accurate value of C for many types of orifices and orifice flow conditions. Several sources of these are listed in the bibliography.

EXAMPLE 8.1

If z_2 in Fig. 8.1 is 3.0 m and the orifice is a square, sharp-edged opening with a side of 30 mm, determine the discharge Q through this orifice if (a) $z_1 = 33.0$ m and (b) $z_1 = 4.0$ m.

SOLUTION

(a) The head on the orifice is

$$h = 33.0 - 3.0$$

$$= 30.0 \text{ m}$$

Therefore, from Table 8.1,

$$C = 0.598$$

Hence, from Eq. (8.8),

$$Q = 0.598 \times 0.03^2 \times \sqrt{2 \times 9.81 \times 30.0}$$
$$= 0.013 \text{ m}^3/\text{s}$$
$$= 13.0 \text{ L/s}$$

(b) In this case the head on the orifice is

$$h = 4.0 - 3.0$$
$$= 1.0 \text{ m}$$

From Table 8.1 and Eq. (8.8), this produces

$$C = 0.603$$

and

$$Q = 0.603 \times 0.03^2 \times \sqrt{2 \times 9.81 \times 1.0}$$
$$= 0.0024 \text{ m}^3/\text{s}$$
$$= 2.4 \text{ L/s}$$

It should be noted that, in both instances, if the values of C had been considered to be

$$C = 0.60$$

the values of Q would for all practical purposes have been those calculated above.

Submerged Orifices

When an orifice has liquid standing above it at both sides, as shown in Fig. 8.3, it is said to be submerged. Flow conditions are then generally as indicated in Fig. 8.3, and application of Bernoulli's theorem between points 1 and 2 will lead to the following expression.

$$0 + 0 + z_1 = \frac{v_2^2}{2g} + \frac{p_2}{\gamma} + z_2 \tag{8.9}$$

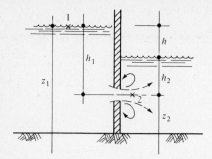

FIGURE 8.3

But

$$\frac{p_2}{\gamma} = h_2 \tag{8.10}$$

Substituting this value in Eq. (8.9) and rearranging yields

$$\frac{{v_2}^2}{2g} = z_1 - z_2 - h_2 \tag{8.11}$$

From Fig. 8.3 it is evident that

$$z_1 - z_2 - h_2 = h \tag{8.12}$$

and, substituting in Eq. (8.11),

$$v = \sqrt{2gh} \tag{8.13}$$

which is the same form as Eq. (8.6); in this case, however, h is the difference in water levels upstream and downstream of the orifice.

Considerations similar to those discussed earlier regarding friction losses and contraction of the jet lead to the general form of the equation for flow through a submerged orifice. This equation will have the same form as Eq. (8.8), or

$$Q = CA\sqrt{2gh}$$

where C is the coefficient of the orifice for submerged flow and h is the difference in head on the orifice.

For many practical applications the value of C for submerged flow through sharp-edged orifices can be taken to be

$$C = 0.63 \tag{8.14}$$

EXAMPLE 8.2

In Fig. 8.3, $z_1 = 4.0$ m, $z_2 = 1.0$ m and $h_2 = 2.0$ m. Determine the capacity of
flow when the orifice has a diameter of 30 mm.

SOLUTION

From the figure,

$$h = 4.0 - 2.0 - 1.0$$

$$= 1.0 \text{ m}$$

Assume C to be 0.63. Then, from Eq. (8.8),

$$Q = 0.63 \times \frac{0.03^2 \times \pi}{4} \times \sqrt{2 \times 9.81 \times 1.0}$$

$$= 0.002 \text{ m}^3/\text{s}$$

$$= 2.0 \text{ L/s}$$

Types of Orifices

In practice many types of orifices are in use. In civil engineering applications many structures behave like orifices, and the principles of orifice flow discussed above are applied to solving problems related to them.

Various types and shapes with their orifice coefficient are shown in Table 8.2. The values for the orifice coefficients shown in this table are average values and are applicable under average conditions. For more detailed and accurate values of C, reference must be made to literature describing orifice flow experiments.

Of specific interest to the civil engineer are orifices such as those related to flow through gates in a dam, flow through culverts, and specialized flow measurement devices. Some of these applications of orifice flow are discussed in detail in future applicable chapters.

Trajectory of a Jet

A jet discharging from an orifice takes on the shape shown in Fig. 8.4. At the vena contracta, for a very short time the flow is horizontal and then past this point the gravitational attraction pulls the jet downward.

Neglecting friction between the jet and the air, which will be minimal in any case, and considering the jet placed in a set of coordinates with the origin

TABLE 8.2   Typical Orifices

Type of Orifice		C	Comments
1. Circular, sharp-edged		0.60–0.65	
2. Square, sharp-edged		0.60–0.61	
3. Rectangular, sharp-edged		0.62–0.68	For use with certain vertical, sliding, and other similar gates in canals and spillways
4. Short tube		0.80–0.83	Values of C applicable only if tube length is sufficiently short to neglect friction loss, but allows full tube flow at exit
5. Short tube		0.60–0.65	Values of C applicable if jet is free of tube wall throughout

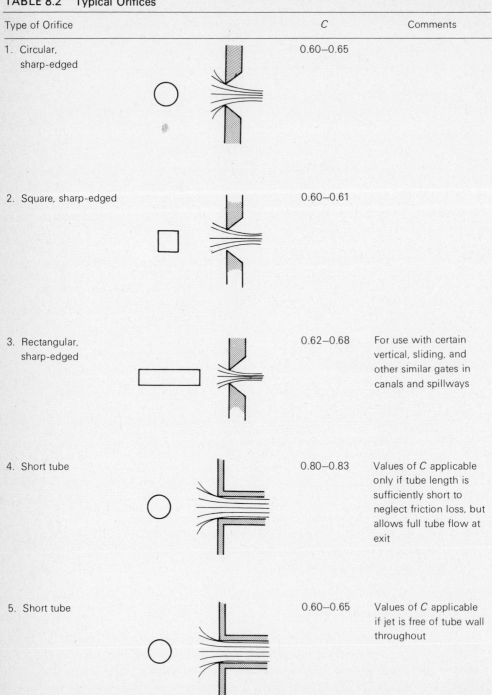

TABLE 8.2   (*Continued*)

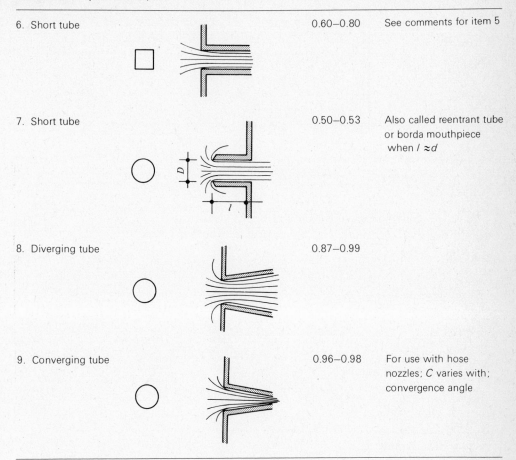

6. Short tube		0.60–0.80	See comments for item 5
7. Short tube		0.50–0.53	Also called reentrant tube or borda mouthpiece when $l \approx d$
8. Diverging tube		0.87–0.99	
9. Converging tube		0.96–0.98	For use with hose nozzles; C varies with; convergence angle

at the vena contracta and the x and y axes horizontal and vertical, respectively, the trajectory of the jet can be studied.

If a particle a in the jet, with horizontal velocity v, reaches a certain point in the jet in the time t, then this particle has traveled since leaving the vena contracta, or origin, a horizontal distance

$$x = vt \qquad (8.15)$$

During this same period of time t the gravitational force makes the particle act like a falling body, and it will have traveled a vertical distance

$$y = \frac{gt^2}{2} \qquad (8.16)$$

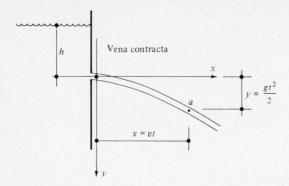

FIGURE 8.4

Solving Eqs. (8.15) and (8.16) for t yields

$$t = \frac{x}{v} \tag{8.17}$$

and

$$t = \sqrt{\frac{2y}{g}} \tag{8.18}$$

from which follows

$$\frac{x}{v} = \sqrt{\frac{2y}{g}} \tag{8.19}$$

or, rearranging,

$$x^2 = \frac{2v^2}{g} y \tag{8.20}$$

which is the equation of a parabola. The actual trajectory of a jet will take on the shape of a parabola, unless undue and unusual conditions exist or air friction is unduly large. Often the flow of a pipe discharging horizontally into the air is measured using this principle.

EXAMPLE 8.3

In Fig. 8.4, $x = 1.25$ m and $y = 0.2$ m. Determine the head of water h above the centerline of the orifice if the orifice diameter is 100 mm. (Assume that $C = 0.63$.)

SOLUTION

From Eq. (8.20),

$$1.25^2 = \frac{2 \times v^2}{9.81} \times 0.2$$

from which

$$v = 6.19 \text{ m/s}$$

and

$$Q = \frac{0.100^2 \times \pi}{4} \times 6.19$$

$$= 0.049 \text{ m}^3/\text{s}$$

$$= 49 \text{ L/s}$$

Therefore, from Eq. (8.8),

$$0.049 \text{ m}^3/\text{s} = 0.63 \times \frac{0.100^2 \times \pi}{4} \sqrt{2gh}$$

or

$$h = 5.0 \text{ m}$$

8.2 FLOW OVER WEIRS

Weirs

Any partial obstruction in the flow in an open channel will be accompanied by an increased velocity of flow at the point of obstruction. In the broadest terms such an obstruction is a *weir*. Waterfalls, rapids, and chutes in a river in this context are all special forms of weirs.

The theory related to the flow over weirs therefore has significance in practical civil engineering. Flow over spillways of dams, road sections when culverts operate under flood conditions, discharge and distribution weirs in water and wastewater treatment plants are all everyday applications of weir flow. Further, the measurement of flowing water in an open channel is very often accomplished by special forms of weirs.

Types of Weirs

Weirs are classified according to the shape of their crest and whether they partially obstruct the width of flow through the channel. Several forms and types of weirs are shown in Fig. 8.5.

Sharp-crested weirs, which are generally in the shape shown in Fig. 8.5*a*, have uses related primarily to the measurement of flow. They can be used in such a way as to maintain the entire width of the channel, as shown in Fig. 8.5*h* or they can reduce the width of the channel in a manner as shown in Fig. 8.5*e*, *f*, and *g*. In the latter instance they are called contracted or notched weirs. Applications of sharp-crested weir flow are also related to overflow spillways, where the spillway crest is a thin membrane, such as a concrete wall or stop logs. In such cases, if the flow over the weir is sufficiently large, the spillway will operate hydraulically much like a sharp-crested weir.

Various forms of broad-crested weirs are in operation wherever dams or spillways are constructed. The forms shown in Fig. 8.5*b* and *d* are very common types. The ogee weir shown in Fig. 8.5*c* is a special type of broad-crested weir designed specifically to suit the actual form of the water stream leaving the weir. It has special applications for flow over dams.

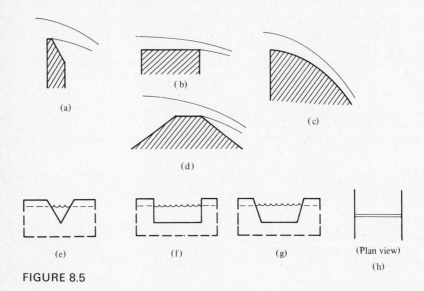

FIGURE 8.5

Weir Flow

The flow over a sharp-crested weir is shown schematically in Fig. 8.6. In relation to the crest of the weir, *H* is the total energy head. Figure 8.6 shows an idealized outline of the stream of water flowing over the weir, as well as the path

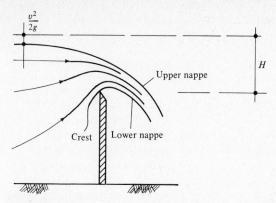

FIGURE 8.6

lines of water particles. Obviously, at the point where the water crosses the weir crest its velocity is considerably more than some distance upstream of the weir. The velocity at a point where the effect of the weir is negligible is called the *approach velocity*.

A more idealized section of the flow over the crest could be as shown in Fig. 8.7. Here it is assumed that for a very short distance across the crest the upper and lower nappe, as well as the path lines of the water particles, are essentially parallel and horizontal. Application of Bernoulli's equation between point 1 on the surface of the water upstream of the weir and point 2 in the stream over the weir will lead to

$$h_1 = \frac{v_2{}^2}{2g} + y \tag{8.21}$$

if one remembers that v_1, p_1, and p_2 are all zero in this particular application. Rearranging Eq. (8.22) yields

$$v_2{}^2 = 2g(h_1 - y) \tag{8.22}$$

or, generally, when applied to any particle in the stream

$$v = \sqrt{2g(h - y)} \tag{8.23}$$

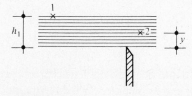

FIGURE 8.7

If the weir in question has a width l, perpendicular to the plane of the page, then the cross-sectional area of the flow over this weir can be expressed as

$$A = l \times h \tag{8.24}$$

and substituting this value of A and the value of v from Eq. (8.23) into Eq. (6.10) will yield

$$Q = l \times h \times \sqrt{2g(h - y)} \tag{8.25}$$

where Q is the total flow over the weir.

A study of Fig. 8.6 and comparison of it with Fig. 8.7 indicate that the derivation of Eq. (8.26) ignores several obvious discrepancies between the two figures. Also, the velocity is not constant throughout the cross section; in fact, experiments have shown that it varies considerably over the area of flow. As in all real flows, friction losses occur due to turbulence created by the constriction of flow and, further, the approach velocity v_1 is never zero.

It is therefore necessary to adjust Eq. (8.25) to make it conform to actual flow conditions. In practice, this is done through the application of a weir coefficient, C. The general form of the equation for flow over a weir then becomes

$$Q = ClH^{3/2} \tag{8.26}$$

where H = total upstream energy head on the weir crest
$\quad l$ = weir length
$\quad C$ = a weir coefficient

If the flow per unit width of weir is called q, Eq. (8.26) can be rewritten as

$$q = CH^{3/2} \tag{8.27}$$

In both equations it must be remembered that H is the total upstream energy head in relation to the weir crest. H must therefore include not only the difference in elevation between the crest and the upstream water level but also the velocity head due to the approach velocity. Moreover, as will be indicated later in this text, measurements of H must be made at specific locations upstream of the weir to avoid errors due to the contraction of the water surface as the weir's effect is felt by the approaching water.

Weir Coefficients

The weir coefficient C must take into account the required adjustments for the various assumptions and discrepancies included in the development of the weir flow formulas of Eqs. (8.26) and (8.27). Some of the more obvious considerations are the type and shape of the weir, the conditions of approach of the flow towards the weir, and the relationship between the head H on the weir and the overall height of the weir.

To allow for these variances several weir coefficient formulas have been developed experimentally. Many of the coefficients are developed directly from experiments or measurements on existing weirs. Values of C so derived are theoretically applicable to other weirs only if the latter represent reasonably accurately the conditions under which the coefficients were developed.

EXAMPLE 8.4

The weir in Fig. 8.6 spans a rectangular channel 2.50 m in width. The difference in elevation between the weir crest and the upstream water surface is 0.25 m, and the weir itself is 0.50 m high. Determine the flow in this channel, assuming the weir coefficient to be 1.9.

SOLUTION

In Eq. (8.27) the value of H must include the velocity head of the flow of the water toward the weir because H is the difference in energy head between the upstream point 1 and the point where the water crosses the weir crest. This velocity head cannot be determined without knowing the value of the capacity of flow.

The simplest form of solution for this problem is to assume a velocity of approach and calculating Q using this assumed v. Subsequently the assumed v can be adjusted until a value for Q is obtained, that corresponds to the assumed v.

In this example assume the velocity of approach to be zero. Then H in Eq. (8.27) will be equal to the difference in elevation between the weir crest and the upstream water surface, or

$$H = 0.25 \text{ m}$$

and, from Eq. (8.27),

$$q = 1.9 \times (0.25)^{3/2}$$
$$= 0.24 \text{ m}^3/\text{s per linear meter}$$

The area of flow upstream of the weir would therefore be

$$A = (0.25 + 0.50) \times 1.0$$
$$= 0.75 \text{ m}^2 \text{ per linear meter}$$

and hence the velocity v is

$$v = 0.24/0.75$$
$$= 0.32 \text{ m/s}$$

The approach velocity head will be

$$\frac{v^2}{2g} = \frac{(0.32)^2}{2 \times 9.81}$$

$$= 0.005 \text{ m}$$

A more accurate value of the head H on the weir is found to be

$$H = 0.25 + 0.005$$

$$= 0.255 \text{ m}$$

Substituting this value of H in Eq. (8.27) yields

$$q = 1.9 \times (0.255)^{3/2}$$

$$= 0.245 \text{ m}^3/\text{s per linear meter}$$

Further adjustment of the value of H leads to an accurate value

$$H = 0.26 \text{ m}$$

and

$$q = 0.25 \text{ m}^3/\text{s per linear meter}$$

or a capacity of flow in the channel of

$$Q = 0.25 \text{ m}^3/\text{s per linear meter} \times 2.50 \text{ m}$$

$$= 0.625 \text{ m}^3/\text{s}$$

$$= 625 \text{ L/s}$$

It should be pointed out that in this instance the adjustment for velocity head makes only a slight difference in the actual calculated values of Q. This will be the case in many situations; consequently, the velocity of approach often can be neglected in practice.

8.3   FLOW NETS

Flow Lines and Equipotential Lines

The section of a stream of flowing fluid shown in Fig. 8.8 has steady flow, and the head loss from point 1 to point 2 is h. Since this head loss is due to only to the flow of the fluid, it can be assumed that the head loss takes place at a constant rate along the distance from point 1 to point 2. Therefore, the energy gradient will be a straight line, and since the cross-sectional area is constant,

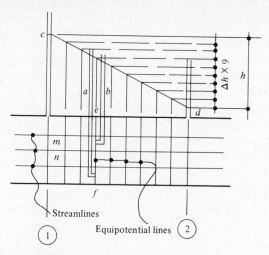

FIGURE 8.8

the velocity is also constant. The hydraulic gradient then will be parallel to the energy gradient and take on the form shown in Fig. 8.8 by the line *cd*.

The vertical lines drawn on the figure, such as the line *ef*, represent points in the stream with equal potential, or equal energy. In other words, piezometers *a* and *b* inserted in the stream along the line *ef* will show a rise of water to a height equal to the elevation of the hydraulic gradient at that point. Any such line is called an equipotential line.

The line *mn* and others dividing the stream in Fig. 8.8 into equal longitudinal sections with equal cross-sectional areas are called flow lines.

A combination of flow lines and equipotential lines is called a flow net. In the special case of the stream of Fig. 8.8, it is obvious that the flow lines and equipotential lines are at right angles to each other and form perfect squares. Mathematical proofs indicate that this orthogonality must exist with any flow net, even the general case shown in Fig. 8.9.

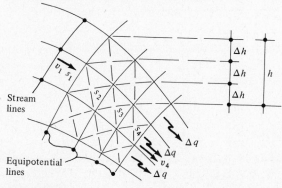

FIGURE 8.9

Analysis of Flow by Means of Flow Nets

Assume that the section of a stream of fluid in Fig. 8.9 is one unit in width, and has a capacity of flow q. With the flow net, drawn to consist of near squares, it is evident that each portion between two flow lines carries a flow Δq equal to one-third of the total flow q.

Since the flow is steady, it follows that Δq is constant. In the case of the flow net in Fig. 8.9, the streamlines converge, so the distance s_1 will be larger than s_4. Therefore, the cross-sectional area at point 4 will be less than the cross-sectional area at point 1. Correspondingly, then, the velocity at point 4 will be larger than the velocity at 1, and the following mathematical relationship for the various points along the stream can be deduced.

$$\frac{s_1}{v_1} = \frac{s_2}{v_2} = \frac{s_3}{v_3} = \cdots = \frac{s_n}{v_n} \tag{8.28}$$

Consequently, if a flow net is drawn to scale and measurements of sides of the various squares s can be made, relationships between the velocities at various points in the stream can be calculated. The general form of Eq. (8.28) can be represented as follows:

$$v_m = \frac{s_m \times v_n}{s_n} \tag{8.29}$$

EXAMPLE 8.5

A sharp-edged gate controls the flow in a rectangular channel, 2.5 m in width. The gate spans the entire width of the channel. For the flow conditions and other dimensions shown in Fig. 8.10, determine the exact shape of the pressure diagram on the gate. Determine also the total pressure on this gate. Assume the orifice coefficient $C=0.6$.

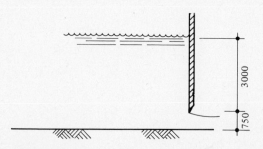

FIGURE 8.10

SOLUTION

Assuming the approach velocity head to be negligible, an approximate value of the flow per unit width through the gate can be found from Eq. (8.8), or

$$q = 0.6 \times (2.5 \times 0.75) \times \sqrt{2 \times g \times 3.0}$$
$$= 8.63 \text{ m}^3/\text{s}$$

and

$$q = \frac{8.63}{2.5}$$
$$= 3.45 \text{ m}^3/\text{s}$$

Hence the approximate approach velocity is

$$v_{app} = \frac{3.45}{3.75}$$
$$= 0.92 \text{ m/s}$$

This approach velocity produces a very small velocity head (0.04 m) and the values calculated above for Q and q can be accepted as sufficiently accurate.

The flow net relating to this problem is shown in Fig. 8.11. This flow net has been drawn by trial-and-error methods, freely sketching the

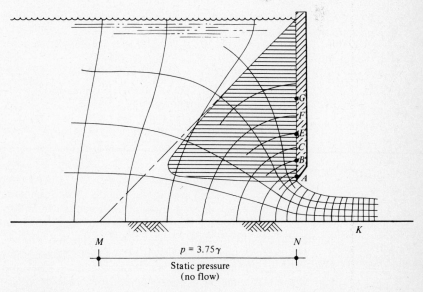

FIGURE 8.11

flow lines and equipotential lines in such a way as to ensure that they suit the requirements of these lines as set out above. In other words, as far as possible the flow lines and equipotential lines form squares. Flow-net sketching is aided by keeping in mind certain given conditions, such as in this instance:

1. One obvious flow line is immediately adjacent to and parallel to the gate.
2. One other flow line is along the bottom of the channel.
3. All equipotential lines must therefore begin and end at right angles to the gate and the channel bottom.

The shape of the upper water surface has been suggested from the flow-line construction. In the vicinity of the gate the flow-net squares have been further subdivided for additional clarity and accuracy.

At point K, downstream of the gate, where the water surface has become essentially parallel to the channel bottom and uniform flow has been reestablished, the depth of the water can be measured to be 0.42 m. Therefore the downstream velocity v_K is

$$v_K = \frac{3.45}{0.42}$$

$$= 8.21 \text{ m/s}$$

At point K the sides of the squares, s_K, measure to be 0.10 m long. If these squares were further subdivided, as in the area adjacent to the gate, values for s_K would be

$$s_K = \frac{0.1}{2}$$

$$= 0.05 \text{ m}$$

Therefore, from Eq. (8.29), values for the velocity at points A, $B, \ldots, G$ can be found as follows:

$$v_A = \frac{0.05 \times 8.21}{s_A}$$

Measuring the squares at points $A, B, \ldots, G$ produces values for $s_A, s_B, \ldots, s_G$, as shown in the table below. Since the squares in the flow net are not exact squares, due to the curvature of the flow lines and the equipotential lines, it is common practice to measure all four sides of a square and use the average value of these as the true value of s.

The dashed line OM in Fig. 8.11 represents the pressure diagram MNO that would result if there was no flow and the pressure was static. Part of the static energy is converted to kinetic energy when the water flows through the gate. This kinetic energy, in the form of velocity head, must be taken into account when determining the actual pressure on the gate. Hence

$$h_{net} = h - \frac{v^2}{2g}$$

Values for h, $v^2/2g$, and h_{net} are also shown on the table. The resulting pressure diagram is shown in Fig. 8.11.

Point	S_x, m	v_x, m/s	$\dfrac{v_x^2}{2g}$, m	h, m	h_{net}, m	Pressure, kPa
A	0.11	3.73	0.71	2.95	2.24	22
B	0.13	3.15	0.51	2.85	2.34	23
C	0.18	2.28	0.26	2.70	2.44	24
D	0.23	1.78	0.16	2.48	2.32	23
E	0.29	1.41	0.10	2.24	2.14	21
F	0.39	1.05	0.06	1.92	1.86	18
G	0.53	0.77	0.03	1.42	1.39	14

Applications of Flow-Net Analysis

Most problems related to two-dimensional flow can be solved and analyzed by means of flow nets. Often, however, more simple and more accurate solutions can be found by other means, many of which are discussed in this text.

Even though they are time-consuming and require a considerable amount of experience and patience, flow nets often function as aids in visualizing details of flow phenomena, such as the conditions of flow illustrated in Fig. 8.11 and other situations related to flow around obstacles or bends.

Probably the most widely used application of flow nets relates to solving problems of flow through soils, such as flows into wells and seepage under dams and under and through earth embankments. This type of flow-net use is discussed in detail in a future chapter.

8.4 SUMMARY OF THE CHAPTER IN BRITISH UNITS

Flow Through Orifices

The flow through orifices is governed by Eq. (8.8), or

$$Q = CA\sqrt{2gh}$$

In the British System of units the symbols in the above equation have the following meaning:

Q = rate of flow, in cubic feet per second
A = cross-sectional area of the orifice, in square feet
h = head on the orifice, in feet
C = the orifice coefficient

When the orifice is discharging freely into the air, the head on the orifice, h, is the vertical distance from the water surface to the centerline of the orifice, as shown in Fig. 8.1. When the orifice is submerged, h is the vertical distance between the upstream and the downstream water surfaces, as shown in Fig. 8.3. Values of the orifice coefficient C are shown in Table 8.2.

EXAMPLE 8.6

A sharp-edged circular orifice with a diameter of 1 inch is subjected to a head of 3.5 ft. Determine the rate of flow through this orifice.

SOLUTION

From Table 8.2, $C = 0.63$. The cross-sectional area of the orifice is

$$A = (1/12)^2 \times \pi/4$$

$$= 0.0055 \text{ ft}^2$$

Hence, from Eq. (8.8),

$$Q = 0.63 \times 0.0055 \times \sqrt{2 \times 32.2 \times 3.5}$$

$$= 0.05 \text{ cfs}$$

EXAMPLE 8.7

In Fig. 8.3, $z_1 = 12$ ft, $z_2 = 3$ ft, and $h = 6.0$ ft. The orifice is sharp edged, 2.0 ft wide and 6 in. high. Determine the rate of flow through this orifice.

SOLUTION

From Table 8.2, $C = 0.65$. Hence, from Eq. (8.8),

$$Q = 0.65 \times 2.0 \times 0.5 \times \sqrt{2 \times 32.2 \times 3.0}$$

$$= 9.0 \text{ cfs}$$

Flow over Weirs

The formula developed for flow over weirs, earlier in this chapter for the SI system of units is Eq. (8.27), or

$$Q = ClH^{3/2}$$

In the British System of units weir flow is also governed by this formula, and then the symbols have the following meanings.

Q = rate of flow, in cubic feet per second
l = length of the weir, in feet
H = head on the weir, in feet
C = the weir coefficient

As pointed out earlier, the head H on the weir must take into account the approach velocity if this velocity is significant. In many instances of weir flow in civil engineering applications the approach velocity is often too slow to produce a high value of the velocity and can be ignored.

The Weir Coefficient

The weir coefficient in Eq. (8.27) must be adjusted for use with the formula in the British System of units. The generally accepted conversion is as follows:

$$C_{Brit} = 1.81 C_{SI} \qquad (8.30)$$

or

$$C_{SI} = 0.552 C_{Brit} \qquad (8.31)$$

SELECTED PROBLEMS IN SI UNITS

8.1 Determine the rate of flow and velocity through a rectangular, sharp-crested orifice, 450 mm wide and 150 mm high, if the orifice is discharging freely into the air and the head on the orifice is (a) 1.5 m, (b) 6.7 m, and (c) 12.8 m.

8.2 Determine the head on a 50-mm-diameter Borda tube, discharging freely into the air, if Q is (a) 3.5 L/s, (b) 1.5 L/s, and (c) 12.5 L/s.

8.3 A 150-mm pipe discharges water horizontally in the air, as shown in Fig. P8.3. Determine the rate of flow in this pipe if the dimensions of the ordinates of the jet are measured by a tool as shown and (a) $x = 420$ mm, $y = 130$ mm; (b) $x = 315$ mm, $y = 430$ mm.

8.4 The orifice in Fig. P8.4 is sharp-edged and has a diameter of 75 mm. Determine the rate of flow through this orifice if (a) $z_1 = 3.4$ m, $z_2 = 1.5$ m; (b) $z_1 = 10.8$ m, $z_2 = 0.5$ m.

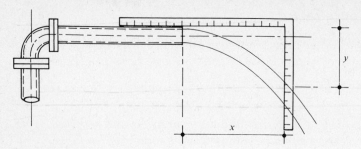

FIGURE P8.3

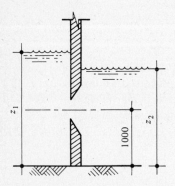

FIGURE P8.4

8.5 A noncontracting, sharp-crested weir spans a rectangular channel that is 1.75 m wide. The weir crest is 1.5 m above the channel bottom, and the rate of flow is 950 L/s. Determine the depth of flow upstream of the weir if the weir coefficient $C = 1.85$.

8.6 The weir in Fig. P8.6 spans a rectangular channel, 1.5 m wide. The rate of flow over the weir is 1500 L/s. By means of a flow net, determine the exact shape of the pressure diagram and the total pressure on this weir ($C = 1.85$).

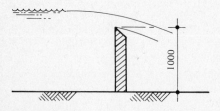

FIGURE P8.6

Civil Engineering Applications of Fluid Flow: Analysis of Pipe Flow and Pipe Networks

9.1 DEFINITION OF PIPE FLOW

In earlier chapters the terms *pressure flow* and *gravity flow* were discussed. Pressure flow was described as a flow condition in which the fluid moves through a closed conduit as a result of a source of energy, generally external to the conduit proper, such as the energy supplied by a pump or an external pressure head.

On the other hand, gravity flow is the condition in which flow takes place due to the energy within the conduit and the flowing fluid—namely, the force of gravity. Although gravity flow can take place in any type of conduit, pressure flow can exist only in a closed conduit, flowing full. Since most practical forms of closed conduits are pipes, it has become common practice to refer to pressure flow as pipe flow, and this practice will be followed in this text.

9.2 PIPE-FLOW FORMULAS

In a previous chapter, the historical and technical development of several fluid flow formulas were discussed. Two of these formulas, developed for use with the flow of water, are the Hazen-Williams and the Manning formulas. Even though it was pointed out in the previous discussions, it should be emphasized again that these formulas are not applicable universally to all fluid-flow problems. The Hazen-Williams and the Manning formulas, in fact, are applicable only to incompressible fluids with a relative density of 1.0. In other words, they should be used only for the flow of water or other liquids with similar properties. However, it is precisely because of their usefulness in water flow that they have become universally accepted for use by civil engineers.

Common practice in North America is to use the Hazen–Williams formula exclusively for applications to pipe flow and the Manning formula for gravity flow. Both formulas, however, are equally applicable to either flow condition. In some European practices it is common to find the Manning formula used for pressure-flow as well as for gravity-flow problems.

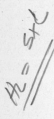

The Hazen-Williams Formula

The Hazen–Williams formula is generally expressed in the following forms:

$$v = 0.849 C R^{0.63} s^{0.54} \tag{9.1}$$

$$Q = 0.849 C A R^{0.63} s^{0.54} \tag{9.2}$$

where v = velocity of flow, in m/s
  C = Hazen–Williams pipe coefficient
  A = cross-sectional area of the flow stream, in m^2
  R = hydraulic radius of the flow stream, in m
  s = slope of the energy gradient or, if the flow is uniform, the hydraulic gradient

Equation (9.2), when solved for s, yields

$$s = \left(\frac{Q}{0.849 C A R^{0.63}} \right)^{1/0.54}$$

or

$$s = K Q^{1.852} \tag{9.3}$$

where

$$K = \left(\frac{1}{0.849 C A R^{0.63}} \right)^{1.852} \tag{9.4}$$

The Hazen-Williams Pipe Coefficient, (C)

Values for the Hazen–Williams pipe coefficient C have been determined from laboratory and field investigations for a number of pipe materials, both

TABLE 9.1 Values of C in the Hazen-Williams Formula and of n in the Manning Formula

Pipe Material	C	n
Asbestos Cement	130–140	0.011–0.015
Cast iron		
Plain	90–100	0.014–0.016
Cement-lined	100–130	0.012–0.015
Concrete		
New	120–130	0.011–0.015
Old	100–120	0.012–0.014
Plastic	140–150	0.009–0.010

in new condition and at various stages of aging. Since the formula has been used primarily in the waterworks industry, only a limited number of values have been published, the most commonly used of which are shown in Table 9.1.

It should be pointed out that the values for C listed in Table 9.1 are applicable only to flow conditions normally encountered in the waterworks industry and could vary with the size of conduit and the velocity of flow.

The ranges of the values shown for C might appear large in some instances. But using conservative averages of these values will, in most civil engineering applications, provide results well within the usually required accuracy of projects of this type. Further, it should also be remembered that prolonged use of any pipe, even with clean water, will result in the formation of a coating on the pipe walls, which will naturally increase the resistance to flow and consequently change the actual value of the pipe coefficient C.

The Manning Formula

The Manning formula is most commonly expressed in the forms

$$v = \frac{R^{2/3}s^{1/2}}{n} \tag{9.5}$$

and

$$Q = \frac{AR^{2/3}s^{1/2}}{n} \tag{9.6}$$

where v = velocity of flow, in m/s

n = the Manning roughness coefficient

A = cross-sectional area of the flow stream, in m^2

R = hydraulic radius of the flow stream, in m

s = slope of the energy gradient or, if the flow is uniform, the hydraulic gradient

Q = capacity of flow, in m³/s

Equation (9.6), when solved for s, yields

$$s = \left(\frac{Qn}{AR^{2/3}}\right)^2$$

or

$$s = KQ^2 \tag{9.7}$$

where

$$K = \left(\frac{n}{AR^{2/3}}\right)^2 \tag{9.8}$$

The Manning Roughness Coefficient (n)

Numerous values for the Manning roughness coefficient, n have been established experimentally. Since the Manning formula is used extensively for gravity flow, the majority of the published values are for applications in this context and apply to natural or constructed open channels. These coefficients will be discussed in the following chapters.

Values for n applicable to pipe-flow conditions are also listed in Table 9.1. Comments made earlier with regard to the values for the Hazen–Williams coefficient C apply equally here. It should be noted that the roughness coefficient n appears in the denominator of the Manning formula. Hence conservative values will be the higher of the ranges shown in the table.

9.3   APPLICATION OF THE PIPE-FLOW FORMULAS

Evidently, both the Hazen–Williams and the Manning formulas are complex, and in many instances no easy or ready solutions to flow problems can be found from them. This is especially so because both A and R are functions of d, the pipe diameter. Furthermore, R is expressed to a different exponent than A, further complicating solutions. For instance, if it were required to find the size of pipe required to deliver a specified capacity at a specified energy gradient, recourse would have to be made to cumbersome and time-consuming solutions involving work by trial and error.

In the past, solutions to both equations were somewhat simplified by graphical means or by using specially designed slide rules, each of which has its limitations in accuracy and convenience. Since the development of the pocket electronic calculator, mathematical solutions to problems requiring the use of these formulas can be accomplished rapidly and accurately. Both the graphical and mathematical solutions are discussed below.

Graphical Solutions (Nomographs)

Nomographs that aid in the solution of the Hazen–Williams and the Manning formulas have been developed and are shown in Figs. 9.1 and 9.2, respectively. Their use is illustrated in the examples below.

Although these nomographs greatly reduce the work involved, they do have their limitations. In many instances accuracy is hard to obtain—for example, when reading larger values of Q versus small pipe diameters, or vice versa. Further, most of the nomographs are constructed for only one value of C or n. Consequently, either a large number of nomographs (one for each value of C or n) must be used or additional calculations are necessary. All of these operations present many opportunities for errors and miscalculations.

Mathematical Solutions

The generalized forms of the Hazen–Williams and the Manning formulas, developed in Section 9.2, can be expressed as

$$s = KQ^m \tag{9.9}$$

or

$$Q = \left(\frac{s}{K}\right)^{1/m} \tag{9.10}$$

and

$$K = \frac{s}{Q_m} \tag{9.11}$$

where m has a value of 1.852 for the Hazen–Williams formula and 2.0 for the Manning formula.

A study of Eqs. (9.4) and (9.8) reveals that K is a function of A, R, and (depending on which formula is used), C or n. In other words, K is a function of the pipe diameter and the roughness coefficient. K is therefore dependent only on the size and type of pipe used.

Tables 9.2 and 9.3 list values for K for various pipe diameters in common use and for different values of C and n. Tables 9.2 and 9.3 have been developed on the basis of commercial pipe sizes presently available. Even though the industry will in the near future convert to the SI system of measurements, it is expected that pipes will continue to be manufactured using existing molds and

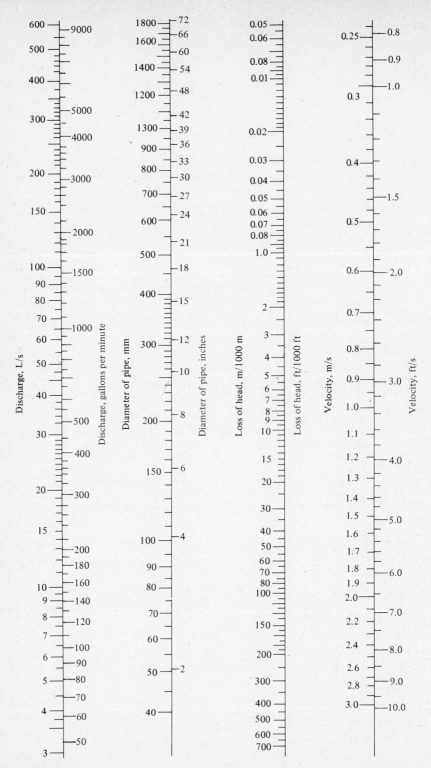

FIGURE 9.1

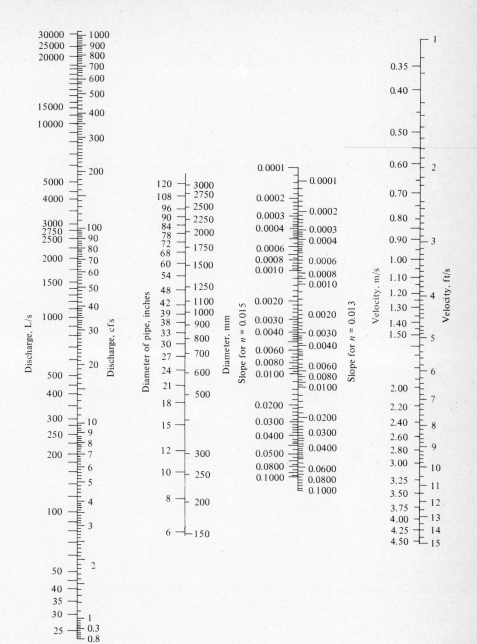

FIGURE 9.2

TABLE 9.2 K and K' Values for Solving the Hazen-Williams Formula, in SI Units (Based on Commercial Pipe Sizes)

Nominal Diameter mm	Nominal Diameter in.	Actual, mm	Area, m²	C 80	C 90	C 100	C 110	C 120	C 130	C 140	C 150	K'
100	4	102	0.0081	219.1	176.1	144.9	121.5	103.4	89.14	77.71	68.39	733 800
150	6	152	0.0182	30.40	24.45	20.11	16.86	14.35	12.37	10.79	9.492	101 800
200	8	203	0.0324	7.489	6.022	4.954	4.153	3.535	3.048	2.657	2.338	25 080
250	10	254	0.0507	2.526	2.031	1.671	1.401	1.192	1.028	0.8961	0.7887	8459
300	12	305	0.0730	0.1039	0.8358	0.6876	0.5764	0.4906	0.4230	0.3688	0.3245	3481
400	15	381	0.1141	0.3506	0.2819	0.2319	0.1944	0.1655	0.1427	0.1244	0.1095	1174
450	18	457	0.1642	0.1443	0.1160	0.0954	0.0800	0.0681	0.0587	0.0512	0.0450	483.0
550	21	533	0.2235	0.0681	0.0548	0.0450	0.0378	0.0321	0.0277	0.0242	0.0213	228.0
600	24	610	0.2029	0.0355	0.0286	0.0235	0.0197	0.0168	0.0145	0.0126	0.0111	119.0
700	27	686	0.3695	0.0200	0.0161	0.0132	0.0111	9.450	8.149	7.104	6.252	67.03
750	30	762	0.4562	0.0120	9.638	7.929	6.646	5.657	4.878	4.252	3.742	40.12
850	33	838	0.5520	7.535	6.059	4.985	4.178	3.556	3.066	2.673	2.353	25.22
900	36	914	0.6570	4.932	3.966	3.263	2.735	2.328	2.007	1.750	1.540	16.51
1000	39	991	0.7710	3.340	2.685	2.209	1.852	1.576	1.359	1.185	1.043	11.18
1050	42	1067	0.8942	2.328	1.872	1.540	1.291	1.099	0.9474	0.8259	0.7268	7.792
1200	48	1219	1.1679	1.215	0.9768	0.8037	0.6737	0.5743	0.4944	0.4310	0.3793	4.066
1400	54	1372	1.4782	0.6846	0.5504	0.4529	0.3796	0.3231	0.2786	0.2419	0.2137	2.291
1500	60	1524	1.8249	0.4098	0.3295	0.2711	0.2272	0.1934	0.1668	0.1454	0.1279	1.371

Note: Values in italics are $K \times 10^3$.

$$s = K \times Q^{1.852} = K' \left(\frac{Q}{C}\right)^{1.852} \quad \text{where } Q \text{ is in } m^3/s.$$

TABLE 9.3 K and K' Values for Solving the Manning Formula, in SI Units (Based on Commercial Pipe Sizes)

Nominal mm	Nominal in.	Actual mm	Area m²	n=0.009	0.010	0.012	0.013	0.015	0.018	0.020	0.024	K'
100	4	102	0.0081	164.9	203.6	293.2	344.1	458.1	659.7	814.4	1173.0	2 038 000
150	6	152	0.0182	18.97	23.42	33.73	39.58	52.70	75.89	93.69	134.9	234 400
200	8	203	0.0324	4.090	5.050	7.272	8.534	11.36	16.36	20.20	29.09	50 540
250	10	254	0.0507	1.244	1.536	2.212	2.596	3.456	4.977	6.145	8.848	15 370
300	12	305	0.0730	0.4706	0.5809	0.8366	0.9818	1.307	1.882	2.324	3.346	5814
400	15	381	0.1141	0.1431	0.1767	0.2545	0.2987	0.3976	0.5726	0.7069	1.018	1769
450	18	457	0.1642	0.0541	0.0668	0.0962	0.1129	0.1504	0.2165	0.2673	0.3850	668.9
550	21	533	0.2235	0.0238	0.0294	0.0423	0.0496	0.0661	0.0952	0.1175	0.1692	294.0
600	24	610	0.2920	0.0117	0.0144	0.0208	0.0244	0.0324	0.0467	0.0576	0.0830	144.2
700	27	686	0.3695	*6.228*	*7.688*	0.0111	0.0130	0.0173	0.0249	0.0308	0.0443	76.94
750	30	762	0.4562	*3.550*	*4.383*	*6.312*	*7.408*	*9.862*	0.0142	0.0175	0.0252	43.87
850	33	838	0.5520	*2.136*	*2.637*	*3.797*	*4.456*	*5.932*	*8.542*	0.0106	0.0152	26.39
900	36	914	0.6570	*1.343*	*1.658*	*2.387*	*2.801*	*3.730*	*5.371*	*6.631*	*9.548*	15.59
1000	39	991	0.7710	*0.8761*	*1.082*	*1.558*	*1.828*	*2.434*	*3.505*	*4.327*	*6.230*	10.83
1050	42	1067	0.8942	*0.5901*	*0.7285*	*1.049*	*1.231*	*1.639*	*2.360*	*2.914*	*4.196*	7.291
1200	48	1219	1.1679	*0.2895*	*0.3574*	*0.5147*	*0.6040*	*0.8041*	*1.158*	*1.430*	*2.059*	3.577
1400	54	1372	1.4782	*0.1545*	*0.1907*	*0.2746*	*0.3223*	*0.4291*	*0.6178*	*0.7628*	*1.098*	1.908
1500	60	1524	1.8249	*0.0881*	*0.1087*	*0.1566*	*0.1837*	*0.2446*	*0.3522*	*0.4349*	*0.6262*	1.088
1700	66	1676	2.2081	*0.0530*	*0.0654*	*0.0942*	*0.1105*	*0.1471*	*0.2119*	*0.2616*	*0.3767*	0.6545
1800	72	1829	2.6278	*0.0333*	*0.0411*	*0.0592*	*0.0695*	*0.0925*	*0.1332*	*0.1645*	*0.2368*	0.4115
2150	84	2134	3.5768	*0.0146*	*0.0181*	*0.0260*	*0.0305*	*0.0407*	*0.0586*	*0.0723*	*0.1041*	0.1808
2450	96	2438	4.6717	*0.0072*	*0.0089*	*0.0128*	*0.0150*	*0.0200*	*0.0287*	*0.0355*	*0.0511*	0.0887
2750	108	2743	5.9126	*0.0038*	*0.0047*	*0.0068*	*0.0080*	*0.0106*	*0.0153*	*0.0189*	*0.0272*	0.0473

Note: Values in italics are K × 10⁻³

$$s = K \times Q^2 = K' \left(\frac{Q}{C}\right)^{1/2} \quad \text{where } Q \text{ is in m}^3/\text{s.}$$

forms. It is expected that the present pipes will simply be renamed with a nominal size or diameter expressed in millimeters. All indications are that these nominal sizes will be as shown in Tables 9.2 and 9.3. For instance, a pipe manufactured with a diameter of 12 inches, will have an actual diameter of 305 mm and will be called a 300-mm pipe.

The values for K and K' in Tables 9.2 and 9.3 are based on these actual dimensions as opposed to theoretical or nominal diameters. Unless specifically stated otherwise in this text, any further reference to pipe diameters will be the nominal diameter, and calculations will be performed on the basis of actual sizes of the pipes.

If the pipe coefficients or roughness coefficients are left out of the calculated values for K, Eqs. (9.9) and (9.10) will read

$$s = K' \left(\frac{Q}{C} \right)^{1.85} \tag{9.12}$$

$$Q = C \left(\frac{s}{K'} \right)^{0.54} \tag{9.13}$$

for the Hazen–Williams formula, and

$$s = K'(Qn)^2 \tag{9.14}$$

$$Q = \frac{1}{n} \left(\frac{s}{K'} \right)^{1/2} \tag{9.15}$$

for the Manning formula.

Values for K' are also shown in Tables 9.2 and 9.3.

EXAMPLE 9.1

A cast-iron water pipe, 400 mm in diameter, carries water at a rate of 0.125 m³/s. Determine, by means of the Hazen-Williams formula, the slope of the hydraulic gradient of this pipe and the velocity of flow.

SOLUTION

(a) *Graphical solution:* On the nomograph of Fig. 9.1, line up the known values, $d = 381$, the actual diameter of a nominal 400-mm pipe, and $Q = 0.125$, and find

$$s = 0.0045 \text{ m/m}$$

and

$$v = 1.09 \text{ m/s}$$

(b) *Mathematical solution:* From Table 9.2, for $d=400$ and $C=100$, find $K=0.232$ and $A=0.114$. Hence

$$s=0.232 \times 0.125^{1.85}=0.005 \text{ m/m}$$

and

$$v=\frac{Q}{A}=\frac{0.125}{0.114}=1.09 \text{ m/s}$$

EXAMPLE 9.2

An asbestos cement water pipe ($C=140$) with a diameter of 300 mm has a slope of the hydraulic gradient of 0.0025 m/m. Determine, using the Hazen-Williams formula, the capacity of the pipe and the velocity of flow.

SOLUTION

(a) *Graphical solution:* From the nomograph in Fig. 9.1, with $d=305$ mm, the actual diameter of a 300-mm pipe, and $s=0.0025$, find $Q=0.048$ m³/s and $v=0.66$ m/s.

However, it must be remembered that the nomograph, as indicated, was constructed for $C=100$, whereas the pipe in question has a $C=140$. Consequently the values of Q and v need to be corrected.

A study of Eqs. (9.1) and (9.2) indicates that Q and v are directly related to C. Hence the actual values of Q and v for this pipe will be as follows:

$$Q=0.048 \times \frac{140}{100}=0.067 \text{ m}^3/\text{s}$$

$$v=0.66 \times \frac{140}{100}=0.92 \text{ m/s}$$

(b) *Mathematical solution:* From Table 9.2, for $d=300$ and $C=140$, find $K=0.369$ and $A=0.730$. Hence

$$Q=\left(\frac{0.0025}{0.369}\right)^{0.54}=0.067 \text{ m}^3/\text{s}$$

$$v=\frac{0.067}{0.730}=0.92 \text{ m/s}$$

EXAMPLE 9.3

A concrete pipe ($C=120$) with a diameter of 200 mm carries water at the rate of 75 L/s. Determine the slope of the hydraulic gradient.

SOLUTION

(a) *Graphical solution:* From the nomograph in Fig. 9.1, with $d=203$, the actual diameter of a 200-mm pipe and $Q=75$ L/s, find

$$s=0.040 \text{ m/m}$$

Since the nomograph provides values only for pipes with $C=100$, a correction for $C=120$ is necessary.

A study of Eqs. (9.1) and (9.2) indicates that Q, v, and s are related to the ratios of C values, to the power of 1.852. Hence, the actual value s for this pipe will be

$$s=0.040 \times \left(\frac{100}{120}\right)^{1.852} =0.029 \text{ m/m}$$

(b) *Mathematical solution:* From Table 9.2 for $d=200$ and $C=120$, find $K=3.54$; hence

$$s=3.54 \times 0.075^{1.85}=0.029 \text{ m/m}$$

EXAMPLE 9.4

Determine the diameter of a PVC pipe ($C=150$) required to carry 100 L/s of water, with a maximum hydraulic gradient of 0.002 m/m.

SOLUTION

(a) *Graphical solution:* In order to enter the nomograph of Fig. 9.1, a corrected value of s corresponding to $C=150$ must be used, or a correction factor of

$$\left(\frac{100}{150}\right)^{1.852} =0.47$$

must be applied. Hence, the corrected value of s will be

$$s=\frac{0.002}{0.47}=0.0042 \text{ m/m}$$

Entering the nomograph with $s=0.0042$ and $Q=100$ L/s, d is found to be approximately equal to 350 mm. Since a 350-mm pipe is not available commercially, the next larger pipe available would be chosen— say, with $d=400$ mm. Therefore, in order to determine the actual value of the hydraulic gradient with this pipe, one could proceed as in Example 9.3. For $C=100$, $Q=100$ L/s, and $d=381$, the actual diameter of a 400-mm pipe, from the nomograph:

$$s=0.003 \text{ m/m}$$

and corrected for $C=150$, this will be

$$s=0.003 \times 0.47 = 0.0014 \text{ m/m}$$

(b) *Mathematical solution:* From Eq. (9.11) find

$$K=\frac{0.002}{0.10^{1.852}}=0.142$$

In Table 9.2, for $C=150$, this value of K is found between $d=300$ and $d=400$. Again, the next available larger pipe would be chosen, or $d=400$ mm. For such a pipe, $K=0.1095$ and

$$s=0.1095 \times 0.10^{1.852}=0.0015 \text{ m/m}$$

EXAMPLE 9.5

A concrete pipe, 400 mm in diameter, carries water at a rate of 125 L/s. Determine by means of the Manning formula the slope of the hydraulic gradient of this pipe and the velocity of flow. Assume $n=0.013$.

SOLUTION

(a) *Graphical solution:* On the nomograph of Fig. 9.2, line up the known values of $d=381$ mm, the actual diameter of a 400-mm pipe, and $Q=125$ L/s and find for $n=0.013$

$$s=0.0047 \text{ m/m}$$

and

$$v=1.09 \text{ m/s}$$

(b) *Mathematical solution:* From Table 9.3 for $d=400$ mm and $n=0.013$, find $K=0.2994$ and $A=0.114$. Hence, from Eq. (9.7)

$$s=0.299 \times 0.125^2$$
$$=0.0047 \text{ m/m}$$

and

$$v=\frac{0.125}{0.114}$$
$$=1.10 \text{ m/s}$$

EXAMPLE 9.6

Determine the diameter of the PVC pipe required to carry 100 L/s of water, with a maximum hydraulic gradient of 0.002 m/m. Assume $n=0.009$. Use the Manning formula.

SOLUTION

(a) *Graphical solution:* In order to enter the nomograph of Fig. 9.2, which is constructed only for use with pipes with $n=0.013$, it is necessary to correct the value of s to correspond with the actual value of $n=0.009$. This can be done by applying a correction factor as follows:

$$\left(\frac{0.013}{0.009}\right)^2 = 2.09$$

In other words, the required slope with which to enter the nomograph is

$$s=0.002 \times 2.09$$

$$=0.00417 \, \text{m/m}$$

Then, from the nomograph, for $s=0.00417$ m/m and $Q=100$ L/s, d is found to be approximately equal to 365 mm. This size pipe is not available, and the next larger commercial size will be chosen, or $d=400$ mm. The actual hydraulic gradient at which this pipe will deliver 100 L/s can now be calculated as in previous examples. For $n=0.013$ and $d=381$ mm, the actual diameter of a 400-mm pipe, from the nomograph,

$$s=0.0029$$

Corrected for $n=0.009$, this will yield

$$s=\frac{0.0029}{2.09}$$

$$=0.0014 \, \text{m/m}$$

(b) *Mathematical solution:* From Eq. (9.11) find

$$K=\frac{0.002}{0.10^2}$$

$$=0.20$$

From Table 9.3 for $n=0.009$, this value of K is found between $d=300$ mm and $d=400$ mm. Hence the required pipe size is 400 mm. The actual hydraulic gradient at which this pipe will deliver 100 L/s is found using Eq. (9.8) and the value for $K=0.143$ found in Table 9.3

for $d = 400$ mm and $n = 0.009$, or

$$s = 0.143 \times 0.10^2$$
$$= 0.0014 \text{ m/m}$$

9.4 HYDRAULIC GRADIENTS IN PIPE FLOW

Hydraulic Gradients and Graphical Solutions to Pipe-Flow Problems

Hydraulic-gradient calculations and constructions are an invaluable tool in assessing the performance of any system of pipes or pipe networks operating under pressure. A properly constructed profile of the pipe, the energy, and the hydraulic gradient provides quick and easy-to-read information as to the quantity and pressure of the fluid that can be supplied at any point along the pipeline.

Quantity and pressure of delivery are two extremely important criteria in the assessment and evaluation of many civil engineering hydraulics problems. In a municipal water supply system, for instance, it is important to know not only whether a certain quantity of water can be delivered to a bathroom but also whether this amount can be delivered to a bathroom on, say, the third floor of a building. Similarly, in the case of firefighting requirements, it must be known whether firefighting water can be brought to bear to the roof of a structure without the use of a special pump from the fire department. A complete analysis of such system often includes the drawing of pressure contour maps of the system for various conditions of flow.

Hydraulic Gradients in Horizontal Pipelines

According to Bernoulli's theorem of the conservation of energy, for the pipe in Fig. 9.3, the total energy head at point 1 is E_1 and equal to the total energy head at point 2, E_2, taking into account the energy losses between points 1 and 2, or

$$E_1 = E_2 + H_l \tag{9.16}$$

where H_l is the energy loss between 1 and 2.

Since in this instance the capacity of flow Q and the pipe diameter are constant, the velocity heads at points 1 and 2 will be equal to each other. Also, because the pipe in Fig. 9.3 is horizontal, the elevation heads are equal.

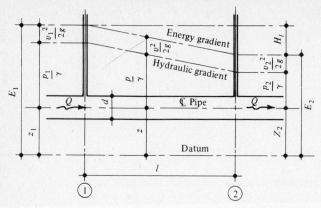

FIGURE 9.3

Therefore, Eq. (9.16) can be rewritten as

$$\frac{p_1}{\gamma} = \frac{p_2}{\gamma} + H_l \qquad (9.17)$$

or

$$\frac{p_1}{\gamma} - \frac{p_2}{\gamma} = H_l \qquad (9.18)$$

Since it is assumed that the energy loss is constant along the length of the pipe, it follows that the energy gradient is a straight line and from the foregoing it is evident that the energy gradient and the hydraulic gradient are parallel and a vertical distance, equal to the velocity head, apart. Both the energy and hydraulic gradients are shown in Fig. 9.3.

From the above and from careful observation of Fig. 9.3 it follows that H_l, the energy-head loss in the pipe, is in this instance also equal to the pressure-head loss.

In the Hazen–Williams and Manning formulas, s represents the slope of the energy gradient, or the energy-head loss per unit length of pipe. Consequently,

$$s \times l = H_l \qquad (9.19)$$

which is also the pressure-head loss in the pipe.

Therefore, in Fig. 9.3 the vertical distance between the centerline of the pipe and the hydraulic gradient at any point along the line represents the pressure head in the pipe at that point.

EXAMPLE 9.7

A concrete pipe, with $C = 120$ and a diameter of 250 mm, carries water at the rate of 100 L/s, from point 1 to point 2. The distance from point 1 to point 2 is 300 m and the pressure at point 1 is 300 kPa. Determine the pressure at point 2 and at point 3, 225 m from point 1.

SOLUTION

Using the Hazen-Williams equation to find the value of H_l yields

$$s = 1.192 \times 0.10^{1.852}$$

$$= 0.017 \text{ m/m}$$

and, from Eq. (9.19),

$$H_l = 0.017 \text{ m/m} \times 300 \text{ m}$$

$$= 5.03 \text{ m}$$

Hence from Eq. (9.18)

$$\frac{300 \text{ kPa}}{\gamma} - \frac{p_2}{\gamma} = 5.03 \text{ m}$$

or

$$\frac{p_2}{\gamma} = 25.55 \text{ m}$$

and

$$p_2 = 251 \text{ kPa}$$

The pressure at point 3 can be found in a similar manner, as follows:

$$H_l = 225 \text{ m} \times 0.017 \text{ m/m}$$

$$= 3.82 \text{ m}$$

$$\frac{p_2}{\gamma} = 26.76 \text{ m}$$

$$p_2 = 263 \text{ kPa}$$

Figure 9.4 represents the pipeline in question and the graphical solution to finding the pressure at point 3.

At point 3 the vertical distance between the pipe and the hydraulic gradient measures 26.76 m, and consequently

$$p_3 = 263 \text{ kPa}$$

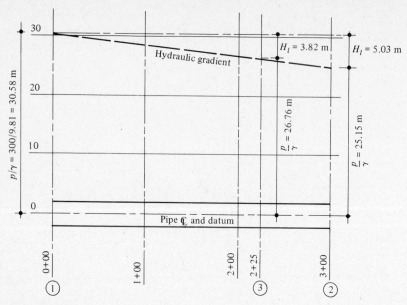

FIGURE 9.4

Hydraulic Gradients in Sloping Pipes

Many pipe installations in civil engineering are not installed horizontally. Municipal water mains, for instance, are installed essentially parallel to the ground surface, buried at a depth sufficient to protect the lines from frost or other damage. The use of hydraulic gradients and graphical solutions to such pipe problems is best illustrated by means of an example.

EXAMPLE 9.8

If the pipe in Fig. 9.5 is cast iron, with $C = 100$, has a diameter of 250 mm, and carries 150 L/s of water, determine the pressure at point 2. The pipe length from point 1 to point 2 is 450 m, $p_1 = 275$ kPa, and the elevations of points 1 and 2 are 235.6 m and 222.5 m, respectively.

SOLUTION

Since the pipe diameter and the flow capacity are constant, the velocity-head terms in Bernoulli's equation will again cancel out and the conservation of energy equation reads

$$\frac{p_1}{\gamma} + z_1 = \frac{p_2}{\gamma} + z_2 + H_l$$

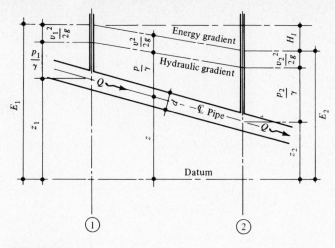

FIGURE 9.5

From Eq. (9.3) and Eq. (9.19)

$$H_l = 1.671 \times 0.15^{1.852} \times 450$$

$$= 22.4 \text{ m}$$

Hence

$$\frac{p_2}{\gamma} = \frac{275}{\gamma} + 235.6 - 222.5 - 22.4$$

$$= 18.7 \text{ m}$$

and

$$p_2 = 184 \text{ kPa}$$

EXAMPLE 9.9

An asbestos cement pipeline, 3000 m long, 400 mm in diameter, and with $C = 140$, supplies water to a municipality at the rate of 17 500 m³/day. The point at which the water is delivered to the municipality is 30 m higher than the point where the pipeline begins. The water must be delivered at a pressure of 150 kPa, determine the pressure required at the beginning of the pipeline.

SOLUTION

Consider the datum to be at the beginning of the pipeline, then $z_1 = 0$ and $z_2 = 30$ m.

In order to calculate H_l, using the Hazen-Williams equation, Q must be expressed in m^3/s, or

$$Q = \frac{17\,500 \text{ m}^3/\text{day}}{24 \text{ h/day} \times 3600 \text{ s/h}}$$

$$= 0.203 \text{ m}^3/\text{s}$$

Consequently,

$$H_l = 0.1244 \times 0.203^{1.852} \times 3000$$

$$= 19.4 \text{ m}$$

Since Q and the pipe size are constant, the velocity heads will also be constant and Bernoulli's equation reduces to

$$\frac{p_1}{\gamma} + 0 = \frac{150}{\gamma} + 30.0 + 19.4$$

or

$$\frac{p_1}{\gamma} = 64.7 \text{ m}$$

and

$$p_1 = 635 \text{ kPa}$$

Pipelines Connecting Two Reservoirs

Figure 9.6 represents a schematic arrangement of the general case of a pipeline connecting two reservoirs, where $d_1, d_2, \ldots$, and $l_1, l_2, \ldots$ represent the respective pipe diameters and lengths.

Essentially, all single pipelines—and indeed all simple pipe systems, as shown later in this chapter—can be represented in a manner similar to that shown in Fig. 9.6. If the pressures at the beginning and at the end of the pipeline are known, for instance, imaginary reservoirs can be visualized at each end of the line, with water-surface elevations equal to the respective pressure heads.

Visualizing pipeline problems in this manner often simplifies the solutions. Bernoulli's equation written between a point on the surface of one reservoir to a point on the surface of the other reservoir is greatly simplified when it is remembered that the pressure at these points is atmospheric, and hence

$$p_1 = p_2 = 0$$

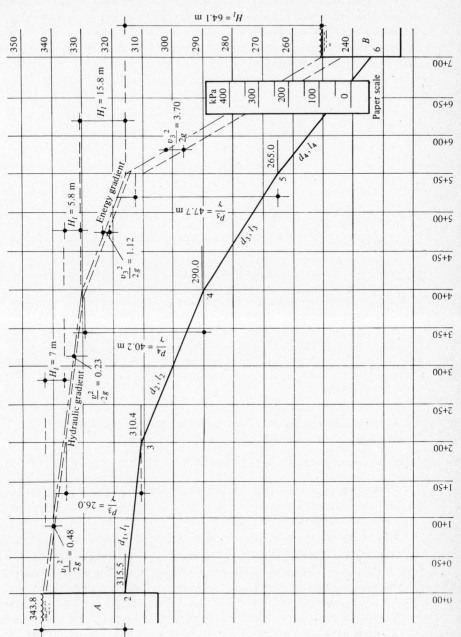

FIGURE 9.6

295

Furthermore, the velocity of the water at the surface of the reservoir is vertical and, since the cross-sectional area of the reservoir is very large in relation to the flow, it follows that the velocity of the points on the reservoir surface must be small. The velocity-head terms in Bernoulli's equation, being equal to $v^2/2g$, will become even smaller and can be considered negligible.

Consequently, the Bernoulli equation reduces to

$$0 + 0 + z_1 = 0 + 0 + z_2 + H_l$$

or

$$H_l = z_1 - z_2 \qquad (9.20)$$

EXAMPLE 9.10

Determine the elevation of the water surface of reservoir B in Fig. 9.6 if the flow of water from reservoir A to B is 150 L/s and if the elevation of reservoir A water surface is 343.8 m. All pipes have a C of 120, and pipe diameters and lengths are as follows: $d_1 = 250$ mm, $d_2 = 300$ mm, $d_3 = 200$ mm, $d_4 = 150$ mm, $l_1 = 200$ m, $l_2 = 200$ m, $l_3 = 150$ m, and $l_4 = 150$ m. The elevations of points 2, 3, 4, 5, and 6 in Fig. 9.6 are, respectively, 315.5 m, 310.4 m, 290.0 m, 265.0 m, and 235.0 m. Draw the energy and hydraulic gradients and determine the pressure at points 2, 3, 4, 5, and 6.

SOLUTION

As mentioned earlier, in problems of this nature, with long pipelines and small flows, entrance and fitting losses can be neglected. Consequently, they will not be taken into account here.

From Eq. (9.20),

$$H_l = z_1 - z_2$$

or

$$z_2 = z_1 - H_l$$

The head loss in this instance is equal to the sum of the head losses in each pipe. Individual head losses can be calculated using the Hazen-Williams equation and are found to be

$$H_{l_1} = 7.1 \text{ m}$$

$$H_{l_2} = 5.8 \text{ m}$$

$$H_{l_3} = 15.8 \text{ m}$$

$$H_{l_4} = 64.1 \text{ m}$$

$$\overline{H_l = 92.8 \text{ m}}$$

Hence

$$z_2 = 343.8 \text{ m} - 92.8 \text{ m}$$

$$= 251.0 \text{ m}$$

which is the elevation of the water surface of reservoir B.

Hydraulic and energy gradients for this pipeline are shown in Fig. 9.6. From this figure, pressure heads at points 2, 3, 4, 5, and 6 can be read directly. Consequently, pressures at these points are

$$p_2 = 278 \text{ kPa}$$

$$p_3 = 255 \text{ kPa}$$

$$p_4 = 394 \text{ kPa}$$

$$p_5 = 467 \text{ kPa}$$

$$p_6 = 121 \text{ kPa}$$

It must be noted that, in the above work, allowance must be made for velocity head. This was accomplished by assuming that the hydraulic gradient takes on the form shown at each point where a change in pipe diameter occurs. Essentially this is similar to assuming that the velocity head at each point is equal to the average of the velocity heads in the pipes before and after the point.

The pressures at these points can also be directly calculated by application of the Bernoulli equation. The resulting calculations are shown in tabular form below.

	Upstream Point			Downstream Point			
Pipe Section	z	p/γ	H_l	z	Ave. $v^2/2g$	p/γ	p
1 to 3	343.8	0	7.1	310.4	0.35	26.0	255
3 to 4	310.4	16.0	5.8	290.0	0.70	40.2	349
4 to 5	290.0	40.2	15.8	265.0	2.40	47.7	467
5 to 6	265.0	47.7	64.1	235.0	3.70	12.3	121

Pressure Contours

In many civil engineering applications it is often important to know the pressure distribution along the pipeline. In pipe networks this can best be accomplished by the construction of pressure contour lines. Later in this chapter pressure contours in pipe networks are discussed in more detail.

At this point the concept and methods of finding points of specific pressures along a pipeline will be demonstrated. In the previous example, pressures in the pipeline varied from a low of 120 kPa, to a high of approximately 470 kPa. In order to demonstrate the use of hydraulic gradients in establishing points along the pipeline with specific pressure, assume that it is necessary to locate points where the pressures are multiples of 50 kPa. In other words, points where the pressures are 150, 200, 250, ..., 450 kPa.

To perform this work mathematically would be time-consuming and tedious. With the help of the hydraulic gradient as already constructed in Fig. 9.6 the work can be greatly simplified. First it is necessary to construct on a strip of paper, a scale, indicating pressures in multiples of 50 kPa, ranging from 0 to, in this instance, 450 kPa, and to the same scale as the vertical scale of Fig. 9.6. If this specially constructed scale is placed with the zero point on the pipe, then the point where the hydraulic gradient intersects the scale will be the pressure in the pipeline at that location. In order to find the points with pressures equal to 150, 200, ..., 450 kPa, it is therefore only necessary to slide this scale along the pipeline, keeping the zero point on the pipe and maintaining the scale in a vertical position, and noting all the points where the hydraulic gradient intersects the scale at a multiple of 50 kPa. Figure 9.6 shows the point as well as the scale located at the point where the pressure is 150 kPa.

Hydraulic Gradients in Pipelines with Pumps

When a pump, located in a pipeline between two reservoirs, is used to lift the water from one reservoir to the other, the pump will add extra energy to the system and boost the hydraulic gradient accordingly. Such a case is illustrated in the following example.

EXAMPLE 9.11

The pump in Fig. 9.7 must deliver 250 L/s of water to the elevated storage tank, as shown. Determine the power required and draw the hydraulic gradient.

SOLUTION

Writing Bernoulli's equation, from a point on the surface of the supply reservoir to a point of the surface of the elevated storage tank, yields

$$0 + 0 + 125.0 + H_u = 0 + 0 + 185.0 + H_l$$

It will be remembered that at both reservoir water surfaces, the pressure is atmospheric, or $p = 0$, and as well that the velocities of the dropping and rising reservoir water levels is so small that their velocity heads can be ignored.

The value of H_l can be found by means of Eq. (9.19) and Table 9.2.

For the suction line,

$$H_{l\,\text{suction}} = 0.836 \times 0.25^{1.852} \times 25$$

$$= 1.60 \text{ m}$$

For the discharge pipe;

$$H_{l\,\text{disch.}} = 0.688 \times 0.25^{1.852} \times 250$$

$$= 13.2 \text{ m}$$

Hence the total head loss is

$$H_l = 1.60 + 13.20$$

$$= 14.8 \text{ m}$$

Substituting this value of H_l in the Bernoulli equation leads to

$$H_u = 185.0 + 14.80 - 125.0$$

$$= 74.8 \text{ m}$$

Since H_u is the energy that must be added by the pump, it follows that the power supplied by the pump is as follows:

$$\text{Power} = 74.8 \times 0.25 \times 9.81$$

$$= 183 \text{ kW}$$

The corresponding hydraulic gradient is shown in Fig. 9.7.

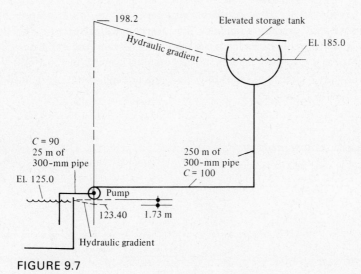

FIGURE 9.7

Multiple Reservoirs in Pipeline Systems

As mentioned earlier, solutions involving pressure pipe flow and energy losses can often be simplified by representing pressure heads at selected locations in the system, with water-surface elevations of imaginary reservoirs. In this manner, for instance, a pump in a pipe system can be replaced by a reservoir, with the water-surface elevation at the level of the hydraulic gradient supplied by the pump.

In complex distribution systems, such as municipal or industrial water supply systems, often more than one such reservoir can be readily envisaged. The system shown schematically in Fig. 9.8 could, for instance, represent an actual reservoir at point 3, an elevation of the hydraulic gradient at which water must be delivered at point 2, and at point 1 a pump capable of supplying the necessary flows and pressure heads. It must be pointed out that the pipelines shown in Fig. 9.8 are schematic representations of possible complex distribution systems.

For the system of pipes and reservoirs in Fig. 9.8, a number of problems could be set. The most common type of these problems are:

Knowing all pipes, the flow in one pipe, and the elevations of the water levels at two reservoirs, determine the elevation of the water surface of the third reservoir.

Knowing all pipes and the elevation of all reservoir water surfaces, determine the flow in each pipe.

Knowing two pipes, the length of the third pipe, and the reservoir water levels, determine the required diameter of the third pipe.

Solutions to these problems can generally be found readily by studying the hydraulic gradients between the reservoir water surfaces and the point A, the junction of the pipes from the reservoirs. If a piezometer is assumed to be at point A and the water level in it is assumed to be as shown in Fig. 9.8, these conclusions can be drawn regarding the flows in each of the pipes of the system:

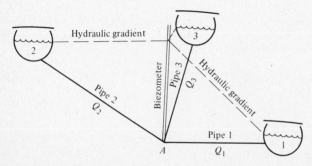

FIGURE 9.8

Water must flow from reservoir 3 to the junction.
Water must flow from the junction to reservoirs 2 and 1.

It can therefore be said that with the water level or hydraulic gradient at the junction as shown, the flow in pipe 3 must be large enough to equal the sum of the flows in pipes 1 and 2, or

$$Q_1 + Q_2 = Q_3 \tag{9.21}$$

Obviously, if the water level in the piezometer is at a different elevation, say lower than the elevation of reservoir 2, the flows from reservoirs 2 and 3 must supply the flow to reservoir 1. A new equation similar to Eq. (9.21) can then be written expressing the relationship between the flows in the pipes.

Solutions to problems related to pipe systems with multiple reservoirs are demonstrated in the following examples.

EXAMPLE 9.12

Figure 9.9*a* shows a schematic sketch of a water distribution system, where reservoir 1 represents a pump providing the pressure head at this point in the system to supply certain quantities of water to points 2 and 3. The latter points are represented by reservoirs 2 and 3, respectively. Determine the flow of water delivered to points 2 and 3.

SOLUTION

The solution of this problem involves finding values for more unknowns than the available equations. Hence a solution by trial and error is indicated.

To facilitate this solution a piezometer is assumed to be located at the junction of the pipes, the point *A*. At this point a water level, or hydraulic gradient, in the piezometer will be assumed. With this assumed water level, the hydraulic gradient and hence the flows in each of the pipes can be calculated. These flows must meet the requirements of an appropriate equation similar to Eq. (9.21). If such an equation is not satisfied, a new elevation of the hydraulic gradient at point *A* is assumed and flows are again checked with the requirements of an applicable form of Eq. (9.21). This process is repeated until agreement is reached in the flows from or to the reservoirs.

As a first assumption, the water level in the piezometer is set at *A*, with an elevation of 139.0 m, as shown in Fig. 9.9*a*. The hydraulic gradients from point *A* to the reservoirs can then be drawn as shown and the following conclusions can be made

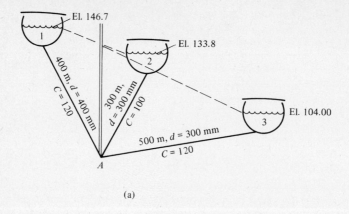

(a)

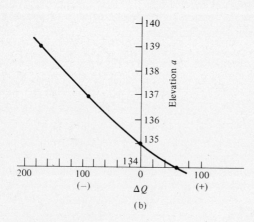

(b)

FIGURE 9.9

The slope of the hydraulic gradient is from reservoir 1 to the junction at A. Therefore the flow will be in this direction.

From the junction at A the slope of the hydraulic gradient is to reservoirs 2 and 3. Hence water will flow from A to 2 and 3.

Therefore at the junction the flow from 1 must equal the sum of the flows to 2 and 3, or Eq. (9.21) must take on the form

$$Q_1 = Q_2 + Q_3$$

or

$$Q_1 - Q_2 - Q_3 = 0$$

With the assumed elevation $a = 139.0$, the flows in each pipe can be calculated by means of the Hazen-Williams formula and Table 9.2. These calculations are shown in tabular form below.

Pipe	Reservoir Elevation	a	H_l	Length	s	K	Q, L/s
1	146.7	139.0	7.7	400	0.0193	0.166	312
2	133.8	139.0	5.2	300	0.0173	0.688	137
3	104.0	139.0	35.0	500	0.0700	0.491	349

Hence

$$Q_1 - Q_2 - Q_3 = 312 - 137 - 349$$
$$= -174 \neq 0$$

Obviously, the assumed elevation at the junction does not result in agreement for Eq. (9.21). The flow in pipe 1 is too small to supply the flows in pipes 2 and 3. Therefore the value of Q_1 must increase. This can be accomplished by providing a steeper hydraulic gradient from reservoir 1 to the junction—in other words, by assuming a new value for a that is lower than the previous assumption.

Below are shown in tabular form the calculations for two further trials, with values of a of 137.0 and 134.0.

Pipe	Reservoir Elevation	a	H_l	Length	s	K	Q, L/s
1	146.7	137.0	9.7	400	0.024	0.166	354
2	133.8	137.0	3.2	300	0.011	0.688	105
3	104.0	137.0	33.0	500	0.066	0.491	338

Hence

$$Q_1 - Q_2 - Q_3 = 354 - 105 - 338$$
$$= -89 \neq 0$$

Pipe	Reservoir Elevation	a	H_l	Length	s	K	Q L/s
1	146.7	134.0	12.7	400	0.032	0.136	409
2	133.8	134.0	0.2	300	0.0007	0.688	24
3	104.0	134.0	30.0	500	0.060	0.491	321

Hence

$$Q_1 - Q_2 - Q_3 = 409 - 24 - 321$$
$$= 64 \neq 0$$

The last elevation at the junction A is too low and further adjustments must be made. It is not always easy to predict accurate values of a. Therefore, in order to avoid further guesses a graph of three values and the resulting errors above is constructed as shown in Fig. 9.9b. From this graph a more accurate value of a can now be read to be 135.0.
Calculations for this value of a are shown below.

Pipe	Reservoir Elevation	a	H_I	Length	s	K	Q, L/s
1	146.7	135.0	11.7	400	0.029	0.166	392
2	133.8	135.0	1.2	300	0.004	0.688	62
3	104.0	135.0	31.0	500	0.062	0.491	327

$$Q_1 - Q_2 - Q_3 = 392 - 62 - 327$$
$$= 3$$

Further refinement is seldom necessary in civil engineering applications. Hence the flows in the pipes can be stated to be

$Q_1 = 392$ L/s, flowing from reservoir 1 to the junction at A

$Q_2 = 62$ L/s, flowing from the junction at A to reservoir 2

$Q_3 = 327$ L/s, flowing from the junction at A to reservoir 3

EXAMPLE 9.13

A pump must deliver 500 L/s of water to a water distribution system in such a way that the following pressures are maintained at three different points in the system: $p_1 = 350$ kPa, $p_2 = 450$ kPa, and $p_3 = 390$ kPa. The pump supplies the water to a common junction, point A, through a pipeline 600 mm in diameter and 1000 m long. Points 2, 3, and 4 are connected to the junction at A, by means of pipes as follows:

To point 2, 800 m of 300-mm pipe
To point 3, 700 m of 200-mm pipe
To point 4, 500 m of 250-mm pipe

Determine the hydraulic gradient or pressure head required at the pump and the amount of water supplied to each of the three points. Assume that for all pipes $C = 120$.

SOLUTION

This problem can be represented by four reservoirs, as shown schematically in Fig. 9.10. In this figure, reservoir 1 represents the pump with unknown elevation of the hydraulic gradient. At points 2, 3, and 4

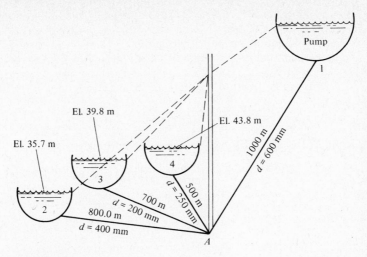

FIGURE 9.10

are shown reservoirs with water-level elevations corresponding to the required pressures at these points. These elevations are 35.7 m for reservoir 2, 43.8 m for reservoir 3, and 39.8 m for reservoir 4.

Again an imaginary piezometer is located at the junction, at A. If the hydraulic gradient at this point is such that flows in the three pipes to the reservoirs 2, 3, and 4 equal the required 500 L/s, the required hydraulic gradient at the pump can be readily found.

In order to determine the elevation of the hydraulic gradient at A, a procedure similar to that in the previous example can be used. The resulting calculations for this work are summarized as follows:

For the first trial elevation at $A = 60.0$ m,

Pipe	Res. Elev.	Elev at A	H_I	Length	s	K	Q, L/s
1	35.7	60.0	24.3	800	0.030	0.166	400
2	43.8	60.0	16.2	700	0.023	3.54	66
3	39.8	60.0	20.2	500	0.040	1.19	16
						Total Q:	482

For the second trial elevation at $A = 62.0$ m,

Pipe	Res. Elev.	Elev. at A	H_I	Length	s	K	Q, L/s
1	35.7	62.0	26.3	800	0.033	0.166	417
2	43.8	62.0	18.2	700	0.026	3.54	70
3	39.8	62.0	22.2	500	0.044	1.19	17
						Total Q:	504

This total of the three Q values of the pipes to the points 1, 2, and 3 is sufficiently close to the required 500 L/s. Therefore, the hydraulic gradient at the pump or the elevation of the reservoir P must be such that 500 L/s will flow to the junction with the hydraulic gradient at A equal to 62.0 m.

From the pump to the junction the head loss with $Q = 500$ L/s is

$$H_1 = 1000.0 \times 0.017 \times 0.5^{1.852}$$

$$= 4.71 \text{ m}$$

Hence the elevation of the hydraulic gradient at the pump must be

$$\text{El. hyd. gradient} = 4.71 + 62.0$$

$$= 66.7 \text{ m}$$

The power output by the pump to supply this flow will be

$$\text{Power} = 66.7 \times 0.5 \times 9.81 \quad = Hyd.\,grad \times Dia \times grad$$

$$= 330 \text{ kW} \qquad Power = \frac{P}{g} \times Y \times Dia$$

9.5 EQUIVALENT PIPES

Use and Definition of Equivalent Pipes

The pipeline systems in Figs. 9.11 and 9.12a cannot always be totally defined hydraulically. That is to say that for certain problems relating to the flow in these systems no direct solution is available. For instance for the system with the three pipes in series in Fig. 9.11 if pressures and elevations at A and D were given and the flow capacity were to be calculated, a solution by trial and error would be indicated. This is understandable since, even though the total head loss from A to D is known, the individual head losses in each section of pipe cannot be determined. Consequently it is impossible to determine the slope s of the energy gradient directly in order to find Q from the flow equations. Similarly, in the case of the pipeloop in Fig. 9.12a, it is impossible to predict or calculate directly the amount of the total Q that will flow in each leg of the loop.

Obviously, the problem related to the three pipes in series would be easily solved, if there was only one pipe, with constant diameter and C coefficient from

| $d = 300$ mm | $d = 250$ mm | $d = 400$ mm | |
| $C = 100$ | B $C = 110$ | C $C = 120$ | D |

A

| 100.0 m | 150.0 m | 100.0 m |

FIGURE 9.11

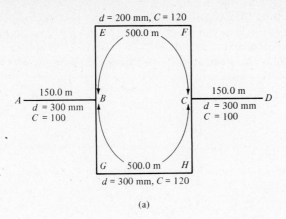

(a)

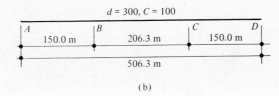

(b)

FIGURE 9.12

A to *D*. The concept of equivalent pipes is used in such instances and often simplifies the work involved in solving problems related to complex pipe systems.

By definition, any two or more pipes that have an equal total head loss for the same flow capacity are hydraulically equivalent. In other words, if pipe *X* has a head loss of *y* with a flow of *Q*, then pipe *Z* will be equivalent to pipe *X* if it also has a head loss *y* with a flow *Q*. Essentially, any two pipes that behave identically from a hydraulic point of view are equivalent.

EXAMPLE 9.14

A cast-iron pipe, $C = 100$, is 300 m long and 250 mm in diameter. Determine four other pipes that are equivalent to this pipe.

SOLUTION

Assume that the pipe in question has a $Q = 100$ L/s, then, from Eq. (9.19)

$$H_l = 1.671 \times 0.10^{1.852} \times 300$$

$$= 7.05 \text{ m}$$

According to the definition of equivalent pipes, any other pipe that with a $Q=100$ L/s has a head loss of 7.05 m is hydraulically equivalent to the pipe in question.

For instance, a cast-iron pipe, $C=100$, and a diameter of 400 mm, will with a $Q=100$ L/s have an

$$s=0.2319 \times 0.10^{1.852}$$

$$=0.00326 \text{ m/m}$$

In order for this pipe to be equivalent to the 250-mm pipe in the question it must have a total head loss of 7.05 m. Consequently, the 400-mm pipe must have a length of

$$l=\frac{7.05}{0.00326}$$

$$=2163 \text{ m}$$

It is evident that innumerable other pipes equivalent to the one in question can be found. Equivalent pipes can be created by changing the diameter, the C coefficient, and the length of the pipe. Most often it is convenient to alter only the diameter and make the length suit the requirements of equal head loss for equal capacity.

Other pipes equivalent to the one in the question are found in tabular form below.

Diam.	C	Q	K	s	l
250	140	100	0.8961	0.013	560
300	120	100	0.4906	0.007	1022
150	100	100	20.11	0.283	25

EXAMPLE 9.15

The pipeline in Fig. 9.11 is horizontal and the pipes have dimensions and C coefficients as shown. If $p_A=750$ kPa and $p_D=375$ kPa, determine the flow from A to D.

SOLUTION

An inspection of the Bernoulli equation reveals that there are more than one unknown values. The velocity heads at A and D cannot be determined without knowing the value of Q and H_l from A to D cannot be calculated. But, since $Z_A=Z_D$,

$$H_I = \frac{v_A{}^2}{2g} - \frac{v_D{}^2}{2g} + \frac{750}{\gamma} - \frac{375}{\gamma}$$

$$= \frac{v_A{}^2 - v_B{}^2}{2g} + 38.23 \text{ m}$$

If from A to D there was only one pipe, then $v_A = v_D$ and a value for H_I could be found. With the use of the Hazen-Williams flow formula, Q could then be calculated.

Therefore in order to solve this problem, replace the three pipes, AB, BC, and CD, with one imaginary pipe that is hydraulically equivalent to the three.

In order to accomplish this, assume that the flow in the pipes is 100 L/s, then the H_I in each of the three sections of the pipeline and in the total pipeline can be calculated. These calculations are shown in tabular form below.

Pipe	Diam.	Length	C	K	s	H_I ($Q=100$)
AB	300	100	100	0.6876	0.0097	0.97
BC	250	150	110	1.671	0.0234	3.52
CD	400	100	120	0.1655	0.0023	0.23
Total H_I:						4.72

The pipe equivalent to this pipeline must have a head loss of 4.72 m when the flow $Q=100$ L/s. Numerous pipes will satisfy this requirement. However, in order to reduce the work a 300-mm pipe with $C=100$ is chosen, since s for this pipe has already been calculated.

This pipe in order to be equivalent to the three pipes of the system must have a length

$$l = \frac{4.72}{0.0097}$$

$$= 487 \text{ m}$$

The system of three pipes in series has now been replaced with one pipe, 300 mm in diameter, $C=100$, and 487 m long. Hydraulically this pipe will perform in an identical manner as the original system. Application of Bernoulli's theorem now yields since $v_A = v_D$,

$$H_I = 38.23 \text{ m}$$

and

$$s = \frac{38.23}{487}$$

$$= 0.0785 \text{ m/m}$$

Consequently, using the Hazen-Williams equation,

$$Q = \left(\frac{0.0785}{0.6876}\right)^{0.54}$$

$$= 0.310 \ m^3/s, \ or \ 310 \ L/s$$

Values for each section of pipe related to the actual flow through them can now be calculated from the Hazen-Williams equation. They are

$$H_{I, AB} = 100.0 \times 0.688 \times 0.310^{1.852}$$

$$= 7.86 \ m$$

$$H_{I, BC} = 150.0 \times 1.670 \times 0.310^{1.852}$$

$$= 28.63 \ m$$

$$H_{I, CD} = 100.0 \times 0.166 \times 0.310^{1.852}$$

$$= 1.90 \ m$$

Equivalent Pipes for Pipes in Series

To replace a system of pipes in series with one equivalent pipe, it is necessary to assume a flow Q through the system, as was done in the previous example. With the assumed Q, head losses for each section of the system are determined and summed to make up the total head loss of the system for the assumed Q.

The equivalent pipe is then calculated, having the same head loss as the system with the assumed Q. Although the choice of pipe diameter is generally immaterial, the following observations should be kept in mind.

1. In order to save time and reduce possibilities of error the equivalent pipe chosen should be the same as one of the pipes of the system or as a pipe to be used later in the problem.
2. Because a certain amount of rounding off is done during the work, replacing short sections of pipe with very long equivalent pipes, or vice versa, could introduce considerable error. Subsequent balancing of the head losses in the actual system might prove difficult.

Equivalent Pipes for Pipes in Parallel

When two pipes are arranged in parallel, or form part of a loop as is the case of pipes $BEFC$ and $BGHC$ in Fig. 9.12a, the head losses in each leg of the loop take place simultaneously—as opposed to consecutively, as is the case for pipes in series. The head loss in each leg is equal to the head loss in the other leg, even though the flow in the legs could be quite different.

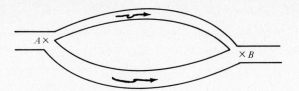

FIGURE 9.13

In order to visualize the latter, consider the loop of pipes shown schematically in Fig. 9.13. Writing Bernoulli's equation for the top part of the loop only will yield

$$\frac{v_A^2}{2g} + \frac{p_A}{\gamma} + z_A = \frac{v_B^2}{2g} + \frac{p_B}{\gamma} + z_B + H_{l(\text{top})} \tag{9.22}$$

Careful consideration of the top leg of the loop leads to the conclusion that the velocity head at A must be equal to the velocity head at B. Therefore, Eq. (9.22) reduces to

$$\frac{p_A}{\gamma} + z_A = \frac{p_B}{\gamma} + z_B + H_{l(\text{top})} \tag{9.23}$$

Similar reasoning will lead to a similar expression for the bottom leg of the loop; namely,

$$\frac{p_A}{\gamma} + z_A = \frac{p_B}{\gamma} + z_b + H_{l(\text{btm})} \tag{9.24}$$

Again, from study of Fig. 9.13, it is evident that p_A, p_B, z_A, and z_B have the same value whether they refer to the bottom or top leg of the loop. Consequently, it must follow that

$$H_{l(\text{top})} = H_{l(\text{btm})} \tag{9.25}$$

Using this principle, any pipe loop can be readily replaced with one equivalent pipe, as is demonstrated in the following example.

EXAMPLE 9.16

The pipe system in Fig. 9.12a is horizontal and the pipes have dimensions and C coefficients as shown. If the pressures at points A and D are 750 and 375 kPa, respectively, determine the total flow in each pipe.

SOLUTION

As in the previous example, application of the Bernoulli equation, and noting that the velocity heads at A and D must be equal to each other, leads to

$$\frac{p_A}{\gamma} = \frac{p_D}{\gamma} + H_{l(AD)}$$

or

$$H_{l(AD)} = 38.23 \text{ m}$$

In this instance even more so than in the previous example, finding the flow in each pipe is impossible, since there is no indication as to the actual split of the flow in each of the legs of the loop BC.

If the pipe system can be replaced with one consisting of, say, three pipes in series as in the previous example, a solution following the method used there can be found. In order to achieve this, the loop BC must be replaced with one equivalent pipe.

Assume that the head loss from C to B is 10 m; then, as previously mentioned,

$$H_{l(BC\,top)} = H_{l(BC\,btm)} = 10 \text{ m}$$

Using the assumed head loss a value for Q in each leg can now be calculated, as tabulated below.

Pipe	Dia.	Length	C	K	s	Q
BEFC	200	500	120	3.535	0.020	0.0612
BGHC	300	500	120	0.4906	0.020	0.1776
Total Q						0.2388

Therefore if $H_l = 10$ m the total Q through the loop is 238.8 L/s. Any pipe that will produce a head loss of 10 m with a $Q = 238.8$ L/s is hydraulically equivalent to the loop.

Since pipes AB and CD have a diameter of 300 mm and a C coefficient of 100, using a similar pipe will reduce the entire system to one pipe and will render the solution simple.

A 300-mm pipe, with $C = 100$ and $Q = 238.8$ L/s, has

$$s = 0.6876 \times 0.2388^{1.852}$$

$$= 0.0485 \text{ m/m}$$

Hence the length required to accomplish a head loss of 10 m is

$$l = \frac{10}{0.0485}$$

$$= 206.3 \text{ m}$$

The pipe system has now been reduced to the simple system shown in Fig. 9.12b, and

$$s = \frac{38.23}{506.3}$$

$$= 0.0755 \text{ m/m}$$

Hence

$$Q = \left(\frac{0.0755}{0.6876}\right)^{0.54}$$

$$= 0.303 \text{ m}^3/\text{s}$$
$$= 303 \text{ L/s}$$

In order to determine the flow through each leg of the loop, consider the calculation table above. It is indicated there that if the flow in the system is 238.8 L/s, the flow in the top leg of the loop is 61.2 L/s or 25.6 percent of the total. Similarly, the bottom section of the loop takes 177.6 L/s or 74.4 percent of the total flow. The actual flows will be split on the basis of the same percentages or proportions. Hence

$$Q_{(BEFC)} = 303 \text{ L/s} \times 25.6\%$$

$$= 77.6 \text{ L/s}$$

and

$$Q_{(BGHC)} = 291 \text{ L/s} \times 74.4\%$$

$$= 225.4 \text{ L/s}$$

In order to check the accuracy of the calculation, verify that the head losses in each leg are equal; that is,

$$H_{l(\text{top})} = 3.535 \times 0.0776^{1.852} \times 500$$

$$= 15.5 \text{ m}$$

and

$$H_{l(\text{btm})} = 0.4906 \times 0.2254^{1.852} \times 500$$

$$= 15.5 \text{ m}$$

Pipe Fittings

Energy losses at bends and at changes in pipe diameters were discussed in a previous chapter. At that time it was pointed out that in the case of long pipelines the minor losses are generally insignificant and can safely be ignored

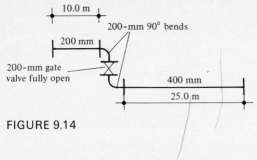

FIGURE 9.14

without appreciably sacrificing accuracy. However, when dealing with short pipe systems containing many forms of minor losses, these losses must be taken into account. In pipe systems these changes in direction or diameter, junctions, and so on are almost always accomplished by means of standard forms of pipe fittings, such as those shown in Fig. 9.14.

A convenient method of allowing for the minor losses due to pipe fittings is by replacing the fittings with lengths of equivalent pipe. This work is facilitated by Fig. 9.15, showing equivalent pipe lengths for various standard pipe fittings. The use of this figure is illustrated in the following examples.

EXAMPLE 9.17

For the pipe system in Fig. 9.14, determine the pressure head required at A if the flow is 200 L/s and the pressure at B is 585 kPa. Consider the system to be horizontal and that all pipes and fittings have $C = 100$.

SOLUTION

By means of the graph of Fig. 9.15 the fittings are all replaced with equivalent pipes. These are

Fitting	Diam.	Eq. Length,	Eq. Pipe Diam.
Elbow	200	5.0	200
Elbow	200	5.0	200
Valve	200	1.4	200
Sudden enlargement	200	4.0	200
Total length		15.4	

Hence the total equivalent length of 200-mm pipe is

$$l_{200} = 15.4 \text{ m} + 10 \text{ m}$$

$$= 25.4 \text{ m}$$

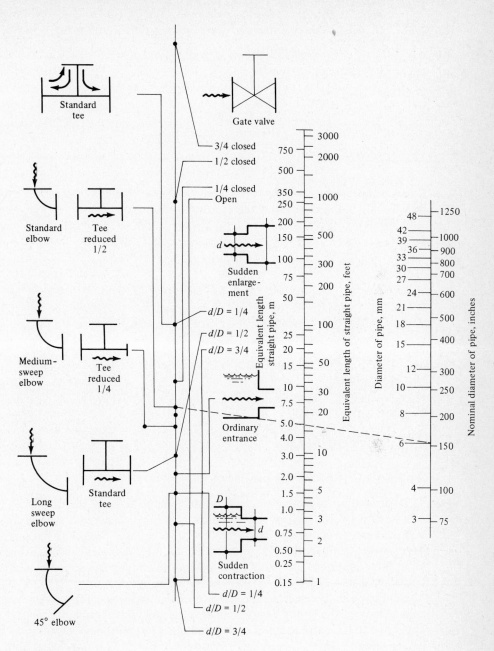

FIGURE 9.15

The head losses in the two portions of the pipe will then be

$$H_{l,200} = 25.4 \times 4.95 \times 0.2^{1.852}$$

$$= 6.38 \text{ m}$$

and

$$H_{l,400} = 25.0 \times 0.232 \times 0.2^{1.852}$$

$$= 0.29 \text{ m}$$

for a total head loss of

$$H_{l,AB} = 6.38 + 0.29$$

$$= 6.67 \text{ m}$$

Therefore the head at A must be equal to

$$H_A = H_B + 6.67$$

$$= 59.63 + 6.67$$

$$= 66.30 \text{ m}$$

and

$$p_A = 650 \text{ kPa}$$

9.6   PIPE NETWORK ANALYSIS

Methods of Analyzing Complex Pipeline Systems

The analysis to determine the flows and energy losses in the various pipes of a large network, such as municipal distribution systems, is a complex but very necessary engineering problem. Several computer programs have been developed to facilitate detailed solutions for the most complex systems. Almost all of these programs are based on a method of analysis developed by Hardy Cross in the mid-1930s.

Other methods, such as the method of sections and electrical analyzers exist. However, the Hardy Cross method is the most commonly used since it is applicable to the most complex systems and lends itself to manual solutions as demonstrated below.

Electrical analyzers are essentially miniature models of the actual pipe network to be studied, but pipes and water are replaced with electrical resistors and electric current. These analyzers are naturally expensive to construct and maintain. Hence they are rapidly giving way to the mathematical modeling capabilities of the computer.

The system analysis by the method of sectioning operates along the same principles as the static analysis of trusses. A section is arbitrarily drawn through the system and energy losses are then calculated along two different routes to the points where the section cuts the pipes. If the energy losses along both routes are equal, the section is located in the right place. If not, the section is moved until equal energy losses in both directions are obtained. This system is practical only for applications to small networks, and even then it is very time-consuming.

The Hardy Cross Method of Pipe Network Analysis

The Hardy Cross method of pipe network analysis is based on making successive corrections to assumed flows in a network until the desired accuracy is reached. In 1936, Hardy Cross developed a formula for calculating successively decreasing values of this correction factor, ΔQ.

The development of the formula for the correction factor is based on two basic principles that must be satisfied in any pipe with an incompressible liquid, such as water. These two principles are as follows:

1. In any closed circuit or loop of pipes, the energy losses caused by flows in the clockwise direction must be equal to the energy losses by flows in the counterclockwise direction.
2. At any point in the system where two or more pipes meet, the inflow to such a junction must equal the outflow from it.

The above can be visualized by considering loop A in Fig. 9.16. In this loop the flow in pipes 1 and 2 are in the same direction, counterclockwise, whereas the flow in pipe 3 is in the opposite direction or clockwise. Hence it is a logical conclusion that the head loss due to Q_1 and Q_2 must equal the head loss due to Q_3. Reasons for this were discussed in some detail earlier in this chapter.

The second principle is directly related to the fact that water is incompressible and any amount entering a junction must also leave that junction.

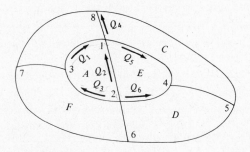

FIGURE 9.16

Thus at joint, or node, 1 in Fig. 9.16 the relationship

$$Q_1 + Q_2 = Q_4 + Q_5 \tag{9.26}$$

must exist.

In solving network analysis problems, no ready available means exist, to enable the determination of the correct flow in each pipe of the system. Hence, assumed flows are employed in all pipes of the system, to start the analysis.

It is reasonably easy to designate flows for all pipes in a network in such a manner that the second condition, the balance of Q's at each node, is met. If the flows in loop A of Fig. 9.16 had been so selected, it would be extremely coincidental if they were also to satisfy the first condition, namely, the balance of the head losses. In other words, the sum of the head losses in pipes 1 and 2 very likely would not be equal to the head loss in pipe 3.

In order to correct this error it is necessary to change the assumed flows in the pipes in such a way that the second condition or principle is still met. Any such correction will bring about a change in the head losses in the pipes. This correction can be readily accomplished by adding or subtracting, as required, a correction factor, ΔQ, to each pipe.

The work involved would be greatly reduced if this correction factor would gradually reduce each time it is applied. In that way only relatively few corrections would need to be made to arrive at a final correct value of the flows in each pipe.

The value of such a correction factor, ΔQ, can be calculated as shown below. Throughout, the sum of all the head losses in a clockwise direction are labeled $\Sigma_c H_l$ and those in a counterclockwise direction are labeled $\Sigma_{cc} H_l$.

Then, from Eq. (9.9), in any loop

$$\Sigma_c H_l = \Sigma_c K Q^m \tag{9.27}$$

and

$$\Sigma_{cc} H_l = \Sigma_{cc} K Q^m \tag{9.28}$$

If the flows in the pipes are the correct flows, the following relationship must be true.

$$\Sigma_c H_l = \Sigma_{cc} H_l \tag{9.29}$$

As mentioned earlier, the likelihood of Eq. (9.29) being satisfied on the first trial is very small. Hence there will be a difference between the energy losses in the clockwise direction and those in the counterclockwise direction. If it assumed that the clockwise energy losses are larger than the counterclockwise ones, or that

$$\Sigma_c H_l > \Sigma_{cc} H_l \tag{9.30}$$

a correction factor ΔQ must be applied. In accordance with Eq. (9.30), this correction factor must be subtracted from the clockwise flows and added to the counterclockwise flows. If ΔQ is the correct value of the correction factor, then the correction will provide the necessary balance in the energy heads in the loop, and the following statement will be true.

$$\Sigma_c K(Q_c - \Delta Q)^m = \Sigma_{cc} K(Q_{cc} + \Delta Q)^m \qquad (9.31)$$

Both sides of Eq. (9.31) are binomials. If it is remembered that m has values of 1.852 and 2.0 for the Hazen–Williams and Manning equations, respectively, it is seen that only the first two terms in each binomial need to be developed. Any terms beyond the first two will be insignificantly small, or inapplicable. Hence Eq. (9.31) can be written as

$$\Sigma_c K(Q_c^m - mQ_c^{m-1}\Delta Q) = \Sigma_{cc} K(Q_{cc}^m + mQ_{cc}^{m-1}\Delta Q) \qquad (9.32)$$

Solving Eq. (9.32) for ΔQ yields

$$\Delta Q = \frac{\Sigma_c K Q_c^m - \Sigma_{cc} K Q_{cc}^m}{m(\Sigma_c K Q_c^{m-1} + \Sigma_{cc} K Q_{cc}^{m-1})} \qquad (9.33)$$

If Eqs. (9.27) and (9.28) are divided by Q_c and Q_{cc}, respectively, the following expressions result.

$$\frac{\Sigma_c H_l}{Q_c} = \Sigma_c K Q_c^{m-1} \qquad (9.34)$$

and

$$\frac{\Sigma_{cc} H_l}{Q_{cc}} = \Sigma_{cc} K Q_{cc}^{m-1} \qquad (9.35)$$

Substituting these values and those from Eqs. (9.27) and (9.28) into Eq. (9.33) yields

$$\Delta Q = \frac{\Sigma_c H_l - \Sigma_{cc} H_l}{m\left(\Sigma_c \dfrac{H_l}{Q} + \Sigma_{cc} \dfrac{H_l}{Q_{cc}}\right)} \qquad (9.36)$$

or, simplified,

$$\Delta Q = \frac{\Sigma_c H_l - \Sigma_{cc} H_l}{m \Sigma \dfrac{H_l}{Q}} \qquad (9.37)$$

Equation (9.37) says that the correction factor ΔQ is equal to the algebraic sum of all the trial head losses in the loop, divided by the product m and the absolute value of the sum of the head losses in all the pipes of the loop, each divided by its respective Q.

Obviously, if a network consisted of only one loop, only one such correction would be required to be made. In a multiloop system, the usual situation, many pipes will be common to two loops. Hence a correction made to a pipe in one loop will likely be undone when this pipe is considered as part of subsequent loop. It will therefore be necessary to make a series of corrections, proceeding from loop to loop until the required correction factor ΔQ is small enough to suit the accuracy requirements of the project.

EXAMPLE 9.18

Figure 9.17*a* represents a closed network of pipes, consisting of four loops. The pipe lengths, diameters, and *C* coefficients are as shown. Also indicated are the flows out of the system at various nodes. Determine the flow in each pipe of the system, to the nearest liter per second.

SOLUTION

This problem will be solved here by means of the Hardy Cross method of pipe network analysis as described above. In view of the complexity of the problem, the solution will be presented in a step-by-step format and in detail.

STEP 1: Determine the total inflow to the system. The inflow must equal the outflow, or

$$Q_{in} = 25 + 40 + 30 + 10 + 20 + 25 + 20 + 20 + 15$$

$$= 205 \text{ L/s}$$

It is important at this stage that a careful check be made of the above equation, because if the initial inflow does not equal the outflow, the system will never be able to be balanced.

STEP 2: Number all system elements. For facility of identification and to maintain orderly progress during the work, number all the loops, pipes, and nodes in the system. A suggested numbering system for this example is shown in Fig. 9.17*b*.

STEP 3: Perform the initial estimated distribution of flow in the pipes. This process should be a rough estimate only, and no attempt should be made to "outguess" the system since at this stage there is no real indication of how much of the flow will go to pipes 5, 1, or 2, for example.

The distribution must however be made in such a way that one of the two requirements or principles set out on page 317 are met. Ensuring that the second principle, the balance of inflow and outflow at each

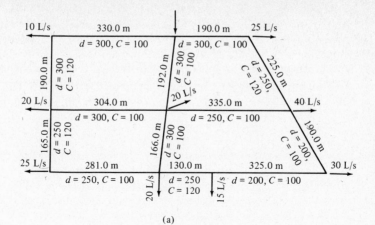

(a)

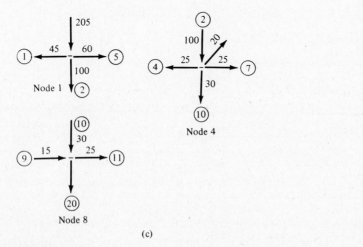

NOTE: Circled numbers
are pipe numbers.

(b)

Node 1

Node 4

Node 8

(c)

FIGURE 9.17

node, is satisfied is relatively easy to accomplish. Figure 9.17*b* shows selected flows in each pipe, chosen in accordance with the above suggestions.

Once the initial distribution is complete, it is recommended that a check be made to ensure that the inflow/outflow requirement is met at each node. Shown in Fig. 9.17*c* is a convenient method for performing this check.

STEP 4: Calculate the clockwise and counterclockwise head losses in loop I. Once the initial guesses of the flows in each of the pipes are complete, the loop must be checked to see whether the clockwise head losses equal the counterclockwise head losses. In order to assist in visualizing this step, loop I has been redrawn in Fig. 9.18*a*. The head losses in each leg are calculated with the aid of the table below and the Hazen-Williams equation. Aside from the normal columns for pipe number, diameter, *K* value, and so forth, this table also includes two other columns, headed "Direction" and "H_l/Q." The latter value is of course required to calculate ΔQ from Eq. (9.37), whereas the former is used to determine whether the head loss is positive or negative, for inclusion in the numerator of this equation.

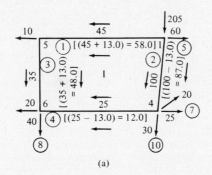

(a)

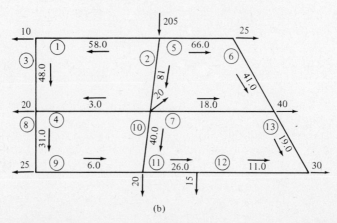

(b)

FIGURE 9.18

Pipe	Dia., mm	Length, m	C	K	Q, m³/s	Direction	S, m/m	H_l, m	H_l/Q
1	300	330	100	0.688	0.045	cc	0.0022	−0.727	16.2
2	300	192	100	0.688	0.100	c	0.0097	1.857	18.6
3	300	190	120	0.491	0.035	cc	0.0098	−0.188	5.4
4	300	304	100	0.688	0.025	c	0.0007	0.226	9.0
		Totals						1.168	49.2

STEP 5: Calculate the correction factor ΔQ. The correction factor ΔQ can now be determined by means of Eq. (9.37), as follows.

$$\Delta Q = \frac{1.168}{1.852 \times 49.2}$$

$$= 0.013 \ m^3/s$$

$$= 13.0 \ L/s$$

STEP 6: Correct the flows in loop I. The correction can now be applied to flows in loop I. In so doing, it must be remembered that in the development of Eq. (9.37) it was assumed that the clockwise head loss was larger than the counterclockwise head loss. This relationship is expressed in Eq. (9.30). As a result of this assumption, the correction factor, ΔQ, was subtracted from the clockwise flows and added to the counterclockwise flows. Since this assumption was maintained throughout the development of Eq. (9.37), corrections must now be made in compliance with this condition. Therefore the corrections to the flows in loop I are as follows:

Pipe	Original Q, L/s	ΔQ, L/s			Corrected Q, L/s
1	45.0	+	13.0	=	58.0
2	100.0	−	13.0	=	87.0
3	35.0	+	13.0	=	48.0
4	25.0	−	13.0	=	12.0

The flow corrections are also shown, in brackets along the pipes, in Fig. 9.18a.

Before completing this step it is again good practice to ensure that no error has been made in the additions and subtractions of the correction factor and that the flows at all the nodes in this loop are still balanced, or

Node 1: $Q_{in} = 205.0 \ L/s$

$Q_{out} = 58.0 + 87.0 + 60.0 = 205.0 \ L/s$

Node 4: $Q_{in} = 87.0$ L/s

$Q_{out} = 12.0 + 20.0 + 30.0 + 25.0 = 87.0$ L/s

Node 5: $Q_{in} = 58.0$ L/s

$Q_{out} = 10.0 + 48.0 = 58.0$ L/s

Node 6: $Q_{in} = 48.0 + 12.0 = 60.0$ L/s

$Q_{out} = 20.0 + 40.0 = 60.0$ L/s

At this stage it is obvious that the above corrections are only the first of several others that might be required, since corrections to loops II and III will alter flows in pipes 2 and 4. Hence the flows in loop *I* will again be unbalanced.

STEP 7: Perform the above work in Steps 4 and 5 for all the other loops in the system. Care must be taken in performing this work to ensure that the flows in all the pipes in each loop are the last corrected values. For instance, in Loop II, pipe 2 will have the corrected flow of 87.0 L/s as corrected in Step 5. Also, in this loop the flow in pipe 2 will be counter-clockwise.

The calculations for Step 7 for the remaining loops—II, III, and IV—are shown in tabular form below.

Loop	Pipe	Dia., mm	Length, m	C	K	Q, m³/s	Dir.	s, m/m	H_l, m	H_l/Q	Corr. Q, m³/s
II	5	300	190	100	0.688	0.060	c	0.0038	0.714	11.9	0.066
	6	250	225	120	1.192	0.035	c	0.0024	0.540	15.4	0.041
	7	250	335	100	1.671	0.025	cc	0.0018	−0.604	24.2	0.019
	2	300	192	100	0.688	0.087	cc	0.0075	−1.435	16.5	0.081
	Totals:								−0.785	68.0	

$$\Delta Q = \frac{-0.785}{1.852 \times 68.0} = -0.006$$

III	4	300	304	100	0.688	0.012	cc	0.0002	−0.058	4.8	0.003
	10	300	166	100	0.688	0.030	c	0.0010	0.173	5.8	0.039
	9	250	281	100	1.671	0.015	cc	0.0007	−0.197	13.1	0.006
	8	250	165	120	1.192	0.040	cc	0.0031	−0.507	12.7	0.031
	Totals:								−0.589	36.4	

$$\Delta Q = \frac{-0.589}{1.852 \times 36.4} = -0.009$$

IV	7	250	335	100	1.671	0.019	c	0.0006	0.217	11.4	0.018
	13	200	190	100	4.954	0.020	c	0.0035	0.672	33.6	0.019
	12	200	325	100	4.954	0.010	cc	0.0010	−0.318	31.8	0.011
	11	250	130	120	1.192	0.025	cc	0.0013	−0.167	6.7	0.026
	10	300	166	100	0.688	0.039	cc	0.0017	−0.281	7.2	0.040
	Totals:								0.123	90.7	

$$\Delta Q = \frac{0.123}{1.852 \times 90.7} = 0.001$$

Although the calculations tabulated above should be self-explanatory, attention is drawn to the addition of the correction factor, when ΔQ is negative as in loops II and III. The corrected value for the flow in pipe 4 in loop III is found to be as follows:

$$\text{Corr. } Q_4 = 0.012 + (-0.009)$$

$$= 0.003$$

Similarly, for pipe 10 in loop III,

$$\text{Corr. } Q_{10} = 0.0300 - (-0.009)$$

$$= 0.039$$

The work performed in this step completes the first "iteration." One iteration consists of the work described in Steps 4, 5, 6, and 7, and includes correcting the flows in all the pipes for each loop of the system.

As mentioned earlier, each time a pipe in a loop is corrected in a subsequent loop, the original loop in which the pipe is located will be unbalanced again. In the example this happens with pipes 2, 4, 7, and 10. Because of this, all the loops in the system—except loop IV, the last one to be corrected—are now unbalanced.

It is therefore necessary to perform further iterations until the required accuracy has been obtained. In the example the required accuracy consists of ensuring that the flows in the pipes are correct to the nearest 1.0 L/s. This accuracy will be achieved when the greatest value of ΔQ in any one iteration does not exceed 1.0 L/s.

The work described above can be simplified somewhat, by arranging the calculations for Steps 4, 5, 6, and 7 in tabular form as shown in Fig. 9.19 and by performing certain repetitious calculations prior to commencing the iterations.

In the table of Fig. 9.19 the following points need explanation and clarification.

1. *Column 6:* The value $L \times K$ is a constant value for each pipe in in the system and therefore needs to be calculated only once.
2. *Column 7:* The value of Q is entered here in liters per second rather than in the cumbersome units of cubic meters per second. Also the value of Q is given a negative sign when the flow is counterclockwise.
3. *Column 8:* Calculation of H_l must, of course, be done using the proper units of cubic meters per second for Q.
4. *Column 9:* If Q is again expressed in units of L/s in calculating H_l/Q, the correction factor ΔQ will also be in L/s.
5. *Column 11:* The corrected value of Q is calculated using algebraic values of both Q and ΔQ. For instance, for pipe 3 in loop I, during the first iteration

$$Q = -35.0 \text{ L/s}$$

$$\Delta Q = 13.0 \text{ L/s}$$

$$Q' = -(35.0 + 13.0) = -48.0 \text{ L/s}$$

PIPE NETWORK
ANALYSIS   NAME:

LINE	LOOP	PIPE	D (mm)	L (m)	C	LK	Q (L/S)	He (m)	He/Q	ΔQ (L/S)	Q' (L/S)	Q (L/S)	He (m)	He/Q	ΔQ (L/S)	Q' (L/S)	Q (L/S)	He (m)	He/Q	ΔQ (L/S)	Q' (L/S)	Q (L/S)
								TRIAL NO.					TRIAL NO.					TRIAL NO.			FINAL	
			3	4	5	6	7	8	9	10	11	12	13	14	15	16	17	18	19	20	21	22
1	I	1	300	330	100	227	-45	-.727	16.2		-58.0	-58	-1.163	20.0		-55.0	-55	-1.055	19.2		-55.0	-55
2		2	300	192	100	132	100	1.857	18.6		67.0	81	1.256	15.5		84.0	83	1.344	16.0		83.0	
3		3	300	190	120	93	-35	-.188	5.4		-48.0	-48	-.336	7.0		-45.0	-45	-.298	6.6		-45.0	-45
4		4	300	304	100	209	25	.226	9.0		12.0	3	.004	1.3		6.0	3	.004	1.3		3.0	
5								1.168	49.2	13.0			-.239	43.8	-3.0			-.005	43.3	0		
6																						
7	II	5	300	190	100	131	60	.714	11.9		66	66	.853	12.9		67.0	67	.877	13.1		67.0	67
8		6	250	225	120	268	35	.540	15.4		44	41	.723	17.6		42.0	42	.756	18.0		42.0	42
9		7	250	335	100	560	-25	-.604	24.2		-19	-18	-.329	18.3		-17.0	-17	-.296	17.4		-17.0	
10		2	300	192	100	132	-87	-1.435	16.5		-81	-84	-1.344	16.0		-83.0	-83	-1.314	15.8		-83.0	-83
11								-.785	68.0	-6.0			-.097	64.8	-1.0			-.023	64.3	0		
12																						
13	III	4	300	304	100	209	-12	-.058	4.8		-3	-6	-.016	2.6		-3.0	-3	-.004	1.5		-4.0	-4
14		10	300	166	100	114	30	.173	5.8		39	40	.294	7.3		43.0	43	.336	7.8		42.0	
15		9	250	281	100	418	-15	-.197	13.1		-6	-6	-.036	6.0		-3.0	-3	-.010	3.3		-4.0	-4
16		8	250	165	120	197	-40	-.507	12.7		-31	-31	-.317	10.2		-28.0	-28	-.262	9.4		-29.0	
17								.589	36.4	-9.0			-.075	20.1	-3.0			.060	22.0	1.0		
18																						
19	IV	7	250	335	100	560	19	.217	11.4		18	17	.296	17.4		17.0	17	.296	17.4		17.0	17
20		13	200	90	100	941	20	.672	33.6		19	19	.611	32.1		19.0	19	.611	32.1		19.0	19
21		12	200	325	100	1610	-10	-.318	31.8		-11	-11	-.380	34.5		-11.0	-11	-.380	34.5		-11.0	-11
22		11	250	130	120	155	-25	-.167	6.7		-26	-26	-.180	6.9		-26.0	-26	-.180	6.9		-26.0	-26
23		10	300	166	100	114	-39	-.281	7.2		-40	-43	-.336	7.8		-43.0	-42	-.321	7.7		-42.0	-42
24								.123	90.7	1.0			.011	64.1	0			.026	98.6	0		
25																						
26																						

FIGURE 9.19

and, for pipe 10 in loop III, first iteration,

$$Q = 30.0 \text{ L/s}$$

$$\Delta Q = -9.0 \text{ L/s}$$

$$Q' = (30.0 - (-9.0)) = 39.0 \text{ L/s}$$

and for pipe 4 in loop III, first iteration

$$Q = -12.0 \text{ L/s}$$

$$\Delta Q = -9.0 \text{ L/s}$$

$$Q' = -(12.0 + (-9.0)) = -3.0 \text{ L/s}$$

The final flows for the pipes in this system are shown in Fig. 9.20.

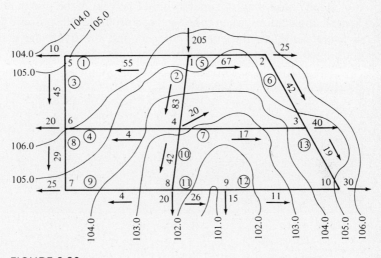

FIGURE 9.20

Pressure Contours in Pipe Networks

A complex pipe network, when analyzed as above in Example 9.17, will be much more defined and easier to assess when pressures can be determined at a glance at any point in the system. Pressure contours—lines connecting points of equal pressure—when drawn on the plan of such a system for a specific design condition provide easy-to-read values of the net pressure at all points in the network. Where pressure contours are closely spaced, they can indicate the need for enlarging certain pipes in order to reduce pressure losses in these locations.

The principles and a method of constructing pressure contours for a single pipe were discussed earlier in this chapter. The development of pressure contours for a pipe network can be accomplished in a similar manner. It is necessary only to construct a profile for each pipe in the system and superimpose on it the energy gradient as found from the network analysis. Locations of points with equal pressure can then be readily performed by means of a specially constructed scale, as demonstrated in the earlier discussion.

In connection with this, it must be pointed out that in municipal water distribution systems or pipe networks, as generally discussed here, the velocities in the pipes are kept relatively low. It is considered to be good practice to maintain velocities less than 3.0 or 4.0 m/s in order to avoid cavitation and water hammer problems in the pipes. A velocity of 3.0 m/s, for instance, corresponds to a velocity head, $v^2/2g$, of less than 0.5 m. The normal pressure heads under which these systems are operated ranges from 35.0 m to 65.0 m. Therefore, the velocity head is approximately only 1 percent of the system pressure. Hence in most applications, when constructing a profile for the purpose of determining pressure contours, the energy gradient and the hydraulic gradient can be considered as one and the same, without undue loss of accuracy.

The following example is included to demonstrate the construction of pressure contours on a pipe network.

EXAMPLE 9.19

Draw pressure contours, at 10-kPa intervals, for the network of Example 9.17. The ground elevation contours of the system are as shown in Fig. 9.20, and the pressure at node 1 is 550 kPa.

SOLUTION

In order to determine the points with equal pressure in the system of Fig. 9.20, it is necessary to first draw a profile of each pipe in the system. Generally, water pipes are installed at a constant depth below the surface of the ground. This depth of installation is for purposes of protecting the pipes from damage resulting from traffic loads, freezing of pipes, or frost heave in the soil surrounding the pipe.

In this example it is assumed that the pipes are buried at a constant depth of 2.0 m, a depth usually accepted in colder climates. The pipe profiles in Fig. 9.21a, b, and c reflect this condition.

The energy gradient for the entire network can be found from the head losses calculated during the last iteration of the Hardy Cross pipe network analysis of Example 9.17. These head losses are shown in column 18 of Fig. 9.19.

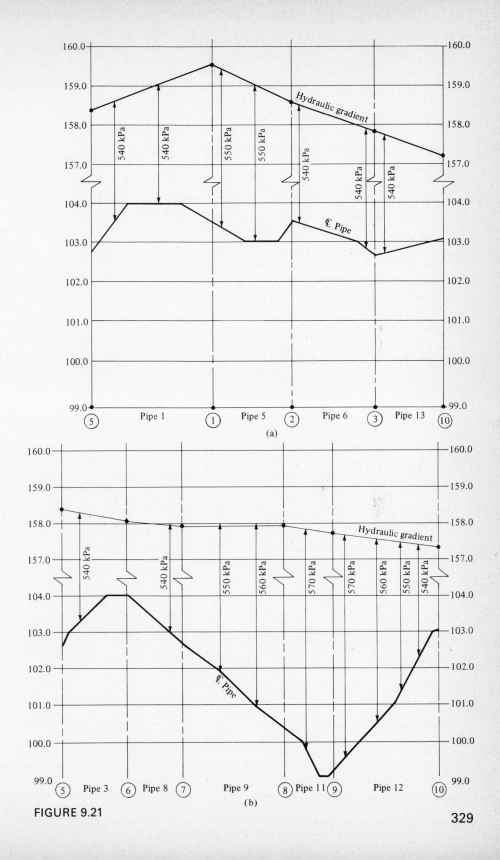

FIGURE 9.21

329

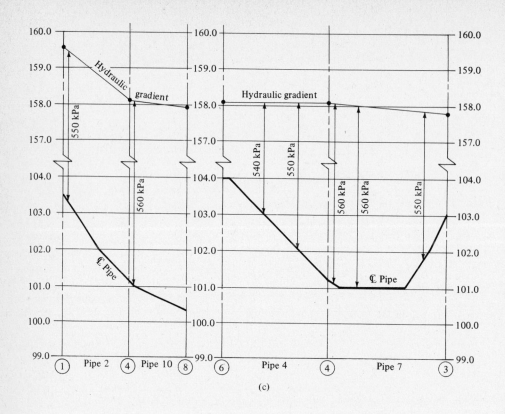

(c)

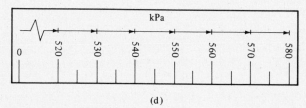

(d)

FIGURE 9.21

The elevation of the energy gradient at node 1 can be found as follows:

$$E_1 = z_1 + \frac{p_1}{\gamma}$$

$$= 103.5 + \frac{550.0}{9.81}$$

$$= 159.5 \text{ m}$$

From this node the energy-gradient elevations of all the other nodes can be found as tabulated at the top of the next page.

Pipe	Node	E	H_l
5	1	159.5	-0.9
	2	158.6	
6			-0.8
	3	157.8	
13			-0.6
	10	157.2	
1	1	159.5	-1.1
	5	158.4	
3			-0.3
	6	158.1	
8			-0.3
	7	157.9	
9			0
	8	157.9	
11			-0.2
	9	157.7	
12			-0.4
	10	157.3	
4	6	158.1	0
	4	158.1	
7			-0.3
	3	157.8	
2	1	159.5	-1.3
	4	158.2	
10			-0.3
	8	157.9	

The corresponding energy gradients are also shown in Fig. 9.21a, b, and c. In constructing these profiles, the vertical scale has been greatly exaggerated in order to facilitate reading values of the pressure heads or pressures. In order to do so, the vertical scale has been broken between elevation 104.0 and 157.0. A gap of 53.0 m or 520 kPa is thus omitted. This has also been done in constructing the special scale shown in Fig. 9.21d.

Following the method discussed earlier in this chapter, the points of equal pressure are located on the profiles by sliding the 0.0 pressure point on the scale along the pipe centerline and reading at the energy gradient points where the scale indicates pressures in multiples of 5 or 10 kPa. This work is shown in Fig. 9.21*a*, *b*, and *c* as well.

Points on the profile with pressures of multiples of 10 kPa are then transferred to the plan of the network and contour lines can be drawn in a manner similar to the drawing of elevation contours. The pressure contours for this system are shown in Fig. 9.22.

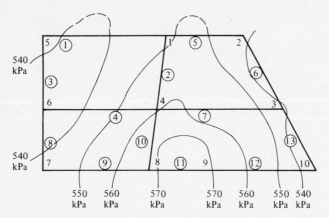

FIGURE 9.22

9.7 WATER-MAIN PIPES

Types of Water Pipes

Probably the first pipe used for transporting water consisted of hollowed-out logs or bamboo. Early civilizations used pipes manufactured of clay and concrete. Romans developed a pipe made of tapered clay sections approximately 500 mm long. Their taper was such that the smaller diameter end would fit into the larger end. In this way they constructed pipelines that were relatively free of leakage. In fact, because of their shape, these pipes would become more watertight as the pressure of the water in them increased.

Pipes used in today's water distribution systems are far removed from these earlier specimens, both in the type of material used and the strength provided. The most commonly utilized pipes are described below, and a very generalized comparison of their properties and usefulness is provided in Table 9.4.

Cast-Iron and Ductile-Iron Pipe (CI, DI)

Ductile-iron pipe is a development resulting from various newer materials, such as asbestos cement and polyvinyl chloride, providing competition to its predecessor, the cast-iron pipe. The manufacture of CI pipe has been virtually

TABLE 9.4   Comparison of Water-Pipe Properties

Property	DI	AC	RC	PVC	PE
Flow characteristics	Good	Good	Good	Very good	Very good
Weight	Relatively heavy	Light	Very heavy	Light	Light
Durability	Good	Good	Very good		
Corrosion:					
Internal	Good if lined	Good	Good	Very good	Very good
External	Good except in acid soils	Good	Good except in alkali soils	Very good	Very good
Strength:					
Flexure	Very good	Good	Good	Weak at high temperature	Very good
Impact	Very good	Good	Very good	Brittle when cold	Good
Handling	Good	Good		Very good	Very good
Tapping for services	Easy	Easy	Difficult	Easy	Easy

stopped since the introduction of the DI pipe on the market. Both CI and DI pipes are centrifugally cast in molds at very high rotational speeds. This high speed of rotation provides centrifugal forces sufficient to ensure that the pipe wall material has a high density and dimensional uniformity. The latter ensures accuracy in adhering to outside and inside diameter tolerances. Ductile iron has a special composition, different from that of cast iron, that renders the material more flexible and able to develop higher tensile stresses. Hence this pipe is considerably thinner in wall thickness, lighter, and easier to handle in the field than the earlier cast iron used to be.

Both CI and DI pipes are subject to internal and external corrosion under certain conditions. Interior corrosion has been significantly reduced by means of lining the pipe with a thin cement layer.

Asbestos Cement Pipe (AC)

Asbestos cement pipe is made of portland cement, a fine aggregate, and asbestos fibers. The asbestos fibers have a high tensile strength and, if properly distributed throughout the mixture, produce a strong pipe material. Due to problems related to the severe health hazard presented by prolonged inhalation of airborne asbestos fibers, special care must be taken in the field when cutting and fitting this pipe.

Reinforced Concrete Pipe (RC)

Reinforced water or pressure pipe is generally manufactured only in the larger sizes, 400 mm and up. Most RC pressure pipes have as an integral part

TABLE 9.5 K and K' Values for Solving the Hazen-Williams Formula, in British Units

Diam. in.	Area, ft²	C								K'
		80	90	100	110	120	130	140	150	
4	0.087	297.7	239.4	196.9	165.1	140.5	121.2	105.6	92.95	0.2629
6	0.196	41.32	33.22	27.33	22.91	19.50	16.82	14.66	12.90	0.7638
8	0.349	10.18	8.184	6.733	5.644	4.804	4.142	3.611	3.178	1.6276
10	0.546	3.433	2.760	2.271	1.904	1.620	1.397	1.218	1.072	2.9270
12	0.786	1.413	1.136	0.9345	0.7833	0.6667	0.5749	0.5012	0.4411	4.7279
15	1.228	0.4765	0.3831	0.3152	0.2642	0.2249	0.1939	0.1600	0.1488	8.5024
18	1.768	0.1961	0.1577	0.1297	0.1087	*92.54*	*79.79*	*69.56*	*61.22*	13.734
21	2.406	*92.55*	*74.41*	*61.22*	*51.32*	*43.68*	*37.66*	*32.83*	*28.89*	20.600
24	3.143	*48.30*	*38.83*	*31.95*	*26.78*	*22.79*	*19.65*	*17.13*	*15.08*	29.267
27	3.978	*27.21*	*21.88*	*18.00*	*15.09*	*12.84*	*11.07*	*9.654*	*8.496*	39.894
30	4.911	*16.29*	*13.10*	*10.78*	*9.033*	*7.688*	*6.629*	*5.779*	*5.086*	52.632
33	5.942	*10.24*	*8.234*	*6.774*	*5.678*	*4.833*	*4.167*	*3.633*	*3.197*	67.626
36	7.071	*6.703*	*5.390*	*4.434*	*3.777*	*3.164*	*2.728*	*2.378*	*2.093*	85.016
39	8.299	*4.539*	*3.650*	*3.003*	*2.577*	*2.142*	*1.847*	*1.610*	*1.417*	104.94
42	9.625	*3.164*	*2.544*	*2.093*	*1.754*	*1.493*	*1.288*	*1.122*	*0.9878*	127.52
48	12.57	*1.652*	*1.329*	*1.093*	*0.9162*	*0.7799*	*0.6724*	*0.5862*	*0.5159*	181.17
54	15.91	*0.9304*	*0.7481*	*0.6155*	*0.5159*	*0.4391*	*0.3786*	*0.3301*	*0.2905*	246.96
60	19.64	*0.5569*	*0.4478*	*0.3684*	*0.3088*	*0.2629*	*0.2266*	*0.1976*	*0.1739*	325.81

Note: Values in italics are $K \times 10^6$. All other K values are $K \times 10^3$.

$$s = KQ^{1.852} = K'\left(\frac{Q}{C}\right)^{1.852} \quad \text{where } Q \text{ is in ft}^3/\text{s}.$$

TABLE 9.6   K and K' Values for Solving the Manning Formula, in British Units

Diam., in.	Area, ft²	n								K'
		0.009	0.010	0.012	0.013	0.015	0.018	0.020	0.024	
4	0.087	132.2	163.2	235.1	275.9	367.3	528.9	653.0	940.3	0.2490
6	0.196	15.21	18.78	27.04	31.74	42.25	60.84	75.12	108.2	0.7342
8	0.349	3.280	4.049	5.830	6.843	9.110	13.12	16.20	23.32	1.5811
10	0.546	0.9976	1.232	1.774	2.081	2.771	3.990	4.927	7.094	2.8667
12	0.786	0.3773	0.4658	0.6707	0.7872	1.048	1.509	1.863	2.683	4.6615
15	1.228	0.1148	0.1417	0.2040	0.2394	0.3188	0.4591	0.5667	0.8161	8.4519
18	1.768	0.0434	0.0536	0.0772	0.0906	0.1206	0.1736	0.2143	0.3086	13.744
21	2.406	0.0191	0.0236	0.0339	0.0398	0.0530	0.0763	0.0942	0.1356	20.732
24	3.143	*9.358*	0.0116	0.0166	0.0195	0.0260	0.0374	0.0462	0.0665	29.599
27	3.978	*4.993*	*6.164*	*8.876*	0.0104	0.0139	0.0200	0.0247	0.0355	40.521
30	4.911	*2.847*	*3.514*	*5.061*	*5.939*	*7.907*	0.0114	0.0141	0.0202	53.666
33	5.942	*1.712*	*2.114*	*3.044*	*3.572*	*4.756*	*6.849*	*8.455*	0.0122	69.196
36	7.071	*1.077*	*1.329*	*1.914*	*2.246*	*2.990*	*4.306*	*5.316*	*7.655*	87.268
39	8.299	*0.7025*	*0.8672*	*1.249*	*1.466*	*1.951*	*2.810*	*3.469*	*4.995*	108.03
42	9.625	*0.4731*	*0.5841*	*0.8411*	*0.9871*	*1.314*	*1.892*	*2.336*	*3.364*	131.64
48	12.57	*0.2321*	*0.2865*	*0.4126*	*0.4843*	*0.6447*	*0.9284*	*1.146*	*1.651*	187.94
54	15.91	*0.1238*	*0.1529*	*0.2202*	*0.2584*	*0.3440*	*0.4854*	*0.6116*	*0.8806*	257.29
60	19.64	*0.0706*	*0.0872*	*0.1255*	*0.1473*	*0.1961*	*0.2824*	*0.3487*	*0.5021*	340.76
66	23.77	*0.0425*	*0.0524*	*0.0755*	*0.0886*	*0.1180*	*0.1699*	*0.2097*	*0.3020*	439.37
72	28.29	*0.0267*	*0.0330*	*0.0475*	*0.0557*	*0.0742*	*0.1068*	*0.1319*	*0.1899*	554.11
84	38.50	*0.0117*	*0.0145*	*0.0209*	*0.0245*	*0.0326*	*0.0469*	*0.0580*	*0.0834*	835.84
96	50.29	*0.0058*	*0.0071*	*0.0102*	*0.0120*	*0.0160*	*0.0230*	*0.0284*	*0.0409*	1193.4
108	63.64	*0.0031*	*0.0038*	*0.0055*	*0.0064*	*0.0085*	*0.0123*	*0.0152*	*0.0218*	1633.7

Note: Values in italics are $K \times 10^6$. All other K values are $K \times 10^3$.

$s = KQ^2 = K'(Qn)^2$, where Q is in ft³/s.

of the pipe wall a thin steel cylinder. This cylinder is surrounded by a cage of reinforcing wire and cast in the concrete of the wall. The cylinder's main purpose is to provide watertightness under high pressure use, when even reinforced concrete will develop tiny tensile cracks capable of allowing water to seep through.

Polyvinyl Chloride Pipe (PVC)

Polyvinyl chloride pipe is a relative newcomer to the market. This pipe is manufactured by extrusion processes, where plastic pellets are heated under pressure and forced through a die to produce the required size of pipe. It is made in sizes up to 300 mm. As is the case with many plastics, the material is susceptible to loss of strength at high temperatures and it becomes relatively brittle at low temperatures.

Polyethylene Pipe (PE)

Polyethylene pipe is manufactured in a manner similar to PVC pipe and exhibits many of that material's characteristics. It is not, however, as affected by cold. The pipe can be joined in the field by butt fusion, at the side of the trench. This is facilitated by the flexibility of the pipe and the longer lengths, up to 11.5 m, in which it is generally manufactured.

9.8   SUMMARY OF THE CHAPTER IN BRITISH UNITS

Pipe-Flow Formulas

The two most commonly used formulas for determining hydraulic values, such as rate of flow or slope of the energy and hydraulic gradients are the Hazen–Williams and the Manning formulas. As pointed out earlier in this chapter, these formulas were developed for use with water as the fluid. In the British System of units, the formulas take on the following forms.

The Hazen–Williams formula is expressed as

$$Q = 1.318 C A R^{0.63} s^{0.54} \tag{9.38}$$

and the Manning formula as

$$Q = \left(\frac{1.486}{n}\right) A R^{2/3} s^{1/2} \tag{9.39}$$

where Q = rate of flow, in cubic feet per second

A = cross-sectional area of the flow, in square feet

R = hydraulic radius, in feet

s = slope of the energy gradient

C, n = pipe roughness coefficients, as shown in Table 9.1

As in the SI System of units, these formulas can be rewritten to read as in Eqs. (9.3) and (9.7), or

$$s = KQ^m$$

where $m = 1.852$ for the Hazen–Williams equation

$\quad n = 2.0$ for the Manning equation

$\quad K$ = a form factor

If the roughness coefficients C or n are omitted from the form factor K, the formulas will read

$$s = K'\left(\frac{Q}{C}\right)^{1.852}$$

for the Hazen–Williams formula, or

$$s = K'(Qn)^2$$

for the Manning formula.

Values for the form factors K and K' to be used in the British System of units are shown in Tables 9.5 and 9.6.

Applications of the Flow Formulas

The Hazen–Williams and Manning flow formulas are used in the same manner as described earlier in this chapter. Their solutions can be accomplished by graphical means, using the nomographs of Figs. 9.1 and 9.2 or by means of Tables 9.5 and 9.6. The following are examples of their use.

EXAMPLE 9.20

A cast-iron water main, 15 inches in diameter, carries water at a rate of 3.5 cfs. Determine, by means of the Hazen-Williams formula, the slope of the hydraulic gradient of this pipe.

SOLUTION

(a) *Graphical solution:* On the nomograph of Fig. 9.1, line up the known values, $d = 15$ in. and $Q = 3.5$ cfs, and find

$$s = 0.003$$

(b) *Mathematical solution:* From Table 9.5, for $d=15$ in. and $C=100$, find $K=0.000315$. Hence

$$s=0.000315 \times 3.5^{1.852}=0.003$$

EXAMPLE 9.21

An asbestos cement pipe $(C=140)$ with a diameter of 12 in. has a slope of the hydraulic gradient of 0.0025. Determine, using the Hazen-Williams equation, the capacity of the pipe and the velocity of flow.

SOLUTION

(a) *Graphical solution:* From the nomograph of Fig. 9.1, with $d=12$ in. and $s=0.0025$, find $Q=1.70$ cfs and $v=2.20$ ft/s.

However, it must be remembered that the nomograph, as indicated, was constructed for $C=100$, whereas the pipe in question has a C value of 140. Consequently the values of Q and v for this pipe will be as follows:

$$Q=1.7 \times \frac{140}{100}=2.38 \text{ cfs}$$

$$v=2.2 \times \frac{140}{100}=3.10 \text{ ft/s}$$

(b) *Mathematical solution:* From Table 9.5 for $d=12$ and $C=140$, find $K=0.00050$, and with $A=0.785$ ft^2,

$$Q= \left(\frac{0.0025}{0.0005}\right)^{0.54}=2.38 \text{ cfs}$$

$$v=\frac{2.38}{0.785}=3.10 \text{ ft/s}$$

EXAMPLE 9.22

A concrete pipe $(n=0.013)$ that is 8 inches in diameter has water flowing through it at a rate of 2.6 cfs. Determine, by means of the Manning equation, the slope of the energy gradient.

SOLUTION

(a) *Graphical solution:* From the nomograph of Fig. 9.2, for $d=8$ in. and $Q=2.6$ cfs, find $s=0.047$.

(b) *Mathematical solution:* From Table 9.6 for $d=8$ in. and $n=0.013$, find $K=0.00684$; hence

$$s=0.00684 \times 2.6^2 = 0.046$$

Hydraulic Gradients

Solving problems related to determining pressures in pipelines is often done graphically, using profiles of the pipelines and their hydraulic gradients. This method is illustrated as it applies to the British System of units in the following example.

EXAMPLE 9.23

A cast-iron pipeline, $C=100$, connects points A, B, C, and D, which have respective elevations of 346.85 ft, 357.23 ft, 361.24 ft, and 351.37 ft. The pipe diameters and lengths are:

from A to B: 6 in. and 450 ft
from B to C: 10 in. and 320 ft
from C to D: 8 in. and 280 ft

The pressure at A is 73 psi and the rate of flow of water is 0.50 cfs. Determine graphically the pressure at a point E, located on the line BC, 100 ft from B.

SOLUTION

Using the Hazen-Williams equation and Table 9.5, the head losses in each section of the pipe can be readily calculated. This work is shown in tabular form below.

Pipe	Dia.	C	l	K	s	H_l
AB	6	100	450	0.0273	0.0413	18.57
BC	10	100	320	0.0023	0.0035	1.11
CD	8	100	280	0.0067	0.0101	2.84

The resulting elevations of the hydraulic gradients have been calculated in the table on the next page. The elevation of the hydraulic gradient at A was determined, remembering that 1 psi is equivalent to a pressure head of 2.31 ft, as pointed out in an earlier chapter.

Pipe	Point	Elevation	H_l
AB	A	515.48	
			18.57
BC	B	496.91	1.11
	C	495.80	
CD			2.84
	D	492.96	

Figure 9.23 shows the corresponding profile and from it the head at the point E can be measured to be 139 ft. Hence

$$p = \frac{139 \text{ ft}}{2.31 \text{ ft/psi}}$$

$$= 60 \text{ psi}$$

It must be pointed out that in this instance the profile in Fig. 9.23 does not take into account the velocity heads in the pipes, and therefore the energy and the hydraulic gradients are shown as one line. This does introduce a slight error. However, considering that the maximum velocity in this system, which occurs in the 6-in. pipe, introduces a maximum value of the velocity head of 0.10 ft, the error can be considered to be negligible.

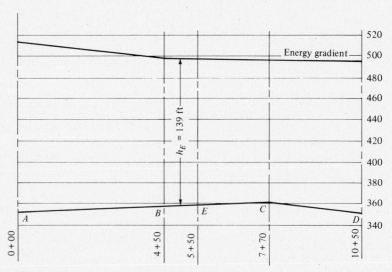

FIGURE 9.23

SELECTED PROBLEMS IN SI UNITS

9.1   For the pipes listed below find the requested values (a) using the Hazen–Williams equation, (b) using the Hazen–Williams nomograph, (c) using the Manning equation, and (d) using the Manning nomograph.

(1)   Dia. = 200 mm; $C = 100$ or $n = 0.10$; $Q = 125$ L/s; $V = ?$; $s = ?$
(2)   Dia. = 400 mm; $C = 120$ or $n = 0.12$; $s = 0.025$; $Q = ?$; $V = ?$
(3) $C = 120$ or $n = 0.012$; $s = 0.15$; $Q = 158.2$ L/s; dia. = ?
(4)   Dia. = 300 mm; $C = 100$ or $n = 0.015$; $v = 2.53$ m/s; $Q = ?$; $s = ?$

9.2   A 150 mm pipe, with $C = 120$, is 1.0 km long from point A to point B and has a flow of 50 L/s. The elevation of point A is 176.456 m and the pressure at point A and B are 375 kPa. Determine the elevation of the pipe at point B.

9.3   A pipe 500 m long, from point A to point B, is horizontal and has a diameter of 450 mm and a C of 130. The pressure at point A is 450 kPa. If the flow of water in this pipe is 200 L/s, determine the pressure at the point B.

9.4   A 700-mm pipe $(C = 120)$ delivers water from point A at an elevation of 146.75 m to a point B, 1500 m away and at an elevation of 132.84 m. The pressures at A and B are 600 kPa and 800 kPa, respectively. Determine the rate of flow in this pipe.

9.5   Determine the rate of flow and the velocity in the pipe in Fig. P9.5.

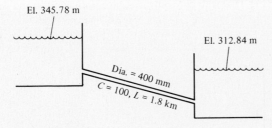

El. 345.78 m

El. 312.84 m

Dia. = 400 mm
$C = 100$, $L = 1.8$ km

FIGURE P9.5

9.6   By means of the Manning formula determine the flow in a 1200-mm pipe, with a roughness coefficient of 0.013, when the slope of the energy gradient is 0.004.

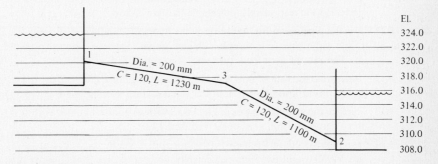

El.
324.0
322.0
320.0
318.0
316.0
314.0
312.0
310.0
308.0

1

Dia. = 200 mm
$C = 120$, $L = 1230$ m

3

Dia. = 200 mm
$C = 120$, $L = 1100$ m

2

FIGURE P9.7

9.7   Determine the rate of flow from reservoir A to reservoir B in Fig. P9.7. Draw the energy gradient and graphically determine the pressure at point 3.

9.8   The pipe shown in plan view in Fig. P9.8 is one of a system of pipes and has Q of 135 L/s, flowing from A to D. The pressure at A is 725 kPa. Determine the points on this pipe where the pressure is a multiple of 25 kPa.

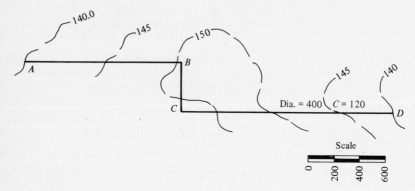

FIGURE P9.8

9.9   Replace the pipe in Fig. P9.9 with one equivalent pipe, with $C = 140$ and a diameter of 300 mm.

Dia. = 300
A $C = 120, L = 400.0$ B
Dia. = 550
$C = 90, L = 250.0$
Dia. = 850.0
C $C = 100, L = 500.0$ D

FIGURE P9.9

9.10   If the pipe AB in Fig. P9.9 is horizontal and the pressures at A and B are 675 kPa and 375 kPa, respectively, determine the rate of flow.

9.11   For the pipeline in Fig. P9.11 draw the hydraulic gradient, determine the elevation of the water surface in reservoir B, and graphically determine the pressure at point M. The rate of flow from reservoir A to reservoir B is 159 L/s.

9.12   Determine the rate of flow and the pressure at points X and Y for the pipe system in Fig. P9.11 if the elevation of the water surface in reservoir B is 143.50 m.

9.13   Replace the system of pipes in Fig. P9.13 with an equivalent pipe with $C = 100$ and a diameter of 300 mm and repeat Problem 9.12.

9.14   The pipe system in Problem 9.13 is horizontal and the pressures at points A and F are 540 kPa and 345 kPa, respectively. Determine the flow in each pipe and the pressure at each pipe junction.

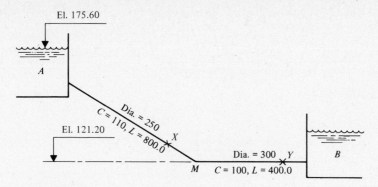

FIGURE P9.11

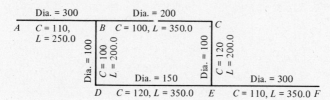

FIGURE P9.13

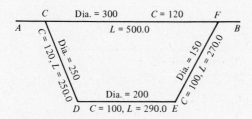

FIGURE P9.14

9.15 The pipe system in Fig. P9.14 is horizontal and the flow from A to B is 175 L/s. Determine the flow in the pipe DC.

9.16 Two reservoirs are connected by a system of DI ($C = 100$) pipes, arranged as follows: Leaving reservoir A, there is a length of 1000 m of 300-mm pipe, which splits into a loop with one leg consisting of a 200-mm pipe, 800 m long and the other leg a 250-mm pipe 1200 m long; these legs join again in a 250-mm pipe, 1500 m long to reach reservoir B. The water levels in reservoirs A and B are 635 m and 585 m, respectively. Determine the rate of flow from reservoir A to reservoir B, as well as the rate of flow in each of the pipes of the loop.

9.17 Replace the pipe system of Fig. P9.17 with an equivalent pipe 250 mm in diameter, with $C = 100$.

9.18 By means of equivalent pipes, determine the flow in each pipe and the pressure at each junction in the pipe system of Fig. P9.18 if the pressures at points A and I are 850 kPa and 335 kPa, respectively.

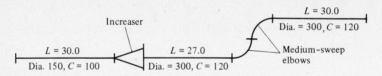

FIGURE P9.17

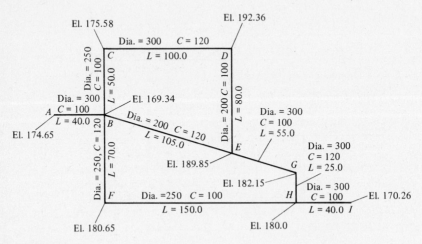

FIGURE P9.18

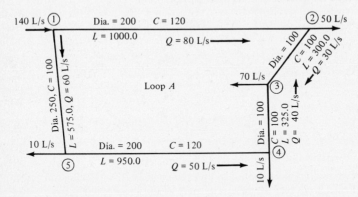

FIGURE P9.19

9.19 Reservoirs A, B, and C have water surface elevations of 132.6 m, 128.4 m and 116.8 m, respectively. They are connected to a water distribution system, which can be represented by three equivalent pipes, as follows: From reservoir A to reservoir C a straight 200-mm DI ($C = 100$) pipe, 3000 m long; from reservoir B a 300-mm AC pipe ($C = 120$), 800 m long, connecting to the pipe from A to C at a point 1200 m from A. Determine the flows to or from each reservoir.

9.22 For loop A in Fig. P9.19 determine the required correction factor to be applied to the assumed flows and correct the flow in each pipe.

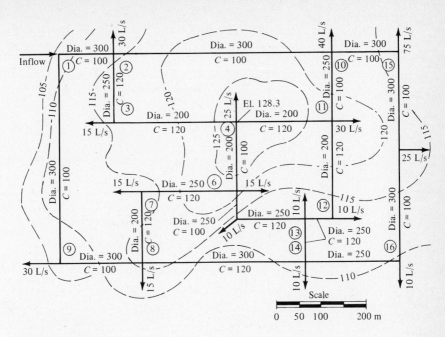

FIGURE P9.23

9.23 For the pipe system in Fig. P9.23 construct pressure contours at 50 kPa intervals when the pressure at point A is 650 kPa and a fire flow of 150 L/s is added to the outflow shown at node 6.

Civil Engineering Applications of Fluid Flow: Uniform Open-Channel Flow

10.1 OPEN CHANNELS

In a previous chapter open-channel flow was defined as the flow that occurs due to the force of gravity only. As a result, it is often referred to as gravity flow. Because of its nature, it follows that open-channel flow takes place only when the flow is not totally confined in the conduit; hence the surface remains free and open to the atmosphere. Therefore open-channel flow differs considerably from pressure or pipe flow. This is inherent in the fact that open-channel flow is not completely confined, but the cross-sectional area of the flow can vary constantly in response to changes created by the conduit roughness, changes in the channel slope, and so forth. As a result, the depth of flow in open channels can be, and often is, variable.

Types of Open Channels

Most open-channel flows occur in drainage structures and facilities. Various forms of open-channel types are shown in Fig. 10.1. In connection with the sewer of Fig. 10.1c, it should be pointed out that even though the flow takes place in a pipe, the flow condition is nevertheless still of the open-channel type since the water surface is unconfined or open to the atmosphere. Only after the flow in such a channel reaches the point where the pipe cross section is 100 percent full are pressure-flow conditions reached.

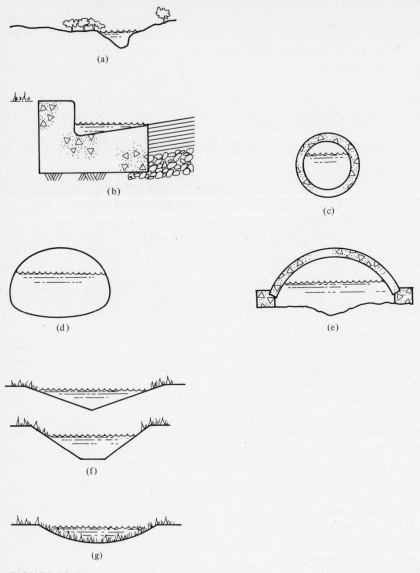

FIGURE 10.1

Ditches and canals usually take on the form of trapezoidal channels. In these instances, side slopes of the channels are usually chosen to be compatible with the soil conditions and/or the lining of the channel walls. Triangular and rectangular channels are special cases of trapezoidal channels. The former represents a trapezoidal cross section with a zero base width, whereas the latter is a trapezoid with vertical sides.

All of the regular geometric cross sections can be readily dealt with for the calculations of flows and energy losses. In the case of the natural streams, where the cross section is often irregular, varies along the length of the channel, and has nonuniform roughness coefficients in any specific cross section, special considerations are necessary. The special conditions related to natural channels are discussed in detail in a subsequent chapter.

Open-Channel Flow

Open-channel flow can occur in all the forms discussed earlier. That is, open-channel flow can be:

1. Steady and uniform, where the capacity of flow and its velocity remain constant, throughout the length of the reach of the channel in question. Such flows are theoretical and are seldom encountered in practice. Nevertheless, many sewers and other drainage structures are designed on the assumption that steady uniform flow prevails.
2. Nonuniform, where—due to changes in cross section, channel slope, roughness coefficient, and other variations or physical channel changes—the velocity of flow varies along the reach. These flows are often encountered and require consideration in the case of large sewers, ditches, and natural streams.
3. Unsteady, where the capacity of flow varies along the length of the reach. Specific applications of this type of flow condition relate to the flow in rivers, where changes in capacity occur, due to passing flood waves and the addition of flow by tributaries.

10.2 OPEN-CHANNEL-FLOW FORMULAS

Manning's Flow Formula

Most of the flow formulas discussed in earlier chapters can be used and are applicable to the open-channel-flow conditions. As a matter of fact, many of these flow formulas were derived from observations and study of open-channel-flow conditions. However, the most commonly used formula in this case is the Manning formula, discussed in Chapter 7. Because of its wide acceptance, a large number of values for the roughness coefficient n have been developed from experimental work.

Manning's flow formula takes on the following three forms.

$$v = \frac{R^{2/3} \times s^{1/2}}{n} \tag{7.5}$$

$$Q = \frac{A \times R^{2/3} \times s^{1/2}}{n} \tag{7.6}$$

$$s = KQ^2 \tag{7.7}$$

where A = cross-sectional area of the flow in the channel, in m^2
    R = hydraulic radius (cross-sectional area/wetted perimeter), in m
    v = velocity of flow, in m/s
    Q = capacity of flow, in m^3/s
    s = slope of the hydraulic gradient
    K = a geometric proportionality coefficient
    n = the roughness coefficient

The use of these formulas and aids in their solution are discussed later in this chapter.

Manning's *n* Coefficient

Values of the *n* coefficient to be used with the Manning formula vary greatly and are dependent upon the materials and conditions of the channel walls and bottom as well as the prevailing flow conditions. The value of *n* ranges from 0.008 for very smooth pipes under laboratory conditions to near 0.20 for natural-stream floodplains with heavy vegetation.

From a study of the Manning formula it is evident that the velocity and capacity of flow are inversely related to the value of *n*; that is, higher values of *n* produce lower values of *v* and *Q*.

Values of Manning's *n* are shown in Table 10.1. Additional values and considerations related to flow in natural channels and rivers are discussed in a later chapter.

Experience and common sense are required in selecting the appropriate value of the *n* coefficient to be used in any particular application of Manning's formula. Generally, tables show three values for *n*, labeled "good," "fair," and "poor." These descriptions relate to the condition of the channel lining. Presumably, all channel linings start off in the "good" condition. However, depending on the type, use, and other circumstances, they can easily and often rapidly deteriorate. Calculations of channel capacities based on the "good" condition generally will apply only during the early stages of the channel structure's life. Often expected flows during this time are smaller than those anticipated toward the end of the design period. It therefore appears to be good practice to choose conservative values of *n*, from the "fair" or even the "poor" condition, when designing such channels.

Energy and Hydraulic Gradients in Open-Channel Flow

Uniform flow occurs in open channels, when the channel cross section is constant and other conditions are such that a uniform velocity is maintained throughout the length of the channel reach. Such flow conditions are represented

TABLE 10.1　Recommended Values of n in Manning's Equation

Type of Channel and Description	Good	Fair	Poor
A. Closed Conduits Flowing Partly Full			
A-1. Metal			
a. Brass, smooth	0.009	0.010*	0.013
b. Steel			
1. Lockbar and welded	0.010	0.012	0.014
2. Riveted and spiral	0.013	0.016	0.017
c. Cast iron			
1. Coated	0.010	0.013	0.014
2. Uncoated	0.011	0.014	0.016
d. Wrought iron			
1. Black	0.012	0.014	0.015
2. Galvanized	0.013	0.016	0.017
e. Corrugated metal			
1. Subdrain	0.017	0.019	0.021
2. Storm drain	0.021	0.024*	0.030
A-2. Nonmetal			
a. Lucite	0.008	0.009	0.010
b. Glass	0.009	0.010	0.013
c. Cement			
1. Neat, surface	0.010	0.011	0.013
2. Mortar	0.011	0.013*	0.015
d. Concrete			
1. Culvert, straight and free of debris	0.010	0.011	0.013
2. Culvert with bends, connections, and some debris	0.011	0.013	0.014
3. Finished	0.011	0.012	0.014
4. Sewer with manholes, inlet, etc., straight	0.013	0.015	0.017
5. Unfinished, steel form	0.012	0.013	0.014
6. Unfinished, smooth wood form	0.012	0.014*	0.016
7. Unfinished, rough wood form	0.015	0.017	0.020
e. Wood			
1. Stave	0.010	0.012	0.014
2. Laminated, treated	0.015	0.017	0.020
f. Clay			
1. Common drainage tile	0.011	0.013*	0.017
2. Vitrified sewer	0.011	0.014	0.017
3. Vitrified sewer with manholes, inlet, etc.	0.013	0.015	0.017
4. Vitrified subdrain with open joint	0.014	0.016	0.018
g. Brickwork			
1. Glazed	0.011	0.013	0.015
2. Lined with cement mortar	0.012	0.015	0.017
h. Sanitary sewers coated with sewage slimes, with bends and connections	0.012	0.013	0.016
i. Paved invert, sewer, smooth bottom	0.016	0.019	0.020

Source: Reproduced from Ven Te Chow, *Open-Channel Hydraulics* (New York: McGraw-Hill, 1959).
*Values recommended for design purposes.

TABLE 10.1   (*Continued*)

Type of Channel and Description	Good	Fair	Poor
A-2. Nonmetal			
j. Rubble masonry, cemented	0.018	0.025	0.030
†k. PVC (polyvinyl chloride)	0.009	0.012	0.013
Plastic	0.010	0.012	0.012
B. Lined or Built-up Channels			
B-1. Metal			
a. Smooth steel surface			
1. Unpainted	0.011	0.012*	0.014
2. Painted	0.012	0.013	0.017
b. Corrugated	0.021	0.025	0.030
B-2. Nonmetal			
a. Cement			
1. Neat, surface	0.010	0.011	0.013
2. Mortar	0.011	0.013	0.015
b. Wood			
1. Planed, untreated	0.010	0.012	0.014
2. Planed, creosoted	0.011	0.012	0.015
3. Unplaned	0.011	0.013	0.015
4. Plank with battens	0.012	0.015	0.018
5. Lined with roofing paper	0.010	0.014	0.017
c. Concrete			
1. Trowel finish	0.011	0.013*	0.015
2. Float finish	0.013	0.015	0.016
3. Finished, with gravel on bottom	0.015	0.017	0.020
4. Unfinished	0.014	0.017	0.020
5. Gunite, good section	0.016	0.019	0.023
6. Gunite, wavy section	0.018	0.022	0.025
7. On good excavated rock	0.017	0.020	
8. On irregular excavated rock	0.022	0.027	
d. Concrete bottom float finished with sides of			
1. Dressed stone in mortar	0.015	0.017	0.020
2. Random stone in mortar	0.017	0.020	0.024
3. Cement rubble masonry, plastered	0.016	0.020	0.024
4. Cement rubble masonry	0.020	0.025	0.030
5. Dry rubble or riprap	0.020	0.030	0.035
e. Gravel bottom with sides of			
1. Formed concrete	0.017	0.020	0.025
2. Random stone in mortar	0.020	0.023	0.026
3. Dry rubble or riprap	0.023	0.033	0.036
f. Brick			
1. Glazed	0.011	0.013*	0.015
2. In cement mortar	0.012	0.015*	0.018
g. Masonry			
1. Cement rubble	0.017	0.025	0.030
2. Dry rubble	0.023	0.032	0.035

†These values were added by the author of this text.
Note: For natural-channel coefficients see Chapter 12.

TABLE 10.1 (*Continued*)

Type of Channel and Description	Good	Fair	Poor
B-2. Nonmetal			
h. Dressed ashlar	0.013	0.015	0.017
i. Asphalt			
1. Smooth	0.013	0.013	
2. Rough	0.016	0.016	
j. Vegetal lining	0.030	. . . . .	0.500
†k. Gabion revetments			
1. 75-mm stones, well placed	0.020	0.025*	0.035
2. 150-mm stones, well placed	0.025	0.030	0.045
†l. 1. Grassed	0.025	0.030	0.045
C. Excavated or Dredged			
a. Earth, straight and uniform			
1. Clean, recently completed	0.016	0.018	0.020
2. Clean, after weathering	0.018	0.022	0.025
3. Gravel, uniform section, clean	0.022	0.025	0.030
4. With short grass, few weeds	0.022	0.027	0.033
b. Earth, winding and sluggish			
1. No vegetation	0.023	0.025	0.030
2. Grass, some weeds	0.025	0.030	0.033
3. Dense weeds or aquatic plants in deep channels	0.030	0.035	0.040
4. Earth bottom and rubble sides	0.028	0.030	0.035
5. Stony bottom and weedy banks	0.025	0.035	0.040
6. Cobble bottom and clean sides	0.030	0.040	0.050
c. Dragline-excavated or dredged			
1. No vegetation	0.025	0.028	0.033
2. Light brush on banks	0.035	0.050	0.060
d. Rock cuts			
1. Smooth and uniform	0.025	0.035	0.040
2. Jagged and irregular	0.035	0.040	0.050
e. Channels not maintained, weeds and brush uncut			
1. Dense weeds, high as flow depth	0.050	0.080	0.120
2. Clean bottom, brush on sides	0.040	0.050	0.080
3. Same, highest stage of flow	0.045	0.070	0.110
4. Dense brush, high stage	0.080	0.100	0.140
D. Natural Streams			
D-1. Minor streams (top width at flood stage <100 ft)			
a. Streams on plain			
1. Clean, straight, full stage, no rifts or or deep pools	0.025	0.030*	0.033
2. Same as above, but more stones and weeds	0.030	0.035	0.040

TABLE 10.1   (*Continued*)

Type of Channel and Description	Good	Fair	Poor
D-1. Minor streams			
3. Clean, winding, some pools and shoals	0.033	0.040	0.045
4. Same as above, but some weeds and stones	0.035	0.045	0.050
5. Same as above, lower stages, more ineffective slopes and sections	0.040	0.048	0.055
6. Same as 4, but more stones	0.045	0.050	0.060
7. Sluggish reaches, weedy, deep pools	0.050	0.070	0.080
8. Very weedy reaches, deep pools, or floodways with heavy stand of timber and underbrush	0.075	0.100	0.150
b. Mountain streams, no vegetation in channel, banks usually steep, trees and brush along banks submerged at high stages			
1. Bottom: gravels, cobbles, and few boulders	0.030	0.040	0.050
2. Bottom: cobbles with large boulders	0.040	0.050	0.070
D-2. Flood plains			
a. Pasture, no brush			
1. Short grass	0.025	0.030	0.035
2. High grass	0.030	0.035	0.050
b. Cultivated areas			
1. No crop	0.020	0.030	0.040
2. Mature row crops	0.025	0.035	0.045
3. Mature field crops	0.030	0.040	0.050
c. Brush			
1. Scattered brush, heavy weeds	0.035	0.050	0.070
2. Light brush and trees, in winter	0.035	0.050	0.060
3. Light brush and trees, in summer	0.040	0.060	0.080
4. Medium to dense brush, in winter	0.045	0.070	0.110
5. Medium to dense brush, in summer	0.070	0.100	0.160
d. Trees			
1. Dense willows, summer, straight	0.110	0.150	0.200
2. Cleared land with tree stumps, no sprouts	0.030	0.040	0.050
3. Same as above, but with heavy growth of sprouts	0.050	0.060	0.080
4. Heavy stand of timber, a few down trees, little undergrowth, flood stage below branches	0.080	0.100	0.120
5. Same as above, but with flood stage reaching branches	0.100	0.120	0.160

TABLE 10.1 (*Continued*)

Type of Channel and Description	Good	Fair	Poor
D-3. Major streams (top width at flood stage >100 ft). The *n* value is less than that for minor streams of similar description, because banks offer less effective resistance.			
a. Regular section with no boulders or brush	0.025		0.060
b. Irregular and rough section	0.035		0.100

schematically in Fig. 10.2. Energy and hydraulic gradients were discussed in previous chapters. Only their applications related to open-channel flow will be discussed here.

Considering Bernoulli's theorem of the conservation of energy, between points 1 and 2 in Fig. 10.2, leads to the expression

$$E_1 = E_2 + H_l \tag{10.1}$$

and s, the slope of the energy gradient, is

$$s = \frac{H_l}{l} \tag{10.2}$$

If the flow is uniform, the cross sections at points 1 and 2 must be constant.

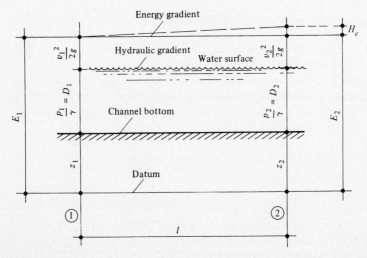

FIGURE 10.2

Consequently, the depth of flow D must also remain constant and

$$D_1 = D_2 \tag{10.3}$$

$$v_1 = v_2 \tag{10.4}$$

From Eq. (10.4) it follows that the velocity head at point 1 must also be equal to the velocity head at 2. Hence, since the depth of flow and the velocity head are constant throughout the flow, a study of Fig. 10.2 indicates that the energy gradient, the water surface and the channel bottom are all parallel. Also, as has already been mentioned in an earlier chapter, since the water surface is a distance equal to the velocity head below the energy gradient, the water surface in fact becomes the hydraulic gradient, and the following relationship ensues

$$s = s_{ws} = s_0 \tag{10.5}$$

where s = slope of the energy gradient

s_{ws} = slope of the hydraulic gradient or water surface

s_0 = slope of the channel bottom or invert

Energy and hydraulic gradients related to non-uniform flow conditions will be discussed later in this text.

10.3   UNIFORM FLOW IN SEWERS, DITCHES, AND CANALS

Application of Manning's Formula

As mentioned earlier, even though in practice true uniform flow seldom exists, it is convenient and often reliably accurate to design sewers, ditches, and canals on the basis of this type of flow condition. Then the relationship between the energy gradient, the hydraulic gradient or water surface, and the channel bottom expressed in Eq. (10.5) is very useful. This is demonstrated in the following examples.

EXAMPLE 10.1

A concrete channel ($n = 0.013$), rectangular in shape and 1.25 m wide, must carry water at a uniform rate of flow of 2000 L/s and a depth of 0.75 m. Determine the required channel bottom slope for this channel.

SOLUTION

A cross section of this channel is shown in Fig. 10.3. From Manning's formula a value for s, the hydraulic gradient, can be found by solving Eq. (7.6) as follows:

$$s = \left(\frac{nQ}{AR^{2/3}}\right)^2$$

In order to find s it is first necessary to determine values of A, the cross-sectional area, and R, the hydraulic radius. From the geometrics of the channel in Fig. 10.3 it follows that

$$A = 1.25 \times 0.75$$
$$= 0.938 \text{ m}^2$$

and the wetted perimeter P is given by

$$P = 0.75 + 1.25 + 0.75$$
$$= 2.75 \text{ m}$$

Hence the hydraulic radius R is

$$R = \frac{0.938}{2.75}$$
$$= 0.341 \text{ m}$$

Substituting these values in the above equation for s yields

$$s = \left(\frac{0.013 \times 2.0}{0.938 \times 0.341^{2/3}}\right)^2$$
$$= 0.003$$

The flow in this instance is uniform and Eq. (10.5) applies. Therefore the slopes of the energy gradient, the water surface, and the channel bottom are identical and

$$s_0 = 0.003$$

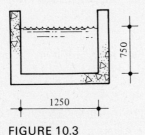

750

1250

FIGURE 10.3

i

EXAMPLE 10.2

A 500-mm asbestos cement sewer pipe ($n=0.012$) has been installed with an invert slope of 0.008. Determine the capacity of flow when this pipe is flowing half full. Assume the flow is uniform.

SOLUTION

In order to find Q from Eq. (10.6), values of A and R must first be determined.

$$A = \frac{0.5^2 \times \pi}{4 \times 2}$$

$$= 0.098 \text{ m}^2$$

From earlier discussions, it should be remembered that for circular pipes, full or half full, the hydraulic radius is equal to one-quarter of the diameter, or

$$R = \frac{0.5}{4}$$

$$= 0.125 \text{ m}$$

Therefore, from Eq. (10.6),

$$Q = \frac{0.098 \times 0.125^{2/3} \times 0.008^{1/2}}{0.012}$$

$$= 0.183 \text{ m}^3/\text{s}$$

$$= 183 \text{ L/s}$$

EXAMPLE 10.3

For the trapezoidal channel shown in Fig. 10.4 determine the slope of the channel if the capacity of flow has to be 4500 L/s. Assume uniform flow and $n=0.025$.

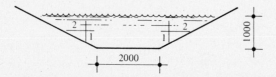

FIGURE 10.4

SOLUTION

The area A, the wetted perimeter P, and the hydraulic radius R are calculated below.

$$\text{Top width} = 2.0 + 2(2 \times 1.0)$$

$$= 6.0 \text{ m}$$

$$A = \frac{2.0 + 6.0}{2} \times 1.0$$

$$= 4.0 \text{ m}^2$$

$$P = 2.0 + 2\sqrt{1.0^2 + 2.0^2}$$

$$= 6.47 \text{ m}$$

and

$$R = \frac{4.0}{6.47}$$

$$= 0.62 \text{ m}$$

Hence, from Eq. (10.6) rearranged,

$$s = \left(\frac{0.025 \times 4.5}{4.0 \times 0.62^{2/3}}\right)^2$$

$$= 0.0015$$

10.4 AIDS IN SOLVING OPEN-CHANNEL-FLOW PROBLEMS

Trapezoidal Channels

Many man-made drainage structures, such as canals, ditches and other open channels have trapezoidal cross sections. Included are the two special cases of trapezoidal channels: rectangular and triangular channels.

Problems to be solved by the engineer in such channels are often related to the finding of the depth of flow D corresponding to a certain cross section, channel bottom slope, and capacity of flow. The nature of the Manning formula is such that a value for D cannot be readily extrapolated from the equation. It is therefore often necessary in such cases to solve for the depth of flow by trial-and-error methods. That is to say, one assumes a D, finding corresponding values of A and R, and then checking the accuracy of the assumed value for D. This is by nature a time-consuming and tedious process.

The introduction of a shape or conveyance factor related to the geometric properties of trapezoidal channels can reduce the amount of work considerably.

Manning's equation, Eq. (7.6), can be expressed as

$$Q = K' \cdot \frac{s^{1/2}}{n}$$

(10.6)

where

$$K' = AR^{2/3}$$

(10.7)

K' is a factor that is related to the geometric properties of the channel cross section. For any specific channel, K' has a constant value and is independent of the channel slope.

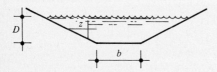

FIGURE 10.5

From a study of Fig. 10.5, the following formulas for the area A, the wetted perimeter P, and the hydraulic radius R can be deduced,

$$A = (b + zD)D$$

(10.8)

$$P = b + 2D(z^2 + 1)^{1/2}$$

(10.9)

where z is the ratio of the side slope of the channel sides, or t/D. If the ratio of the base width b to the depth of flow D is called r, that is, if

$$\frac{b}{D} = r$$

(10.10)

then

$$b = Dr$$

(10.11)

Substitution of this value of b into Eqs. (10.8) and (10.9) leads to

$$A = (r + z)D^2$$

(10.12)

and

$$P = rD + 2D(z^2 + 1)^{1/2}$$

(10.13)

Consequently, the hydraulic radius can be calculated as

$$R = D\frac{(r+z)}{r + 2(z^2 + 1)^{1/2}} \tag{10.14}$$

and, from Eq. (10.7),

$$K' = AR^{2/3} = D^{8/3}\frac{(r+z)^{5/3}}{[r + 2(z^2 + 1)^{1/2}]^{2/3}} \tag{10.15}$$

If the fraction of the right side of Eq. (10.15) is called K, or·

$$K = \frac{(r+z)^{5/3}}{[r + 2(z^2 + 1)^{1/2}]^{2/3}} \tag{10.16}$$

and appropriate substitutions in Eq. (10.6) are made, the following form of the Manning equation, applicable to trapezoidal channels only, results.

$$Q = D^{8/3}K\frac{S^{1/2}}{n} \tag{10.17}$$

Values of K for various values of the ratio b/D and channel side slopes are shown in Table 10.2. If values of K are plotted versus b/D, a nonlinear or curved graph will result. However, the curvature is so slight that straight-line interpolations between the values shown on Table 10.2 produce K values very close to those actually calculated by the use of Eq. (10.16). Use of such interpolated values will therefore generally produce values of Q and v well within the usually · acceptable accuracy required for the design of trapezoidal drainage channels.

In this connection, it must be pointed out that practical designs of drainage channels have inherent in them several approximations and allowances, such as the following:

1. The estimated nature of the roughness coefficient n.
2. The uncertainty and estimated nature of the expected storm runoff flows.
3. The generally accepted practice of allowing a freeboard of approximately 250 mm above the maximum expected water level in order to allow for channel irregularities and waves.

Keeping the above limitations in mind, it would be misleading to determine estimated depths of flow more accurately than to within the nearest 50 or 100 mm.

The use of Table 10.2 and Eq. (10.17) is illustrated in the following examples.

TABLE 10.2 Values for K for Trapezoidal Channels

$\dfrac{b}{D}$	z									
	0.00	0.25	0.50	0.75	1.00	1.50	2.00	3.00	4.00	5.00
0.0	0.000	0.061	0.185	0.336	0.500	0.836	1.170	1.825	2.469	3.109
0.01	0.0003	0.065	0.190	0.343	0.507	0.844	1.178	1.833	2.478	3.117
0.10	0.013	0.104	0.242	0.403	0.573	0.914	1.250	1.907	2.553	3.192
0.20	0.040	0.153	0.305	0.473	0.647	0.993	1.332	1.997	2.636	3.276
0.30	0.077	0.208	0.371	0.546	0.724	1.074	1.414	2.074	2.720	3.360
0.40	0.121	0.268	0.440	0.621	0.802	1.156	1.497	2.158	2.805	3.445
0.50	0.171	0.331	0.511	0.697	0.882	1.238	1.581	2.242	2.889	3.530
1.00	0.481	0.688	0.898	1.102	1.297	1.664	2.010	2.672	3.319	3.958
1.50	0.853	1.090	1.319	1.533	1.734	2.105	2.451	3.112	4.711	5.344
2.00	1.260	1.518	1.759	1.980	2.184	2.557	2.902	3.559	4.200	4.834
3.00	2.134	2.419	2.676	2.905	3.112	3.484	3.825	4.472	5.103	5.730
4.00	3.053	3.354	3.620	3.854	4.062	4.431	4.767	5.402	6.023	6.641
5.00	3.995	4.309	4.581	4.817	5.025	5.391	5.722	6.346	6.956	7.565
6.00	4.953	5.275	5.551	5.788	5.997	6.359	6.685	7.298	7.898	8.499
8.00	6.894	7.227	7.509	7.749	7.956	8.313	8.631	9.224	9.805	10.389
10.00	8.855	9.196	9.482	9.722	9.928	10.281	10.591	11.169	11.734	12.301
15.00	13.799	14.150	14.441	14.682	14.886	15.230	15.529	16.077	16.609	17.145
20.00	18.769	19.126	19.419	19.660	19.863	20.202	20.492	21.020	21.530	22.042

$$Q = D^{8/3} K \frac{s^{1/2}}{n}$$

EXAMPLE 10.4

Solve the problem of Example 10.3 by means of Eq. (10.17) and Table 10.2.

SOLUTION

From Fig. 10.4,

$$z = 2.0$$

$$\frac{b}{D} = \frac{2.0}{1.0}$$

$$= 2.0$$

From Table 10.2,

$$K = 2.902$$

Hence, from Eq. (10.17) rearranged,

$$s = \left(\frac{4.5 \times 0.025}{1.0^{8/3} \times 2.902} \right)^2$$

$$= 0.0015$$

EXAMPLE 10.5

Determine the capacity of flow of a trapezoidal channel with side slopes of 1.5 horizontal to 1.0 vertical, a base width of 1.5 m, and a depth of 1.25 m. The channel bottom slope is 0.001 and the roughness coefficient $n = 0.020$.

SOLUTION

Using the given information,

$$z = 1.50$$

$$\frac{b}{D} = \frac{1.50}{1.25}$$

$$= 1.20$$

Interpolation between b/D values of 1.0 and 1.5 from Table 10.2 yields

$$K = 1.664 + \frac{2.105 - 1.664}{5} \times 2$$

$$= 1.84$$

and, from Eq. (10.17),

$$Q = 1.25^{8/3} \times 1.84 \times \frac{0.001^{1/2}}{0.020}$$

$$= 5.27 \text{ m}^3/\text{s}$$

Solving this problem without the help of Table 10.2 would have required the following steps.

From Eq. (10.8),

$$A = (1.50 + 1.50 \times 1.25)1.25$$

$$= 4.219 \text{ m}^2$$

and, from Eq. (10.9),

$$P = 1.5 + 2 \times 1.25(1.50^2 + 1.0)^{1/2}$$

$$= 6.007 \text{ m}$$

Therefore

$$R = \frac{4.219}{6.007}$$

$$= 0.702 \text{ m}$$

Consequently, by substitution in the Manning equation,

$$Q = 4.219 \times 0.702^{2/3} \times \frac{0.001^{1/2}}{0.020}$$

$$= 5.27 \text{ m}^3/\text{s}$$

EXAMPLE 10.6

Solve the problem of Example 10.1 by using Eq. (10.17) and Table 10.2.

SOLUTION

From Fig. 10.3,

$$z = 0$$

and

$$\frac{b}{D} = \frac{1.25}{0.75}$$

$$= 1.67$$

Interpolation between appropriate values from Table 10.2 yields

$$K = 0.853 + (1.26 - 0.853) \times \frac{1.7}{5}$$

$$= 0.991$$

and, from Eq. (10.17) rearranged,

$$s = \left(\frac{2.0 \times 0.013}{0.75^{8/3} \times 0.991} \right)^2$$

$$= 0.003$$

EXAMPLE 10.7

A triangular channel has side slopes of 1 horizontal to 1 vertical and is constructed of concrete with a roughness coefficient $n = 0.013$. The channel has a slope of $s = 0.0035$. Determine the capacity of flow when the depth of flow is 0.78 m.

SOLUTION

Since a triangular channel has no bottom width,

$$\frac{b}{D} = 0$$

Consequently, from Table 10.2,

$$K = 0.50$$

and hence

$$Q = 0.78^{8/3} \times 0.5 \times \frac{0.0035^{1/2}}{0.013}$$

$$= 1.17 \text{ m}^3/\text{s}$$

EXAMPLE 10.8

A trapezoidal channel is to be constructed in a soil requiring a maximum side slope of 1.50 horizontal to 1.0 vertical. The channel will be grass-lined, and a roughness coefficient $n = 0.025$ is expected. The base width of the channel is to be 1.50 m and the channel bottom slope $s = 0.0075$. Determine the depth of flow when the capacity of flow is 3.75 m³/s.

SOLUTION

Even with the aid of Table 10.2 this problem requires a solution by trial and error. A convenient and orderly method to perform such calculations is by means of a table. The method shown below for solving this problem consists of assuming a value of the depth of flow D, calculating the corresponding Q and comparing the latter with the required Q. These calculations are then repeated until a value of Q is obtained sufficiently close to the actual value.

Solution of Eq. (10.17) requires values of D, K and $s^{1/2}/n$. All these elements are included in the table below. It should be remembered that the value of $s^{1/2}/n$ is constant and could be stored for easy recall in a calculator's memory bank.

Trial D	$\dfrac{b}{D}$	K			$\dfrac{s^{1/2}}{n}$	Q	Comment
		Lower	Higher	Actual			
1.00	0.67	1.238	1.664	1.374	3.644	5.01	Too high
0.75	2.00			2.557		4.33	Too high
0.65	2.31	2.557	3.484	2.840		3.29	Too low
0.70	2.14			2.687		3.78	Too high
0.69	2.17			2.710		3.68	Too low
0.695	2.16			2.710		3.74	OK

It must be pointed out that the above calculations have been carried to a point beyond the usually required accuracy. As mentioned earlier, a statement that the depth of flow will be 0.695 m would be misleading in its indication of the accuracy obtainable.

The practical answer in this instance is arrived at on line 4 of the above table, where the depth of flow is 700 mm. If a 250-mm allowance for freeboard is to be taken into account, the channel should be constructed to a depth of 950 mm.

Asymmetrical Trapezoidal Channels

Obviously Table 10.2 is designed for trapezoidal channels in which both sides have the same slopes. Trapezoidal channels not symmetrical about their vertical centerline, but having a side slope at the left different from that at the right, can nevertheless be solved by means of Table 10.2. In such cases the actual value of K to be used in Eq. (10.17) can be approximated by taking the average value of the K's that would result if these values were calculated for two channels, each with one of the side slopes. The solution of such problems is illustrated in the following example.

EXAMPLE 10.9

A grass-lined trapezoidal channel, with $n=0.027$ and $s=0.0045$, has a bottom width of 3.0 m. The side slope of the left bank is 1.5 to 1 and that of the right bank is 3 to 1. Determine Q if the depth of flow is 1.5 m.

SOLUTION

For this channel,

$$\frac{b}{D}=\frac{3.0}{1.5}$$

$$=2.0$$

For the left side of the channel,

$$z_l=1.5$$

For the right side,

$$z_r=3.0$$

Correspondingly, from Table 10.2 the following K values obtain

$$K_l=2.557$$

and

$$K_r=3.559$$

The average K value to be used in the Manning equation is

$$K_{ave}=\frac{2.557+3.559}{2}$$

$$=3.058$$

and hence

$$Q=1.5^{8/3}\times3.058\times\frac{0.0045^{1/2}}{0.027}$$

$$=22.4 \text{ m}^3/\text{s}$$

Circular Channels: Flowing Full

One of the most common drainage devices is the pipe, or circular channel. It can be easily preconstructed in a variety of sizes and materials. When the flow is such that the entire cross section of the pipe is full, the flow conditions are at the point of change from gravity flow to pressure flow. Solutions of the Manning flow formula when pipes are full have been discussed briefly in an earlier chapter.

The solution of Manning's equation for full pipe flow can be simplified by the use of tables and nomographs. These tables and nomographs, shown

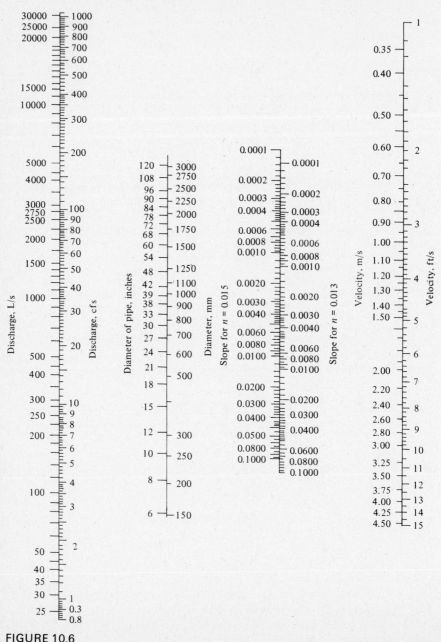

FIGURE 10.6

TABLE 10.3 K and K' Values for Solving the Manning Formula, in SI Units (Based on Commercial Pipe Sizes)

Nominal mm	Nominal in.	Actual mm	Area, m²	0.009	0.010	0.012	0.013	0.015	0.018	0.020	0.024	K'
100	4	102	0.0081	164.9	203.6	293.2	344.1	458.1	659.7	814.4	1173.0	2 038 000
150	6	152	0.0182	18.97	23.42	33.73	39.58	52.70	75.89	93.69	134.9	234 400
200	8	203	0.0324	4.090	5.050	7.272	8.534	11.36	16.36	20.20	29.09	50 540
250	10	254	0.0507	1.244	1.536	2.212	2.596	3.456	4.977	6.145	8.848	15 370
300	12	305	0.0730	0.4706	0.5809	0.8366	0.9818	1.307	1.882	2.324	3.346	5814
400	15	381	0.1141	0.1431	0.1767	0.2545	0.2987	0.3976	0.5726	0.7069	1.018	1769
450	18	457	0.1642	0.0541	0.0668	0.0962	0.1129	0.1504	0.2165	0.2673	0.3850	668.9
550	21	533	0.2235	0.0238	0.0294	0.0423	0.0496	0.0661	0.0952	0.1175	0.1692	294.0
600	24	610	0.2920	0.0117	0.0144	0.0208	0.0244	0.0324	0.0467	0.0576	0.0830	144.2
700	27	686	0.3695	*6.228*	*7.688*	0.0111	0.0130	0.0173	0.0249	0.0308	0.0443	76.94
750	30	762	0.4562	*3.550*	*4.383*	*6.312*	*7.408*	*9.862*	0.0142	0.0175	0.0252	43.87
850	33	838	0.5520	*2.136*	*2.637*	*3.797*	*4.456*	*5.932*	*8.542*	0.0106	0.0152	26.39
900	36	914	0.6570	*1.343*	*1.658*	*2.387*	*2.801*	*3.730*	*5.371*	*6.631*	*9.548*	15.59
1000	39	991	0.7710	*0.8761*	*1.082*	*1.558*	*1.828*	*2.434*	*3.505*	*4.327*	*6.230*	10.83
1050	42	1067	0.8942	*0.5901*	*0.7285*	*1.049*	*1.231*	*1.639*	*2.360*	*2.914*	*4.196*	7.291
1200	48	1219	1.1679	*0.2895*	*0.3574*	*0.5147*	*0.6040*	*0.8041*	*1.158*	*1.430*	*2.059*	3.577
1400	54	1372	1.4782	*0.1545*	*0.1907*	*0.2746*	*0.3223*	*0.4291*	*0.6178*	*0.7628*	*1.098*	1.908
1500	60	1524	1.8249	*0.0881*	*0.1087*	*0.1566*	*0.1837*	*0.2446*	*0.3522*	*0.4349*	*0.6262*	1.088
1700	66	1676	2.2081	*0.0530*	*0.0654*	*0.0942*	*0.1105*	*0.1471*	*0.2119*	*0.2616*	*0.3767*	0.6545
1800	72	1829	2.6278	*0.0333*	*0.0411*	*0.0592*	*0.0695*	*0.0925*	*0.1332*	*0.1645*	*0.2368*	0.4115
2150	84	2134	3.5768	*0.0146*	*0.0181*	*0.0260*	*0.0305*	*0.0407*	*0.0586*	*0.0723*	*0.1041*	0.1808
2450	96	2438	4.6717	*0.0072*	*0.0089*	*0.0128*	*0.0150*	*0.0200*	*0.0287*	*0.0355*	*0.0511*	0.0887
2570	108	2743	5.9126	*0.0038*	*0.0047*	*0.0068*	*0.0080*	*0.0106*	*0.0153*	*0.0189*	*0.0272*	0.0473

Note: Values in italics are $K \times 10^3$.

$s = K \times Q^2 = K'\left(\dfrac{Q}{C}\right)^{1/2}$ where Q is in m³/s.

earlier as Table 9.3 and Fig. 9.2, are here reproduced, for convenience, as Table 10.3 and Fig. 10.6, respectively.

Manning's formula can, as demonstrated in Chapter 9, be rearranged for circular pipes as follows:

$$s = KQ^2 \qquad (10.18)$$

where s = slope of the hydraulic gradient
$\quad Q$ = rate of flow
$\quad K$ = a proportionality factor, listed in Table 10.3, for various values of pipe diameter d and roughness coefficient n.

TABLE 10.4   Values of n in the Manning Formula

Pipe Material	n	Pipe Material	n
Asbestos Cement	0.011–0.015	Concrete	0.011–0.015
Cast iron		New	0.012–0.014
Plain	0.014–0.016	Old	0.009–0.010
Cement-lined	0.012–0.015	Plastic	

Values of n coefficients specifically applicable to pipe materials are listed in Table 10.4. General comments about the choice of n made earlier in this chapter apply in this instance as well. Additionally, long use of pipes in many sewer applications tends to produce a thin layer of deposit on the pipe walls, roughening some and rendering others more smooth. After long use, unless structurally damaged, most pipes of whatever material tend to exhibit relatively similar hydraulic properties, with n somewhere in the vicinity of 0.013.

The following examples are included here to illustrate the use of the nomograph of Fig. 10.6 and Table 10.3.

EXAMPLE 10.10
A 300 mm concrete pipe, with $n = 0.013$, flows full and is installed at a bottom slope of 0.01. Determine the capacity and velocity of flow in this pipe.

SOLUTION

(a) *Mathematical solution:* From Table 10.3, for $D = 300$ mm and $n = 0.013$, $K = 1.069$. Hence, from Eq. (10.18) rearranged,

$$Q = \left(\frac{s}{K}\right)^{1/2}$$
$$= \left(\frac{0.010}{1.069}\right)^{1/2}$$
$$= 0.097 \text{ m}^3/\text{s}$$
$$= 97 \text{ L/s}$$

Consequently,

$$v = \frac{0.097 \text{ m}^3/\text{s}}{0.071 \text{ m}^2}$$

$$= 1.37 \text{ m/s}$$

(b) *Graphical solution:* On Fig. 10.6, lining up $D = 300$ mm and $s = 0.01$ and reading directly,

$$Q = 97 \text{ L/s}$$

and

$$v = 1.37 \text{ m/s}$$

In this instance, it must be pointed out that the monograph in Fig. 10.6 has been specifically designed for pipes with $n = 0.013$. If pipes with n other than 0.013 are used, adjustments to the flows must be made, as is illustrated in some of the following examples.

EXAMPLE 10.11

A 600-mm pipe with $n = 0.015$ has water flowing in it at the rate of 450 L/s. If the pipe is flowing full at this rate, determine the slope of the hydraulic gradient and the pipe.

SOLUTION

(a) *Mathematical solution:* From Table 10.3, for $d = 600$ mm and $n = 0.015$,

$$K = 0.0324$$

and, from Eq. (10.18),

$$s = 0.0324 \times 0.45^2$$

$$= 0.0066$$

(b) *Graphical solution:* Since the pipe in question has $n = 0.017$, the nomograph in Fig. 10.6, which has been constructed for n values of 0.013, cannot be used without making adjustment for this fact.

From the general form of Manning's formula, Eq. (9.6) it is evident that Q is directly related to $1/n$. Therefore the Q to be used to enter Fig. 10.6 must be adjusted as follows:

$$Q' = 450 \times \frac{0.015}{0.013}$$

$$= 519 \text{ L/s}$$

Lining up this value of Q and $d=600$ mm on the nomograph of Fig. 10.6 leads to

$$s=0.0066$$

EXAMPLE 10.12

Determine Q for a pipe with a diameter of 300 mm, a roughness coefficient of 0.010 and a slope of $s=0.02$. Assume the pipe flows full.

SOLUTION

(a) *Mathematical solution:* From Table 10.3, for $d=600$ mm and $n=0.010$ find

$$K=0.581$$

Consequently, from Eq. (10.18) rearranged,

$$Q=\left(\frac{0.02}{0.581}\right)^{1/2}$$

$$=0.186 \text{ m}^3/\text{s}$$

$$=186 \text{ L/s}$$

(b) *Graphical solution:* From the nomograph, for $d=300$ mm and $s=0.020$, find

$$Q'=143 \text{ L/s}$$

or, adjusted for $n=0.010$,

$$Q=143 \times \frac{13}{10}$$

$$=186 \text{ L/s}$$

Circular Channels: Flowing Partially Full

Most applications of gravity flow in pipes or circular channels will consist of pipes flowing only partially full, as shown in Fig. 10.1c. Calculating values of the cross-sectional area and the hydraulic radius in such cases can prove to be complex and time-consuming. Aids in solving problems dealing with partially full pipes have been developed. Two of those aids—one a graphical and the other a mathematical method—are shown in Fig. 10.7 and Table 10.5, respectively.

Both methods are identical in concept and use the relationship between the percentage or fraction of the diameter used in the partial flow and the corresponding ratio of the hydraulic elements, such as the Q and v of the partial section, to those of the full section, so that the following equations ensue

$$K_p = \frac{Q_p}{Q_f} \qquad (10.19)$$

and

$$K'_p = \frac{v_p}{v_f} \qquad (10.20)$$

where K_p and K'_p are values found on the graph of Fig. 10.7 or Table 10.5; Q_p and v_p are the actual Q and velocity at the partial depth; and Q_f and v_f are the corresponding actual capacity and velocity when the pipe is flowing full.

In connection with the phenomena of partial-flow conditions, a study of Fig. 10.7 indicates that at certain depths of flow, the actual capacities and velocities for partial-flow conditions can be larger than those for full-flow condition. This is of course inherent in the nature of Manning's equation, where Q and v are directly related to the hydraulic radius of the pipe. From

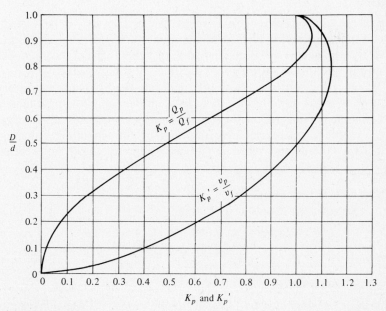

FIGURE 10.7

TABLE 10.5   Circular Pipes: Values for K_p and K_p' Related to Fraction of the Diameter

D/d	0.00	0.01	0.02	0.03	0.04	0.05	0.06	0.07	0.08	0.09
0.0	0.0	0.0002	0.0007	0.0016	0.0030	0.0048	0.0071	0.0098	0.0130	0.0167
	0.0	*0.0890*	*0.1408*	*0.1839*	*0.2221*	*0.2569*	*0.2829*	*0.3194*	*0.3480*	*0.3752*
0.10	0.0209	0.0255	0.0306	0.0361	0.0421	0.0486	0.0555	0.0629	0.0707	0.0789
	0.4012	*0.4260*	*0.4500*	*0.4730*	*0.4953*	*0.5168*	*0.5376*	*0.5578*	*0.5775*	*0.5865*
0.20	0.0876	0.0966	0.1061	0.1160	0.1263	0.1370	0.1480	0.1595	0.1412	0.1834
	0.6151	*0.6331*	*0.6507*	*0.6678*	*0.6844*	*0.7006*	*0.7165*	*0.7320*	*0.7471*	*0.7618*
0.30	0.1958	0.2086	0.2218	0.2352	0.2489	0.2629	0.2772	0.2918	0.3066	0.3217
	0.7761	*0.7902*	*0.8038*	*0.8172*	*0.8302*	*0.8430*	*0.8554*	*0.8675*	*0.8794*	*0.8909*
0.40	0.3370	0.3525	0.3682	0.3842	0.4003	0.4165	0.4330	0.4495	0.4662	0.4831
	0.9022	*0.9133*	*0.9239*	*0.9343*	*0.9445*	*0.9544*	*0.9640*	*0.9734*	*0.9825*	*0.9914*
0.50	0.5000	0.5170	0.5341	0.5513	0.5685	0.5857	0.6030	0.6202	0.6375	0.6547
	1.0000	*1.0084*	*1.0165*	*1.0243*	*1.0319*	*1.0393*	*1.0464*	*1.0533*	*1.0599*	*1.0663*
0.60	0.6718	0.6889	0.7060	0.7229	0.7397	0.7564	0.7729	0.7893	0.8055	0.8215
	1.0724	*1.0783*	*1.0839*	*1.0893*	*1.0944*	*1.0993*	*1.1039*	*1.1083*	*1.1124*	*1.1162*
0.70	0.8372	0.8527	0.8680	0.8829	0.8976	0.9119	0.9258	0.9394	0.9525	0.9652
	1.1200	*1.1231*	*1.1261*	*1.1288*	*1.1313*	*1.1335*	*1.1353*	*1.1369*	*1.1382*	*1.1391*
0.80	0.9775	0.9892	1.0004	1.0110	1.0211	1.0304	1.0391	1.0471	1.0542	1.0605
	1.1397	*1.1400*	*1.1399*	*1.1395*	*1.1387*	*1.1374*	*1.1358*	*1.1337*	*1.1311*	*1.1280*
0.90	1.0658	1.0701	1.0733	1.0752	1.0757	1.0745	1.0714	1.0657	1.0567	1.0420
	1.1243	*1.1200*	*1.1151*	*1.1093*	*1.1027*	*1.0950*	*1.0859*	*1.0751*	*1.0618*	*1.0437*

Note: Numbers in italics are values of K_p'.
$Q_p = K_p Q_f$; $v_p = K_p' v_f$.

earlier discussions of properties of circular sections it is remembered that the hydraulic radius of the full section is equal to the hydraulic radius of the half-full section, or $R = d/4$. Both the figures of Table 10.5 and the graph of Fig. 10.7 substantiate this in showing that the value of K_p' is equal to 1.0.

Once the section is more than half full, the hydraulic radius and consequently the velocity continue to increase until a point at approximately 82 percent of the depth is reached. From this point on, increases in the wetted perimeter exceed increases in area, resulting in corresponding reduction in the hydraulic radius and velocity. The curve representing Q ratios behaves in a similar manner.

EXAMPLE 10.13

A 300-mm sewer, with $n = 0.013$, carries a flow of 90 L/s and has a slope of $s = 0.020$. Determine the depth of flow and the velocity.

SOLUTION

(a) *Mathematical solution:* From Table 10.3, for $d = 300$ and $n = 0.013$, find

$$K = 0.982$$

and

$$A = 0.071 \text{ m}^2$$

Therefore, for full flow, the capacity and velocity of flow will be, according to Eq. (10.18) rearranged,

$$Q_f = \left(\frac{0.02}{0.982}\right)^{1/2}$$

$$= 0.143 \text{ m}^3/\text{s}$$

and

$$v_f = \frac{0.143}{0.071}$$

$$= 1.95 \text{ m/s}$$

Therefore from Eq. (10.19),

$$K_p = \frac{90}{143}$$

$$= 0.63$$

This value in Table 10.5 corresponds to a depth ratio D/d of 0.59, and the depth of flow

$$D = 300 \times 0.63$$

$$= 189 \text{ mm}$$

Also, from Table 10.5, for $K_p = 0.63$,

$$K_p' = 1.06$$

Consequently, from Eq. (10.20),

$$v_p = 1.95 \times 1.06$$

$$= 2.07 \text{ m/s}$$

(b) *Graphical solution:* From the nomograph of Fig. 10.6, for $d = 300$ mm and $s = 0.02$, read

$$Q_f = 143 \text{ L/s}$$

and

$$v_f = 1.95 \text{ m/s}$$

Hence, from Eqs. (10.19) and (10.20),

$$K_p = \frac{90}{143}$$

$$= 0.63$$

Entering Fig. 10.7 with this value of K_p, or Q_p/Q_f, and reading the corresponding value of the fraction of the diameter in use leads to

$$D = 0.63 \times 300 \text{ mm}$$

$$= 189 \text{ mm}$$

Further, a fraction of d of 0.59 corresponds to a value of K_p' or v_p/v_f to be read on the graph and

$$K_p' = 1.06$$

Hence

$$v_p = 1.06 \times 1.93$$

$$= 2.01 \text{ m/s}$$

EXAMPLE 10.14

In a pipe 600 mm in diameter, with $n = 0.015$ and $s = 0.006$, the flow is 375 mm deep. Determine the capacity of flow and the velocity.

SOLUTION

(a) *Mathematical solution:* From Table 10.4, for $d = 600$ mm and $n = 0.015$, find

$$K = 0.0324$$

and

$$A = 0.292$$

Hence

$$Q_f = \left(\frac{0.006}{0.0324}\right)^{1/2}$$

$$= 0.430 \text{ m}^3/\text{s}$$

$$= 430 \text{ L/s}$$

and

$$v_f = \frac{0.430}{0.292}$$

$$= 1.52 \text{ m/s}$$

Since the flow is 375 mm deep, the fraction of the total depth is

$$\frac{d}{D} = \frac{375}{600} = 0.63$$

Therefore, from Table 10.5, read

$$K_p = 0.72$$

and

$$K_p' = 1.09$$

Consequently, from Eqs. (10.19) and (10.20),

$$Q_p = 0.72 \times 430$$

$$= 310 \text{ L/s}$$

and

$$v_p = 1.09 \times 1.52$$

$$= 1.66 \text{ m/s}$$

(b) *Graphical solution:* When full, from the nomograph of Fig. 10.6 and adjusting for $n = 0.015$, it is found that

$$Q_f = 495 \times \frac{13}{15}$$

$$= 430 \text{ L/s}$$

and

$$v_f = 1.75 \times \frac{13}{15}$$

$$= 1.52 \text{ m/s}$$

Also, the fraction of the diameter used in this flow is

$$\frac{d}{D} = \frac{375}{600}$$

$$= 0.63$$

Entering the graph of Fig. 10.7 with this ratio and extending to the appropriate curves yields

$$K_p = 0.72$$

and

$$K'_p = 1.09$$

Consequently,

$$Q_p = 430 \times 0.72$$
$$= 310 \text{ L/s}$$

and

$$v_p = 1.52 \times 1.10$$
$$= 1.66 \text{ m/s}$$

10.5 PRACTICAL DESIGN CONSIDERATIONS

The Design of Sewers

Modern municipal sewage collection systems are almost always constructed of pipes or circular sections. Municipal sewer systems are primarily of two types: those sewers that carry only storm water and those that carry domestic and other wastewater.

Domestic sewage includes wastewater discharged from residential, commercial, and certain types of industrial users. Although such sewage contains over 99 percent water and for all intents and purposes can be considered to have the same hydraulic properties as water, it nevertheless contains a certain amount of solids. Similarly, storm water, particularly in the northern climates in winter and spring, contains a certain amount of sand and gravel, as well as other debris from the roads.

If domestic or storm sewage is allowed to flow predominantly at very low velocities in the pipes, most of these solids carried by it will settle out onto the bottom of the sewer. This will naturally lead to considerable reduction in pipe cross section and reduction in the hydraulic capacity of the pipe. It often results in the pipe becoming sufficiently clogged to create backing up or completely stopping the flow upstream, resulting in flooding of basements and/or roads.

Good sewer design must take this into account and provide for sufficient velocity to maintain the sewage solids in suspension, at least to ensure that solid deposits are such that they do not create flooding and backup conditions.

A velocity of flow that will ensure such conditions is generally referred to as a "self-cleaning velocity."

The problem is of course somewhat aggravated due to the fact that neither domestic nor storm sewage flows at a constant rate. Indeed, many domestic sewers and all storm sewers will experience periods where no flow occurs. In the case of domestic sewage flow, the capacity varies considerably during varying periods of the day. Obviously, during the early morning hours very little flow will exist since most users are asleep. As breakfast time approaches, the flow increases to its first daily peak. This peak attenuates somewhat during the rest of the day. A second, generally larger peak flow takes place around the supper hour. Although the ratios of peak flow to average flow vary somewhat from municipality to municipality, a good rule-of-thumb value of the peak flow being 250 percent of the average daily flow appears reasonable. The latter ratio is often accepted for design purposes, particularly when other, more accurate or detailed information is lacking.

Although most municipalities have their own design standards, some more complex and sophisticated than others, many have a general design requirement that the average flow will take place at a velocity of 1.0 m/s and that this flow must occur when the pipe is flowing approximately half full. This will then leave pipe capacity for peak flows and provide, during such peaks, velocities sufficiently large to remove any deposits that have accumulated during very low flows.

The selection of pipe sizes for municipal sewers based on the above criteria is demonstrated in the following example.

EXAMPLE 10.15

Determine the required diameter for a proposed sanitary sewer that has to carry domestic sewage at an average rate of 125 L/s. The maximum rate of flow during peak period is 275 L/s. Consider using a concrete pipe with $n = 0.013$. The sewer is to be installed on a street with a general slope of 0.8 percent, and it is desirable to install the sewer parallel to the road surface to maintain economy in construction cost. The minimum allowable velocity of flow during average flow conditions is 1.0 m/s. Determine also the depth and the velocity of flow during maximum-flow conditions and when the flow is 25 L/s, 50 L/s, and 75 L/s.

SOLUTION

In order to find an approximate pipe size required, solve for K in Eq. (10.18) using maximum Q and $s = 0.008$.

$$K = \frac{0.008}{(0.275)^2}$$

$$= 0.106$$

From Table 10.3 and for $n=0.013$, this corresponds to pipe diameter between 450 and 550 mm. Since the 450-mm pipe will be too small, a 550-mm pipe will have to be used. Such a pipe has a value of K of

$$K=0.050$$

Hence,

$$Q_f=\left(\frac{0.008}{0.050}\right)^{1/2}$$

$$=400$$

and

$$v_f=\frac{0.400}{0.224}$$

$$=1.79 \text{ m/s}$$

Obviously, this pipe is somewhat larger than that required to carry the maximum flow. However, the next smaller pipe, the 400-mm pipe would be too small, since it has a value

$$K=0.299$$

and hence

$$Q_f=\left(\frac{0.008}{0.299}\right)^{1/2}$$

$$=164 \text{ L/s} < 275 \text{ L/s}$$

The other work required to determine actual velocities and depths of flow is best arranged in table form as shown below. (The K_p' and D/d values in the table were found from Table 10.5.)

Q L/s	K_p	K_p'	v_p' m/s	$\dfrac{D}{d}$	d, mm
25	0.06	0.56	1.00	0.17	94
50	0.12	0.68	1.22	0.24	132
75	0.19	0.77	1.38	0.30	165
125	0.31	0.88	1.58	0.38	209

The proposed sewer, with a diameter of 550 mm and a slope of 0.8 percent, will therefore satisfy the requirements in that the velocity of flow is higher than the required minimum of 1.0 m/s during average flow conditions. Also during average flow the pipe will flow nearly half full.

The Design of Lined Drainage Channels

Many of the everyday drainage channels are constructed with a lining on the bottom and the sides. These linings are used:

To protect the channel bottom and sides against erosion caused by the flowing water.

To enable channel bank side slopes of a steeper nature than those allowed by the natural angle of repose of the soil in which the channel is constructed.

To improve the flow characteristics for the channel by reducing the value of Manning's roughness coefficient.

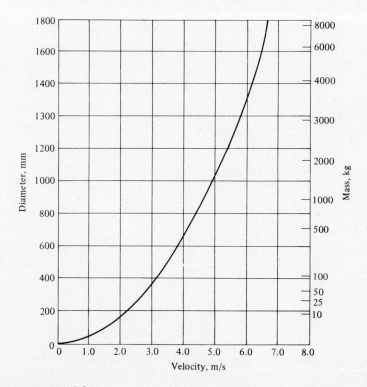

FIGURE 10.8

The most common materials used in channel linings are concrete, asphalt, gabions, riprap, and grass. Plank bottom and channel sides are used for special applications, such as log chutes. Corrugated-metal pipes of half-circular cross section are often used to carry water down the steep sides of embankments.

The recommended n coefficients for the various channel linings are included in Table 10.1.

GABION AND RIPRAP CHANNEL LINING

Gabion baskets are constructed of regularly shaped wire-mesh forms, filled in place with rock. The individual baskets are tied together to form a homogeneous flexible structure.

Riprap consists of stones hand-placed or dumped and spread mechanically on the sides and bottoms of channels. Stone size and mass are determined on the basis of the expected velocities in the channel. Figure 10.8 is a general guide in the selection of riprap size and mass for various velocities. The recommended sizes on this graph represent minimum sizes to be used. Generally, good riprap must consist of a well-graded mixture of different stone sizes, but most stones must be of the size indicated from the graph.

Both gabions and riprap must be supported on well-graded and designed granular filter beds or fiber filter cloths. The filters act as a barrier against the washing out of fine materials in the earth beneath the lining. A typical channel cross section of the type mentioned above is shown in Fig. 10.9.

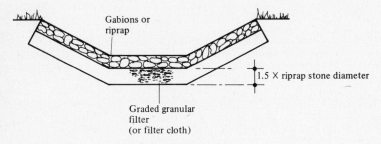

FIGURE 10.9

EXAMPLE 10.16

A trapezoidal channel must carry a flow of 5.5 m³/s. The bottom width of the channel is 2.0 m, and the channel side slopes are 2 horizontal to 2 vertical. The channel is constructed with a bottom slope of 1.0 percent. It is intended to line the channel with riprap. Determine the required depth of this channel and the required minimum size of the riprap stones.

SOLUTION

From Table 10.1 the n coefficient for this channel is found to be 0.033. The depth of flow can be found as in Example 10.8, and consequently the following table of calculations results.

Trial D	z	$\dfrac{b}{D}$	K	$\dfrac{s^{1/2}}{n}$	Q
1.0	2.0	0.50	1.58	3.03	4.80
1.5		0.75	1.80		16.00
1.1		0.55	1.62		6.30
1.05		0.53	1.60		5.50

Hence the depth of flow is

$$D = 1.05 \text{ m}$$

and the channel must be constructed to a depth of 1.30 m to provide for a 250-mm freeboard.

At the design depth of 1.05 m the area of the cross section of flow is found from Eq. (10.12), as follows:

$$A = (0.53 + 2.0) \times (1.05)^2$$

$$= 2.80 \text{ m}^2$$

and therefore the velocity of flow in the channel is

$$v = \frac{5.50}{2.80}$$

$$= 1.96$$

$$\approx 2.0 \text{ m/s}$$

Hence the recommended stone sizes for the riprap, as found from Fig. 10.8, are

$$\text{Min. diam.} = 180 \text{ mm}$$

$$\text{Min. mass} = 10 \text{ kg}$$

CONCRETE CHANNEL LININGS

Concrete channel linings can take on the form of interlocking prefabricated slabs or special shapes placed on the channel bottom and sides, or they can be cast-in-place concrete linings. In either case, care must be taken with these linings to ensure that proper groundwater drainage is provided through the channel sides. Lacking such drainage can result in serious cracking and failure of the lining from hydrostatic pressure or frost heave, or both.

GRASS-LINED CHANNELS

Golf courses, parks, and other areas where aesthetics dictate, or where large flows occur infrequently, are typical areas suited for grass-lined channels. These channels generally take the shape of shallow triangular cross sections with relatively flat side slopes. If the channels are such that a certain amount of water will be standing in the bottom of the channel for long periods of time, the grass in these areas will not grow and serious erosion problems could result. For grass-lined waterways maximum recommended velocities should not exceed 1.0 to 1.5 m/s.

EXAMPLE 10.17

A grass-lined channel, triangular in cross section and with side slopes of 4 horizontal to 1 vertical, must carry a flow of water of 3.0 m³/s. Determine the maximum slope at which this channel can be installed.

SOLUTION

From Table 10.1, the n coefficient of grass-lined channels is found to be 0.027. For this channel,

$$\frac{b}{D} = 0$$

and

$$z = 4.0$$

Assuming the maximum allowable velocity to be 1.5 m/s, the required area can be calculated as follows

$$A = \frac{3.00}{1.50}$$

$$= 2.00 \text{ m}^2$$

and from Eq. (10.12) rearranged, it follows that

$$D = \left(\frac{2.0}{0 + 4.0}\right)^{1/2}$$

$$= 0.71 \text{ m}$$

Therefore, from Table 10.2,

$$K = 2.47$$

and from Eq. (10.17) rearranged,

$$s = \left(\frac{3.0 \times 0.027}{0.71^{8/3} \times 2.47} \right)^2$$
$$= 0.006$$

CURB-AND-GUTTER CHANNELS

All paved roads have a curb-and-gutter type of channel at either edge of the pavement. The design of these drainage structures is generally in accordance with certain standards, varying somewhat from one municipality or authority to another. They generally take on a form similar to those shown in Figs. 10.1b and 10.10. These channels often follow the slope of the road centerline. Their hydraulic design is concerned with obtaining cleansing velocities during storm flows and to ensure that under normal flow the water surface does not extend onto the road pavement. The latter is for the convenience of motorists and pedestrians alike.

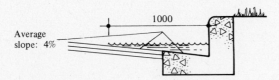

FIGURE 10.10

Design of Unlined Channels

Unlined open channels, when constructed in certain soils, are very susceptible to erosion of the bottom and sides of the channel. In addition, much care must be taken to ensure that the side slopes of the channel banks are compatible with the soil and will remain stable under all flow conditions.

Table 10.6 shows recommended values for maximum velocities and side slopes for channels constructed in a variety of soils. The values recommended in this table are for aged channels and are based on normal flow conditions, without excessive turbulence.

TABLE 10.6   Recommended Maximum Velocities and
Side Slopes for Unlined Channels

Soil Type	Maximum Velocity, m/s	Maximum z (hor./vert.)
Clay and sandy clay	0.6 –0.8	1.0–2.0
Silty sand	0.3 –0.6	1.5–2.0
Fine sand	0.15–0.3	2.0–4.0
Coarse sand	0.2 –0.5	1.5–3.0
Fine gravel	0.4 –0.9	1.5–2.5

10.6   SUMMARY OF THE CHAPTER IN BRITISH UNITS

Theoretical Considerations

The general theoretical discussions in this chapter related to the flow in open channels in the SI System of units applies equally to applications in the British System of units. North American practice generally accepts the Manning equation for use in open-channel flow. This formula, converted for use in the British System of units takes on the forms as outlined earlier, or

$$v = \frac{1.486 R^{2/3} s^{1/2}}{n} \tag{10.21}$$

or

$$Q = \frac{1.486 A R^{2/3} s^{1/2}}{n} \tag{10.22}$$

For circular pipe,

$$s = K Q^2 \tag{10.23}$$

and

$$s = K' Q^2 \tag{10.24}$$

where A = cross-sectional area of the flow in the channel, in square feet
R = hydraulic radius, in feet
v = velocity of flow, in feet per second
Q = capacity of flow, in cubic feet per second
s = slope of the energy gradient

n = channel roughness coefficient

K and K' are form factors.

Values of n are shown in Table 10.1 and values of K and K' are shown in Table 9.7.

Equation (10.17), developed in this chapter for trapezoidal channels, when translated to British units reads as follows:

$$Q = \frac{1.486KD^{8/3}s^{1/2}}{n} \qquad (10.25)$$

with values of the form factor K obtained from Table 10.2.

The Manning Formula Applied in British Units

The aids discussed in this chapter, as related to the SI System of units are also applicable to the solution of problems in the British System of units. This applies specifically to the use of Table 10.2 for trapezoidal channels and Table 10.5 and Fig. 10.7 for circular pipes, flowing full or partially full. The nomograph of Fig. 10.6 is in both systems of units.

Applications of the Manning formula and the aids mentioned above are illustrated in the following examples.

EXAMPLE 10.18

A concrete-lined channel, n=0.013, trapezoidal in shape, has side slopes of 2.0 horizontal, to 1.0 vertical and a bottom width of 3.0 feet. The rate of flow in this channel is 150 cfs, when the flow is uniform and at a depth of 18 inches. Determine the slope of the channel bottom.

SOLUTION

When the flow is uniform, as discussed earlier, the slope of the energy gradient, s, is equal to the slope of the channel bottom. For

$$z = 2.0$$

and

$$\frac{b}{D} = \frac{3.0}{1.5}$$

$$= 2.0$$

Table 10.2 provides a value for the form factor K as follows:

$$K = 2.902 \qquad \text{əd)}$$

Hence, from Eq. (10.25),

$$s = \left(\frac{150 \times 0.013}{1.5^{8/3} \times 1.486 \times 2.902} \right)^2$$

$$= 0.0089$$

EXAMPLE 10.19

A trapezoidal channel must be constructed with side slopes of 3.0 horizontal to 1.0 vertical. The channel will be grass lined, with $n = 0.025$. The bottom width of the channel is to be 2.5 feet and the channel bottom slope is 0.005. Determine the depth of flow, when the rate of flow is 90.0 cfs.

SOLUTION

As in the previous example of this type, this problem is best solved by a trial-and-error method. The method, in this instance, consists of choosing a trial depth and comparing the resulting Q with the required value of Q. The work involved in this operation is shown in tabular form below.

Trial D	$\dfrac{b}{D}$	K Low	K High	Actual	$\dfrac{1.486 s^{1/2}}{n}$	Q	Comment
2.5	1.0			2.672	4.20	129.0	high
1.25	2.0			3.559		27.0	low
2.0	1.25	2.672	3.112	2.892		77.0	low
2.15	1.16	2.672	3.112	2.813		91.0	OK

Note that the value of $D = 2.15$ ft was chosen after its determination from Fig. 10.11.

Allowing a freeboard of 10 to 12 inches, the recommended depth for this channel would be

$$\text{Depth} = 2 \text{ ft } 2 \text{ in.} + 10 \text{ in.}$$

$$= 36 \text{ in., or } 3.0 \text{ ft}$$

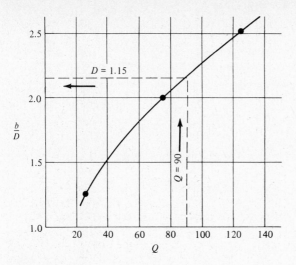

FIGURE 10.11

EXAMPLE 10.20

Determine the required diameter for a proposed sanitary sewer, which must carry domestic sewage at an average rate of 1.5 cfs. The maximum rate of flow, during peak flow, is estimated to be 250 percent of the average flow. The pipe is to be used is PVC, with $n=0.009$ and it is to be installed at a slope of 0.6 percent. The minimum allowable velocity in the pipe during average flow is 2.5 ft/s. Determine also the depth of flow at average and maximum flow conditions.

SOLUTION

The maximum flow this pipe will have to accommodate is

$$Q = 1.5 \times 2.5$$
$$= 3.75 \text{ cfs}$$

Using this Q and the specified slope, the form factor K for this pipe can be calculated from Manning's equation [Eq. (10.23)] as follows:

$$K = \frac{0.006}{(3.75)^2}$$
$$= 0.000427$$

From Table 9.6 for $n=0.009$ the required pipe diameter falls between 10 and 12 inches. Consequently a 12-inch pipe should be chosen.

At a slope of 0.6 percent such a pipe will have a capacity, when flowing full,

$$Q_f = \left(\frac{0.006}{0.0003773} \right)^{1/2}$$

$$= 3.99 \text{ cfs}$$

and

$$v_f = \frac{3.99}{0.785}$$

$$= 5.08 \text{ ft/s}$$

When flowing at maximum Q,

$$K_p = \frac{3.75}{3.99}$$

$$= 0.94$$

Hence, from Table 10.5,

$$\frac{D}{d} = 0.77$$

and

$$D = 12 \text{ in.} \times 0.77$$

$$= 9\tfrac{1}{4} \text{ in.}$$

At average Q,

$$K_p = \frac{1.5}{3.99}$$

$$= 0.38$$

and

$$\frac{D}{d} = 0.43$$

Hence

$$D = 12 \text{ in.} \times 0.43$$

$$= 5\tfrac{1}{8} \text{ in.}$$

During average Q, when $D/d = 0.43$, Table 10.5 shows that

$$K'_p = 0.93$$

Therefore, the velocity at average flow is

$$v_{ave} = 5.0\% \times 0.93$$
$$= 4.7 \text{ ft/s} > 2.5 \text{ ft/s}$$

which is acceptable.

SELECTED PROBLEMS IN SI UNITS

10.1   An earth canal, $n = 0.05$, has a bottom width of 3.0 m and side slopes of 2:1. If the slope of the canal bottom is 0.6 m/km, determine its capacity for depths of flow at 1.0 m, 1.25 m, 1.5 m, 1.75 m and 2.0 m.

10.2   Determine the bottom slope in meters per kilometer for a concrete-lined channel, $n = 0.012$, which is to carry 3.78 m³/s of water, if the channel bottom width is 3.25, the depth of flow is 2.26 m, and the channel side slopes are $\frac{3}{4}$:1.

10.3   A triangular channel has side slopes of 3:1, $n = 0.025$, $s = 0.006$, and $Q = 137$ L/s. Determine the velocity and the depth of flow.

10.4   A rectangular channel has an $n = 0.010$, a bottom width of 5.25 m, and a slope of the channel bottom of 0.001 m/m. Determine the depth of flow when the rate of flow is 32.6 m³/s.

10.5   A trapezoidal channel with $n = 0.035$, side slopes of 3:1, and a bottom slope of 7 m/km has a $Q = 375$ L/s. Determine the depth of flow.

10.6   The channel with a cross-section as shown in Fig. P10.6 has an n coefficient of 0.015 and a bottom slope of 0.0035 m/m. Determine the rate of flow.

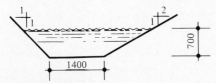

FIGURE P10.6

10.7   A concrete sewer pipe has a Q of 75 L/s. The diameter is 550 mm and $n = 0.015$. Determine the slope at which this pipe should be installed in order to have the velocity fall between 1.0 and 1.3 m/s.

10.8   A 300-mm-diameter storm sewer has a bottom slope of 0.05 m/m and carries a flow of 42 L/s. If $n = 0.015$, determine the depth and the velocity of flow in this pipe.

10.9   A trapezoidal channel has a bottom width of 2.0 m, side slopes of 1.5:1.0 and a bottom slope of 0.003 m/m. If $n = 0.013$, determine the depth of flow when the rate of flow is 145 L/s.

10.10   A concrete sewer pipe 300 m in diameter and installed at a slope of 0.025, carries a flow of 80 L/s. Determine the velocity and the depth of flow. Assume that $n = 0.013$.

10.11   A 200-mm asbestos cement pipe is laid on a slope of $s = 0.0028$ and water flows in it to depth of 165 mm. Determine the capacity of flow ($n = 0.015$).

11

Civil Engineering Applications of Fluid Flow: Nonuniform Open-Channel Flow

11.1 GENERAL CONSIDERATIONS

Steady Nonuniform Flow

In the preceding chapter the flow was uniform in all instances under consideration. That is, the velocity of flow remained constant throughout a certain reach of the stream. It was already pointed out then that many practical flow conditions do not conform to this criterion, but that the velocity of flow, the cross section and even the capacity of flow vary along the length of the reach under study.

In this chapter the flow conditions studied relate to steady, but nonuniform, flow. This type of flow is created by, among others, the following major causes:

1. Changes in the channel cross section.
2. Changes in the channel slope.
3. Certain obstructions, such as dams or gates, in the stream's path.
4. Changes in the discharge—such as in a river, where tributaries enter the main stream.

The first three of the above situations can exist without a change in the rate of flow, or during steady-flow conditions. The last is an example of unsteady flow and will not be discussed here.

Steady nonuniform flow then can take on two forms, retarded flow and accelerated flow, where the flow with a constant capacity either slows down or increases in velocity, depending on the conditions of the channel.

Accelerated and Retarded Flow

An idealized section of a reach of a channel with accelerated and retarded flow conditions is shown in Fig. 11.1. As the flow accelerates, with the rate of flow constant, the depth D must decrease from point 1 to point 2, and a water-surface profile as shown in Fig. 11.1a results. Retarded flow will produce a water-surface profile as shown in Fig. 11.1b.

Significant in each one of the above cases is the fact that now the water surface is a curved line and no longer parallel to the channel bottom and the energy gradient, as was the case for uniform flow. The following points are made in connection with the above observations.

1. The water surface, as will be shown later, can have a concave or convex shape.
2. The energy gradient is not necessarily a straight line; however, as discussed earlier, it is assumed that the energy loss is constant along the length of the reach and it will be so represented and considered to have a slope $s = H_l/l$.
3. As was done in the case of uniform flow, it is here also accepted that the depth of flow D is equal to the pressure head in the energy equation. Obviously, this applies only when the slope of the channel bottom is small. For very steep slopes, allowances for this discrepancy must be made.

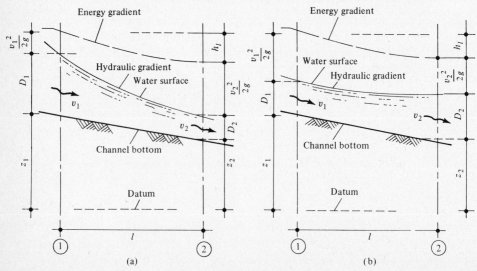

FIGURE 11.1

Total and Specific Energy in Nonuniform Open-Channel Flow

A study of the water-surface profiles and conditions of Fig. 11.1 indicates that the total energy and energy head at any point along these streams can be expressed, as for uniform flow, in the form of Eq. (6.39), or

$$E = \frac{v^2}{2g} + D + z$$

where E = total energy head
$\quad v$ = average velocity of flow of the liquid
$\quad D$ = pressure head or the depth of flow
$\quad z$ = elevation head, related to a datum

Figure 11.2 represents a longitudinal section through an open channel and illustrates the above relationship. In the case of flow under pressure, pipe flow, it is customary to refer to the centerline of the conduit as the point to which the elevation head refers. This is of course logical, since the cross section in these cases is well defined and constant. Hence the centerline is a fixed point in the conduit and a handy reference for elevations.

In open-channel flow no such fixed centerline exists, since the water surface varies in elevation. It is therefore necessary to find a more permanent reference point. This is of course the channel bottom. Much of the calculations related to open-channel flow are performed by using the energy related to the channel bottom rather than a common constant datum. This energy, using the channel bottom as the base, is called the *specific energy head H*. From the above and from a study of Fig. 11.2 it follows that

$$H = \frac{v^2}{2g} + D \tag{11.1}$$

Equation (11.1) is a very useful relationship in studying and comparing energy relationships at various points along a reach of an open channel.

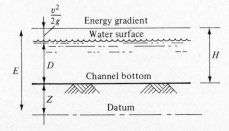

FIGURE 11.2

Alternate Stages of Flow and Critical Depth

If Eq. (11.1) is solved for v, the following expression results.

$$v = (2g(H - D))^{1/2} \tag{11.2}$$

Multiplying both sides of Eq. (11.2) with A, the cross-sectional area of the flow, and remembering that $Q = vA$, produces

$$Q = A(2g(H - D))^{1/2} \tag{11.3}$$

An analysis of these equations indicates that they will yield two real values of D for each value of Q or v.

To illustrate the above relationship between the depth of flow D, the specific energy head H, and the rate of flow Q, consider the simpler case of a rectangular channel with steady flow and graphically represent the relationships of Eq. (11.3).

For a rectangular channel, with a width b and a rate of flow Q, the rate of flow per unit width, q, is

$$q = \frac{Q}{b} \tag{11.4}$$

From Fig. 11.3, the cross-sectional area a per unit width of channel is

$$a = D \times 1.0 \tag{11.5}$$

Hence, Eq. (11.3) can be rearranged for a rectangular channel to read

$$q = D(2g(H - D))^{1/2} \tag{11.6}$$

Plotting values of q—calculated for various values of D from Eq. (11.6), maintaining a constant specific energy head—produces the graph shown in Fig. 11.4a. The graph in question was derived for a rectangular channel, 2.0 m wide, with a constant specific energy head of 3.0 m.

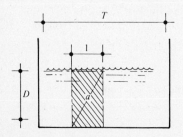

FIGURE 11.3

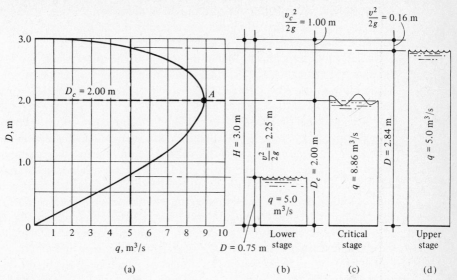

FIGURE 11.4

An analysis of the graph of Fig. 11.4a leads to the following observations:

1. Any value of q corresponds to two values of D. For example, with a constant specific energy head, $H = 3.0$ m, a flow of 5.0 m³/s can take place at a depth of 0.75 m or 2.84 m.
2. The only exception to the above is the point A. At this point only one value of D is found to correspond to one value of q. This value of q is the maximum value q can reach with an energy head of 3.0 m.
3. In the immediate vicinity of the point A, relatively small changes in q will produce large changes in D.

The three observations above are important considerations and lead to definitions used in the study of nonuniform open-channel flow.

When flow in the channel takes place where D and q are represented by the point A on the graph of Fig. 11.4a, the flow is said to take place at *critical depth*. The velocity of flow and the slope when the flow is at critical depth are referred to as the *critical velocity* v_c and the *critical slope* s_c. The depth at which this flow occurs is called the critical depth, D_c. This depth has twofold significance in open-channel flow.

First, if the flow is not taking place at the critical depth, there are always two possible other depths corresponding to a particular q. One of these alternate depths is always above and the other is always below the critical depth, D_c. Thus D_c divides the channel into two regions or stages of flow: the upper stage, where the depth of flow D is larger than D_c, and the lower stage, where D is less than D_c. For instance, in the case of the graph of Fig. 11.4a, when q is 5.0 m³/s, the flow could take place at a depth of 0.75 m at the lower stage or 2.84 m at the upper stage. The actual longitudinal sections of the channel at these flows, together with their respective velocity and energy heads, are shown in Figs. 11.4*b* and 11.4*d* respectively. When the flow is in the upper stage, the velocity is less than the critical velocity and this condition of flow is often called *subcritical flow*. Flow at the lower stage, with velocities larger than critical velocity, is called *supercritical flow*.

Second, when the flow takes place at a depth near critical depth, minor changes in energy head or *q* will produce large changes in depth. Actually, depending on conditions, the depth of flow could be expected to fluctuate between upper and lower stage. Such conditions are in practice accompanied by relatively rough water-surface conditions. Small waves are formed on the surface and these could prove damaging to channel walls and certain types of channel linings. Flow at critical depth is shown schematically in Fig. 11.4*c*.

As will become more evident later, and as can already be deduced from the above discussions, location of critical depth in an open channel is an important consideration in the solving of design and analysis of problems related to open-channel flow.

Figure 11.4*a* indicates that the critical depth D_c is the value of D that produces the maximum value of *q* in Eq. (11.6). The value of D_c can therefore be found by means of differential analysis. In the general case, such analysis indicates that the following condition must be satisfied in order for the flow to be taking place at critical depth.

$$\frac{A^3}{T} = \frac{Q^2}{g} \tag{11.7}$$

where A = cross-sectional area of the flow (see Fig. 11.5)

T = top width of the channel at the water surface (see Fig. 11.5)

Q = rate of flow

g = acceleration due to gravity

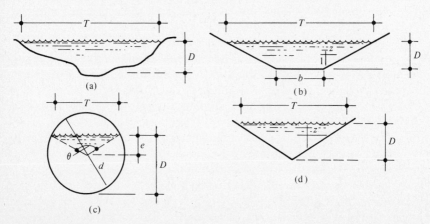

FIGURE 11.5

EXAMPLE 11.1

Determine whether $D_c = 2.0$ m in the channel in Fig. 11.4 satisfies the requirements of Eq. (11.7), and find v_c for this channel. The channel is 4.0 m wide.

SOLUTION

From the graph of Fig. 11.4 when the depth is 2.0 m the corresponding q per unit of channel width is 8.86 m³/s per meter. Therefore

$$Q = 8.86 \times 4.0$$

$$= 35.4 \text{ m}^3/\text{s}$$

and

$$A = 2.0 \times 4.0$$

$$= 8.0 \text{ m}^2$$

Since the channel is rectangular and 4.0 m wide,

$$T = 4.0 \text{ m}$$

Substitution of these values in Eq. (11.7) yields

$$\frac{(8.0)^3}{4.0} = \frac{(35.4)^2}{9.81}$$

$$128 \text{ m}^2 = 128 \text{ m}^2$$

and Eq. (11.7) is satisfied.

The velocity when D is at critical depth is v_c, or

$$v_c = \frac{35.4}{8.0}$$

$$= 4.43 \text{ m/s}$$

The Froude Number and Its Relationship to Critical Depth

The *Froude number* is a dimensionless number related to the velocity and depth of flow. It is generally defined as

$$F_N = \frac{v}{(gD)^{1/2}} \qquad (11.8)$$

Equation (11.1), the expression for the specific energy in a flowing fluid, can be expressed in terms of the Froude number. Dividing this equation by D yields

$$\frac{H}{D} = \frac{v^2}{2gD} + 1 \tag{11.9}$$

or

$$H = D\left(\frac{F_N^2}{2} + 1\right) \tag{11.10}$$

Figure 11.6 is a plot of values of H versus F_N of Eq. (11.10). This graph is for the conditions of Fig. 11.4, namely for a rectangular channel, with a rate of flow per unit width of $q = 8.86\ \mathrm{m^3/s}$. As will be remembered from the discussions earlier, D_c for this channel is at 2.0 m.

At this depth the velocity is critical and equal to

$$v_c = \frac{8.86}{2.0}$$

$$= 4.43\ \mathrm{m/s}$$

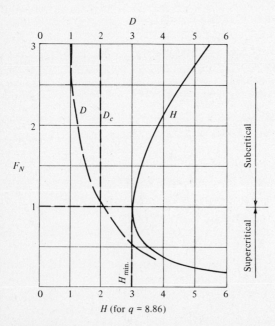

FIGURE 11.6

Hence

$$F_N = \frac{4.43}{(9.81 \times 2.0)^{1/2}}$$

$$= 1.0$$

Reference to Fig. 11.5 indicates that this value of F_N corresponds with a specific energy head of 3.0 m, which is also the minimum energy head, as was to be expected. The graph of Fig. 11.5 further indicates an important conclusion in that when F_N is larger than 1.0, the depth of flow is more than critical depth, and the flow is subcritical. While with F_N less than 1.0, the flow is supercritical.

The Froude number therefore is an important and often convenient tool in determining whether flows are subcritical or supercritical.

Aids in Determining Critical Depth

Solving Eq. (11.7) can be time-consuming, particularly when applied to trapezoidal and circular channels. Methods to facilitate the work involved are presented here.

Rectangular channels are easily solved when the work is done by using a channel section of unit width. In these channels the various factors required for Eq. (11.7) are

$$A = D_c T \tag{11.11}$$

$$\frac{A^3}{T} = D_c{}^3 T^2 \tag{11.12}$$

and if q is the rate of flow per unit width

$$Q = Tq \tag{11.13}$$

Substitution of these values in Eq. (11.7) yields

$$D_c{}^3 T^2 = \frac{T^2 q^2}{g} \tag{11.14}$$

which simplifies to

$$D_c = \left(\frac{q^2}{g}\right)^{1/3} \tag{11.15}$$

For triangular channels, using Fig. 11.5 and a similar reasoning as above, the following ensues.

$$T = 2zD_c \tag{11.16}$$

$$A = zD_c{}^2 \tag{11.17}$$

$$\frac{A^3}{T} = \frac{z^3 D_c{}^6}{2zD_c} \tag{11.18}$$

which, when simplified and substituted into Eq. (11.7), yields

$$D_c = \frac{2}{g} \left(\frac{Q}{z} \right)^{2/5} \tag{11.19}$$

From Fig. 11.6 and earlier work in Chapter 10, the following values for A and T for trapezoidal channels can be found.

$$A = D^2(r + z) \tag{11.20}$$

and

$$T = D(r + 2z) \tag{11.21}$$

where

$$r = \frac{b}{D} \tag{11.22}$$

When the flow is at critical depth, substituting the corresponding values in Eq. (11.7) leads to

$$\frac{D_c{}^6(r + z)^3}{D_c(r + z)} = \frac{Q^2}{g} \tag{11.23}$$

or

$$D_c = \left(\frac{r + 2z}{g(r + z)^3} \right)^{1/5} Q^{2/5} \tag{11.24}$$

which can be simplified to

$$D_c = K_c Q^{2/5} \tag{11.25}$$

where

$$K_c = \left(\frac{r + 2z}{g(r + z)^3} \right)^{1/5} \tag{11.26}$$

TABLE 11.1   Values for K_c for Determining D_c for Trapezoidal Channels

b/D	z 0.25	0.50	0.75	1.00	1.25	1.50	1.75	2.00	3.00	4.00	5.00
0.00	1.270	0.960	0.816	0.728	0.665	0.619	0.582	0.551	0.469	0.418	0.382
0.50	0.753	0.687	0.636	0.596	0.564	0.537	0.514	0.494	0.434	0.394	0.364
1.00	0.601	0.570	0.544	0.521	0.500	0.482	0.466	0.452	0.407	0.374	0.349
1.50	0.520	0.502	0.485	0.470	0.456	0.443	0.431	0.420	0.384	0.357	0.336
2.00	0.477	0.455	0.444	0.432	0.422	0.412	0.403	0.395	0.366	0.343	0.324
2.50	0.430	0.421	0.412	0.404	0.395	0.388	0.380	0.374	0.349	0.330	0.313
3.00	0.401	0.394	0.387	0.380	0.374	0.368	0.362	0.356	0.335	0.318	0.304
3.50	0.378	0.372	0.367	0.361	0.356	0.351	0.346	0.341	0.323	0.308	0.295
4.00	0.359	0.354	0.350	0.345	0.340	0.336	0.332	0.328	0.312	0.299	0.287
4.50	0.343	0.339	0.335	0.331	0.327	0.323	0.320	0.316	0.303	0.291	0.280
5.00	0.329	0.326	0.322	0.319	0.316	0.312	0.309	0.306	0.294	0.283	0.274
6.00	0.307	0.304	0.301	0.299	0.296	0.293	0.291	0.288	0.279	0.270	0.262
7.00	0.289	0.287	0.284	0.282	0.280	0.278	0.276	0.274	0.266	0.258	0.251
8.00	0.274	0.272	0.270	0.268	0.267	0.265	0.263	0.262	0.255	0.248	0.242
9.00	0.262	0.260	0.258	0.257	0.256	0.254	0.252	0.251	0.245	0.240	0.234
10.0	0.251	0.250	0.248	0.247	0.246	0.244	0.243	0.242	0.237	0.232	0.227
15.0	0.214	0.213	0.212	0.212	0.211	0.210	0.209	0.209	0.206	0.203	0.200
20.0	0.191	0.190	0.190	0.189	0.189	0.188	0.188	0.187	0.185	0.183	0.181

$D_c = K_c Q^{2/5}$

TABLE 11.2 Values of K_c for Determining D_c in Circular Channels

D/d	0.00	0.01	0.02	0.03	0.04	0.05	0.06	0.07	0.08	0.09
0.00	0.000	0.244	0.280	0.304	0.323	0.338	0.350	0.362	0.372	0.381
0.10	0.389	0.397	0.404	0.411	0.418	0.424	0.430	0.436	0.441	0.446
0.20	0.451	0.456	0.460	0.465	0.469	0.474	0.478	0.482	0.486	0.489
0.30	0.493	0.497	0.500	0.504	0.507	0.511	0.514	0.517	0.520	0.524
0.40	0.527	0.530	0.533	0.536	0.538	0.541	0.544	0.547	0.550	0.552
0.50	0.555	0.557	0.560	0.563	0.565	0.568	0.570	0.572	0.575	0.577
0.60	0.579	0.581	0.584	0.586	0.588	0.590	0.592	0.594	0.596	0.598
0.70	0.600	0.601	0.603	0.605	0.606	0.608	0.609	0.611	0.612	0.613
0.80	0.614	0.615	0.616	0.617	0.617	0.617	0.617	0.617	0.617	0.616
0.90	0.614	0.612	0.610	0.606	0.602	0.596	0.588	0.576	0.558	0.525

$D_c = K_c Q^{2/5}$

Values for K_c for different side slopes and z and r ratios are shown in Table 11.1.

A similar reasoning will lead to the fact that Eq. (11.26) can be adapted to flow in circular channels. For circular channels,

$$K_c = \left(\frac{1024r(1/r - 1)^{1/2}}{g(\theta - \sin \theta)^3}\right)^{1/5} \tag{11.27}$$

where r is the ratio of the depth of flow to the pipe diameter, or

$$r = \frac{D}{d} \tag{11.28}$$

and other values are as shown in Fig. 11.5c. Values for K_c for circular channels are shown in Table 11.2.

The following examples illustrate the calculations related to critical depth and the use of Tables 11.1 and 11.2.

EXAMPLE 11.2

Determine D_c in a rectangular channel 2.5 m wide, with a steady flow of 15.0 m³/s.

SOLUTION

From the given information,

$$q = \frac{15.0}{2.50}$$

$$= 6.0 \text{ m}^3/\text{s per meter}$$

Therefore, from Eq. (11.13),

$$D_c = \left(\frac{6.0^2}{9.81}\right)^{1/3}$$

$$= 1.54 \text{ m}$$

EXAMPLE 11.3

At a point in rectangular channel, the depth of flow is observed to be 1.15 m. The channel is 2.0 m wide and the rate of flow is 10.0 m³/s. Is the flow in this channel in the upper or lower stage?

SOLUTION

The rate of flow per unit width of channel is

$$q = \frac{10.0}{2.0}$$

$$= 5.0 \text{ m}^3/\text{s per meter}$$

Therefore, from Eq. (11.13),

$$D_c = \left(\frac{5.0^2}{9.81}\right)^{1/3}$$

$$= 1.37 \text{ m}$$

Hence, since the depth of flow of 1.15 m is less than the critical depth of 1.37 m, the flow is in the lower stage.

EXAMPLE 11.4

If the channel in the above problem is constructed of concrete, with $n = 0.015$, at which slope of the channel bottom would flow occur at critical depth, and what would be the critical velocity?

SOLUTION

For this channel to flow at critical depth, the channel slope has to be such that the Manning equation is satisfied for $D = D_c$, or 1.37 m. Hence, from Eq. (10.17),

$$10.0 = 1.37^{8/3} \times K \times \frac{s^{1/2}}{0.015}$$

In this case,

$$\frac{D}{b} = \frac{1.37}{2.0}$$

$$= 0.69$$

and, from Table 10.2,

$$K = \frac{0.481 - 0.171}{50} \times 17 + 0.171$$

$$= 0.276$$

Hence substituting this value of K yields

$$s_c = \left(\frac{10.0 \times 0.015}{1.37^{8/3} \times 1.83} \right)^2$$

$$= 0.0551$$

Also,

$$v_c = \frac{10.0}{2.0 \times 1.37}$$

$$= 3.65 \text{ m/s}$$

EXAMPLE 11.5

A grass-lined trapezoidal channel with $n = 0.035$, a base width of 1.50 m and side slopes of 3 vertical to 1 horizontal must carry a flow of 65.0 m³/s. The channel bottom has a slope of 4.5 m per km. Will the flow in this channel be in the upper or lower stage?

SOLUTION

In order to determine whether the flow is in the upper or lower stage, it is necessary to find the normal depth of flow, D, at which the flow will be 65.0 m³/s, as well as the value of the critical depth D_c.

The normal depth D, can be found as illustrated in Example 10.8. In this instance, $n = 0.035$, $Q = 65.0$ m³/s, and $s = 0.0045$. The work table then will be as follows.

| Trial | b | K | | | $s^{1/2}$ | |
D	D	High	Low	Average	n	Q
2.00	0.75	2.67	2.24	2.46	1.92	30.0
3.00	0.50			2.24		80.5
2.70	0.56	2.67	2.24	2.28		62.1
2.75	0.55			2.28		65.0

Consequently, the normal depth of flow for this channel with $Q = 65.0$ m³/s is 2.75 m.

The value of D_c can be found in a manner similar to the above, by trial and error, and using Table 11.1 for values of K_c and Eq. (11.25). An appropriate work table to facilitate the work involved is as shown below.

Trial D_c	$r=\dfrac{b}{D}$	K_c Low	K_c High	K_c Average	$Q^{2/5}$	D_c
2.00	0.75	0.407	0.434	0.420	5.31	2.23
2.50	0.60			0.429		2.28
2.30	0.65			0.426		2.26
2.25	0.67			0.425		2.26

Therefore D_c for this channel, when the flow is 65.0 m³/s, is 2.25 m.

The flow is obviously in the upper stage, since the normal depth of flow, 2.75 m, is larger than D_c.

A quick check of this value for D_c can be made by means of Eq. (11.7), as follows:

$$A = 2.25^2(0.67 + 3.0)$$

$$= 18.6 \text{ m}^2$$

$$T = 2.25(0.67 + 2 \times 3.0)$$

$$= 15.0 \text{ m}$$

$$\frac{A^3}{T} = \frac{18.6^3}{15.0}$$

$$= 429 \text{ m}^2$$

$$\frac{Q^2}{g} = \frac{65.0^2}{9.81}$$

$$= 430 \text{ m}^2$$

EXAMPLE 11.6

The normal flow in a circular PVC pipe, with $n = 0.010$ and a diameter of 700 mm, is estimated to be 450 L/s. The pipe is installed at a slope of $s = 0.008$. Determine whether the flow in this pipe will be in the upper or lower stage.

SOLUTION

To find the normal depth of flow one of the methods described in the previous chapter can be used. From Table 10.3, for a pipe 700 mm in diameter and with $n = 0.010$,

$$K = 0.0077$$

Hence, from Eq. (10.18),

$$Q_f = \left(\frac{0.008}{0.0077}\right)^{1/2}$$

$$= 1.02 \text{ m}^3/\text{s}$$

Therefore, from Eq. (10.19),

$$K_p = \frac{0.450}{1.020}$$

$$= 0.44$$

and, from Table 10.5,

$$\frac{D}{d} = 0.45$$

Hence the normal depth of flow D is

$$D = 0.45 \times 700$$

$$= 315 \text{ mm}$$

To find D_c, a trial-and-error procedure, using Eq. (11.25) and Table 11.2 can be used. The resulting work table is shown below.

Trial D_c	r	K_c	$Q^{2/5}$	"D_c"
350	0.50	0.555	0.727	400
450	0.64	0.588		427
400	0.57	0.572		416
420	0.60	0.579		421

D_c is 421 mm, which is greater than the normal depth D of 278 mm; hence the flow will be at the lower stage.

Channel Cross-Section Changes

Often a channel cross section is reduced by narrowing of the channel width or by a rise in the channel bottom, or both. Examples of the former are cases of bridge abutments intruding on a river cross section.

In such instances if uniform flow existed upstream of the constriction, flow would change to nonuniform conditions at the constriction. The velocity would increase and the depth of flow would also change. If the constriction of the channel is sufficiently large, the flow upstream will not remain uniform but will require an increase in energy to maintain flow at a steady or constant rate. This is generally accompanied by an increase in the depth of flow upstream, or by "backing up" of the flow.

An analysis of flow conditions related to constrictions in the channel cross section and the relationships between depth of flow and specific energy are demonstrated in the following example.

EXAMPLE 11.7

A rectangular channel 4.0 m wide carries a flow of 32.0 m³/s at a constant depth of 3.18 m. At a point downstream, the channel width has been reduced for a short distance to 3.0 m. Determine the depth and velocity of flow at this constriction. Determine also the maximum constriction that can be introduced in this channel before backing up will occur upstream.

SOLUTION

The plan of this channel in the vicinity of the constriction is shown in Fig. 11.7d. Equation (11.7) expresses the relationship between the specific energy head and the depth of flow.

For the channel in question, upstream of the constriction,

$$q = \frac{32.0}{4.0}$$

$$= 8.0 \text{ m}^3/\text{s per meter}$$

$$D = 3.18 \text{ m}$$

$$v = \frac{8.0}{3.18}$$

$$= 2.52 \text{ m/s}$$

$$H = 3.18 + \frac{2.52^2}{2 \times 9.81}$$

$$= 3.50 \text{ m}$$

At the constriction,

$$q = \frac{32.0}{3.0}$$

$$= 10.67 \text{ m}^3/\text{s per meter}$$

At the constriction, Eq. (11.7) takes on the following form

$$10.67 = D((2 \times 9.81(3.50 - D))^{1/2}$$

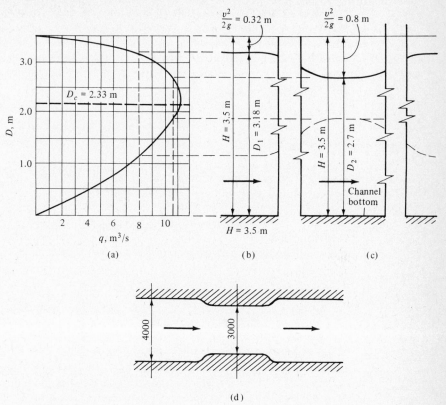

FIGURE 11.7

The flow being steady and uniform before the constriction, there is no change in the specific energy. However, at the constriction itself, due to the turbulence created by forcing the water at the outside of the channel to converge to the center, there will be a certain amount of energy loss. The magnitude of this energy loss will of course depend on the type of constriction, the channel materials, and the rate and velocity of flow. For this discussion the energy loss through the constriction has been assumed to be negligible, a circumstance that could often occur in practice.

To solve the above equation for values of D and to assist the required analysis a graph of D versus q was constructed, as was done previously in Fig. 11.4a. The resulting graph is shown in Fig. 11.7a.

A study of this graph provides the following information regarding the flow upstream of the constriction:

1. The critical depth is $D_c = 2.33$ m.
2. The flow upstream, when at a depth of 3.18 m, is in the upper stage.
3. The corresponding lower-stage depth is 1.18 m.

At the constriction, when $q=10.67$ m³/s per meter, the graph indicates the flow in the upper stage to be at a depth of 2.70 m. Therefore at the constriction the water surface will reduce and generally conform to the shape shown in Fig. 11.7c. The velocity of flow at the constriction will be

$$v = \frac{10.67}{2.70}$$

$$= 3.95 \text{ m/s}$$

If the channel width is reduced further than that shown, the rate of flow per unit width, q, will increase until it reaches the maximum value or, as read from the graph,

$$q_{max} = 11.16 \text{ m}^3\text{/s per meter}$$

A channel width corresponding to this q and the steady flow Q is

$$b = \frac{32.0}{11.16}$$

$$= 2.87 \text{ m}$$

Any further reduction in channel width would result in a further increase in q and a corresponding increase in Q. This can only happen if the graph in Fig. 11.7a is altered to allow q_{max} to move to the right of its present position. In order to do so will require an increase in H and D, or backing up of the flow.

In the above example, if the flow had been at the lower stage upstream of the constriction, the constriction would create a rise in the water surface similar to that shown in dotted line in Fig. 11.7c.

A rise in the channel bottom will also create a change in the water surface. The specific energy is related to the channel bottom and as such a change in it will result in a change in the specific energy head.

In Eq. (11.1) if v is substituted with its value Q/A, the following expression results.

$$H = \frac{Q^2}{2gA^2} + D \tag{11.29}$$

For a rectangular channel, with q the rate of flow per unit width, this equation becomes

$$H = \frac{q^2}{2gD^2} + D \tag{11.30}$$

The behavior of the water surface when the channel bottom is raised or lowered for a short distance in a channel can be readily predicted by means of these expressions, as is demonstrated in the following example.

EXAMPLE 11.8

The channel in Example 11.7 has a rise in the bottom of 0.6 m as shown in Fig. 11.8c. The channel width, rate of flow, and upstream depth of flow remain 4.0 m, 32.0 m³/s, and 3.18 m, respectively. Determine the depth of flow and velocity at the point where the rise in channel bottom occurs. Determine also the maximum rise that can be allowed in the channel bottom without creating backing up conditions upstream.

SOLUTION

Equation (11.30) will also present two values for D for every value of H. Figure 11.8a represents a graphical presentation of the values of D related to values of H. This graph is based on

$$D = 3.18 \text{ m}$$

$$q = \frac{32.0}{4.0}$$

$$= 8.0 \text{ m}^3/\text{s per meter}$$

$$H = \frac{3.26}{D^2} + D$$

Values of D and the velocity head can be read from this graph for values of H. It should be noted that H reaches its minimum value when

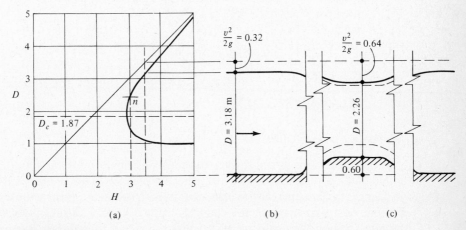

FIGURE 11.8

$D = D_c$. Figure 11.8*b* represents flow conditions upstream of the channel constriction.

When the bottom of the channel is raised by 0.6 m, as shown in Fig. 11.8*c*, the specific energy head H is reduced to

$$H = 3.5 - 0.6$$

$$= 2.90 \text{ m}$$

From observation of the graph of Fig. 11.8*a*, a specific energy head of 2.9 m, corresponds to a depth of flow of 2.26 m. Therefore the new water surface elevation will be at a distance of

$$D' = 0.6 + 2.26$$

$$= 2.86 \text{ m}$$

above the original channel bottom. This represents a net drop in water surface of 0.32 m. An approximation of the water-surface profile is shown in Fig. 11.8.

The reduced depth of flow leads to an increase in velocity and the velocity through the constriction is

$$v = \frac{8.0}{2.26}$$

$$= 3.54 \text{ m/s}$$

Similar reasoning as that used in the previous chapter will lead to the conclusion that the depth of flow cannot be reduced beyond D_c, without the addition of energy, or backing up, upstream. From the graph, D_c occurs when $H = 2.80$ m. Therefore, the maximum amount that the bottom can be raised is

$$H - D_c = 2.80 - 1.87$$

$$= 0.93 \text{ m}$$

This condition is represented in dotted lines in Fig. 11.8*c*.

11.2   GRADUALLY VARYING FLOW

Water-Surface Profiles

In the previous sections, when channel constrictions were discussed it was found that a point is reached where further constriction of the channel will result in changes in the water surface upstream. In both instances, a rise in the channel bottom or a narrowing of the channel width beyond a certain point would require a rise in the upstream water surface in order to maintain steady flow.

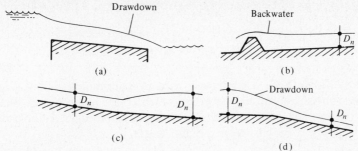

FIGURE 11.9

Some other situations resulting in such changes are shown in Fig. 11.9. In these figures, the notation D_n indicates the *normal depth* of flow, or the depth at which flow is steady and uniform.

The condition depicted in Fig. 11.9a is the case of a channel connecting two reservoirs. If this channel is sufficiently long, uniform flow can occur over a certain distance in the center portion of the channel, where the effects of the entry and discharge conditions are not felt. Also, as will be seen later, in the cases of Figs. 11.9b and c, the conditions could be such that the water-surface profile will change rapidly, accompanied by high turbulence and energy losses.

If the Bernoulli expression is written for points 1 and 2 in Fig. 11.1, the following expression results.

$$\frac{v_1{}^2}{2g} + D_1 + z_1 = \frac{v_2{}^2}{2g} + D_2 + z_2 + H_l \tag{11.31}$$

From the geometry of the channel and the flows in Fig. 11.1, the following deductions can be made.

1. If s is the average slope of the energy gradient, and l is the distance between points 1 and 2, then

$$H_l = s \times l \tag{11.32}$$

2. If s_0 is the slope of the channel bottom, then

$$s_0 \times l = z_1 - z_2 \tag{11.33}$$

Substituting values of H_l and $z_1 - z_2$ from Eqs. (11.32) and (11.33) into Eq. (11.31) and solving for l yields

$$l = \frac{\dfrac{v_2{}^2}{2g} + D_2 - \dfrac{v_1{}^2}{2g} - D_1}{s_0 - s} \tag{11.34}$$

or, using the specific energy-head relationship of Eq. (11.1),

$$l = \frac{H_2 - H_1}{s_0 - s}$$

(11.35)

Equation (11.35) is commonly used in determining actual water-surface profiles. The slope s of the energy gradient in this equation varies all along the length of the reach of the channel when varying flow occurs. As indicated earlier, it is the average slope between points 1 and 2.

Using the Manning equation for determining s in this instance would naturally introduce error into the calculations, since this equation has been developed only for uniform flow. This relationship is illustrated in Fig. 11.10. Here the line ab represents the energy gradient for a reach of the channel with length l. Obviously, the straight line ab, which represents the average slope of the energy gradient, varies considerably from the actual energy gradient. However, if the reach of the channel is shortened, to say the length l', the variance between the average and actual energy gradients is greatly reduced. Obviously, if the reach length were to be made infinitesimally small, the average and actual values of s would become identical and values of s calculated by means of Eq. (11.34) or (11.35) would be the actual slope of the energy gradient.

In practice, channel profiles are analyzed by dividing the channel in a number of convenient reaches and solving Eq. (11.35) for unknown values, usually the reach length, by assuming that s is the slope of the energy gradient if the flow were uniform in the reach. This procedure is illustrated in detail in the following example.

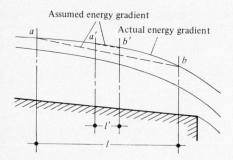

FIGURE 11.10

EXAMPLE 11.9

The flow in a long, rectangular concrete channel, $n = 0.013$, 3.5 m wide and with a bottom slope of 1 m/km, is controlled by a gate as shown in Fig. 11.11. Determine the water-surface profile upstream of this gate.

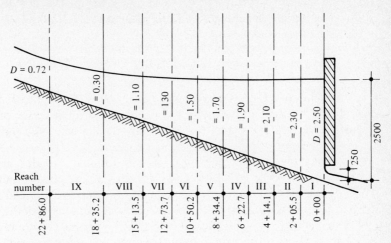

FIGURE 11.11

SOLUTION

The capacity of flow can be determined from Eq. (8.14), or

$$Q = CA(2gh)^{1/2}$$

In this instance,

$$A = 0.25 \times 3.5$$
$$= 0.875 \text{ m}^2$$
$$h = 2.5 - 0.125$$
$$= 2.375 \text{ m}$$

and, from Table 8.2,

$$C = 0.65$$

Hence

$$Q = 0.65 \times 0.875 \times (2 \times 9.81 \times 2.375)^{1/2}$$
$$= 3.9 \text{ m}^3/\text{s}$$

and the flow per unit width is

$$q = \frac{3.9}{3.5}$$
$$= 1.11 \text{ m}^3/\text{s per meter}$$

Because of the gate the flow is retarded, and for a certain distance upstream of the gate flow will not be uniform. However, at a certain point a sufficient distance upstream of the gate, the flow will be essentially uniform again and at a depth D_n. This depth can be determined as illustrated previously and with the aid of Eq. (10.17) and Table 10.2. The calculations are tabulated below.

Trial D_n	$\dfrac{b}{D}$	K	$\dfrac{s^{1/2}}{n}$	Q
0.70	5.0	3.995	2.43	3.75
0.75	4.67	3.68		4.15
0.72	4.86	3.86		3.90

Hence the normal depth of flow in this channel is

$$D_n = 0.72 \text{ m}$$

With the aid of Eq. (11.35), an approximate value of the distance upstream of the gate where the flow will be uniform can be found in the following way. The subscripts D and U will be used in this case to denote values pertaining to the downstream and upstream cross sections, respectively. Then

$$V_D = \frac{3.9}{3.5 \times 2.5}$$

$$= 0.44 \text{ m/s}$$

$$V_U = \frac{3.9}{3.5 \times 0.72}$$

$$= 1.55 \text{ m/s}$$

$$H_D = \frac{0.44^2}{2 \times 9.81} + 2.5$$

$$= 2.51$$

$$H_U = \frac{1.55^2}{2 \times 9.81} + 0.72$$

$$= 0.84$$

The only remaining term to be determined in Eq. (11.35) is s, the slope of the energy gradient. This is accomplished by assuming that the flow in the channel is uniform. In this case, s can be found from Eq. (10.17). When flow is uniform, the depth is constant. Then the constant depth would be the average of the upstream and downstream depths, or

$$D_{\text{ave}} = \frac{2.5 + 0.72}{2}$$

$$= 1.61 \text{ m}$$

Hence, solving Eq. (10.17) yields

$$\frac{b}{D}=\frac{3.5}{1.61}$$

$$=2.17$$

$$K=1.41$$

and

$$s=\left(\frac{3.9\times0.013}{1.61^{8/3}\times1.41}\right)^2$$

$$=0.000102$$

Therefore, from Eq. (11.35),

$$l=\frac{2.51-0.84}{0.001-0.000102}$$

$$=1860\text{ m}$$

As mentioned earlier, and obviously when considering the great disparity between the average depths and actual depths, this procedure introduces a considerable error.

In order to obtain a more accurate estimate of the distance at which the flow becomes uniform, as well as to determine the actual water-surface profile, the channel is divided in a number of shorter reaches. The length of each, corresponding to predetermined depths, can then be calculated as above. Obviously, the shorter the reaches the greater the accuracy. In practice, it is found that, for most cases, sufficient accuracy can be maintained by choosing reaches in such a manner that D_D and D_U vary by no more than 10 percent.

The table below is a summary of the calculations required for this problem.

Reach	D_D	D_U	D_{ave}	v_D	v_U	H_D	H_U	ΔH	$\dfrac{b}{D_{ave}}$	K	s $\times10^{-3}$	(s_0-s) $\times10^{-3}$	l
I	2.50	2.30	2.40	0.44	0.48	2.51	2.31	0.198	1.46	0.82	0.035	0.96	205.5
II	2.30	2.10	2.20	0.48	0.53	2.31	2.11	0.198	1.59	0.85	0.052	0.95	208.6
III	2.10	1.90	2.00	0.53	0.58	2.11	1.92	0.197	1.75	1.06	0.056	0.94	208.6
IV	1.90	1.70	1.80	0.58	0.65	1.92	1.72	0.196	1.94	1.21	0.076	0.92	211.7
V	1.70	1.50	1.60	0.65	0.74	1.72	1.53	0.194	2.19	1.43	0.102	0.90	215.8
VI	1.50	1.30	1.40	0.74	0.85	1.53	1.34	0.191	2.50	1.70	0.147	0.85	223.5
VII	1.30	1.10	1.20	0.85	1.01	1.34	1.15	0.185	2.92	2.06	0.227	0.77	239.8
VIII	1.10	0.90	1.00	1.10	1.23	1.25	0.98	0.174	3.50	2.36	0.458	0.54	321.7
IX	0.90	0.72	0.81	1.23	1.55	0.98	0.84	0.136	4.32	3.35	0.697	0.30	450.8
	Total:												2286.0

The length from the gate to the point where uniform flow is restored is thus estimated to be 2290 m. This distance is approximately 20 percent longer than estimated by approximation earlier.

The water-surface profile can now be plotted as shown in Fig. 11.11. Due to the long horizontal distances involved, the vertical scale has been exaggerated considerably for clarity.

The General Equation for Varying Flow

Equation (11.34) can be arranged and differentiated to take on the form

$$\frac{dD}{dl} = \frac{s_0 - s}{1 - \dfrac{v^2}{gD}} \tag{11.36}$$

or, substituting Eq. (11.8),

$$\frac{dD}{dl} = \frac{s_0 - s}{1 - F_N{}^2} \tag{11.37}$$

If Eqs. (11.36) and (11.37) are visualized in a coordinate system in which values of D are on the ordinate, or y axis, and values of l are on the abscissa, or x axis the ratio dD/dl represents the slope of the tangent to the water surface at any point along the channel.

These relationships therefore indicate whether at any point along the channel the water surface is rising (backwater condition) or dropping (drawdown condition). Immediately the following deductions can be made:

When $dD/dl < 0$, the slope of the water surface is dropping in the downstream direction and the depth decreases downstream.

When $dD/dl = 0$, the slope of the water surface is parallel to the channel bottom and uniform flow exists. This can be readily seen from Eq. (11.36) since, for this condition, $s_0 - s$ must equal zero.

When $dD/dl > 0$, the slope of the water surface rises in the downstream direction and D increases downstream.

When $dD/dl = \infty$, which requires that $1 - F_N{}^2 = 0$, or $F_N = 1.0$, the slope of the water surface must theoretically be vertical. This flow occurs when the flow changes from subcritical to supercritical, or vice versa, as indicated by the value of F_N. A vertical water surface does not occur in reality; however, a very noticeable change in the water surface takes place. This is especially so when the flow changes from below D_c to above D_c. In such instance a phenomenon known as the *hydraulic jump* occurs. The hydraulic jump is discussed later in this chapter.

Types of Water-Surface Profiles

From the foregoing discussion, it is evident that the relationships expressed in Eqs. (11.36) and (11.37) provide a considerable amount of information as to the shape of the water-surface profile in an open channel. A further examination of these formulas yields the following observations.

1. The relationship between the slope of the channel bottom and the slope of the energy gradient determines whether the numerator of the equations is positive or negative.
2. The denominator of the equation is positive when $F_N < 1.0$, and vice versa. In other words, when the flow is supercritical (F_N less than 1) the denominator is positive, and when the flow is subcritical (F_N more than 1) the denominator is negative.

These relationships and their effect on the water-surface profile are demonstrated in the following example.

EXAMPLE 11.10

In the rectangular channel of Example 11.9, at a point 1050 m upstream of the gate, the depth of flow is observed to be 1.50 m. Determine whether the depth increases or decreases downstream from this point.

SOLUTION

In order to determine the slope of the water surface, it is necessary to determine the value of dD/dl, from Eq. (11.36) or (11.37). Hence the following values must be obtained. From Eq. (11.15),

$$D_c = \left(\frac{1.11}{9.81}\right)^{1/3}$$

$$= 0.48 \text{ m}$$

From the work table of Example 11.9,

$$s_0 - s > 0$$

and

$$v = 0.74 \text{ m/s}$$

$$F_N{}^2 = \frac{0.74^2}{9.81 \times 1.5}$$

$$= 0.04 < 1.0$$

Hence

$$\frac{dD}{dl} = \frac{+}{+} > 0$$

and the water surface rises downstream, or D increases.

Determination of the sign of the denominator in this instance could have been simplified by considering that, since the flow in this channel is supercritical, F_N will be less than 1.

In the above example, the actual depth at the point of investigation—of 1.50 m in this instance—is called the observed depth, D_0.

The conditions at which flow in an open channel can take place and the possible relationships between the *observed depth* D_0, the normal depth at which flow is uniform D_n, and the critical depth D_c are illustrated in Fig. 11.12.

It is evident from this figure that there are three regions of channel depths at which flow can be observed:

Region 1, with D_0 greater than D_n and D_c.
Region 2, with D_0 between D_n and D_c.
Region 3, with D_0 less than D_n and D_c.

Further, any channel can be classified by the steepness of the channel-bottom slope. This classification always relates to the critical slope of the channel, as follows:

Steep slope when s_0 is greater than s_c
Critical slope when s_0 is equal to s_c
Mild slope when s_0 is less than s_c

Two additional channel-bottom conditions or slopes exist. These do not really constitute open-channel flow, but gravity flow can take place along them.

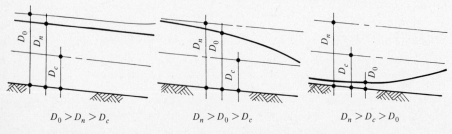

FIGURE 11.12

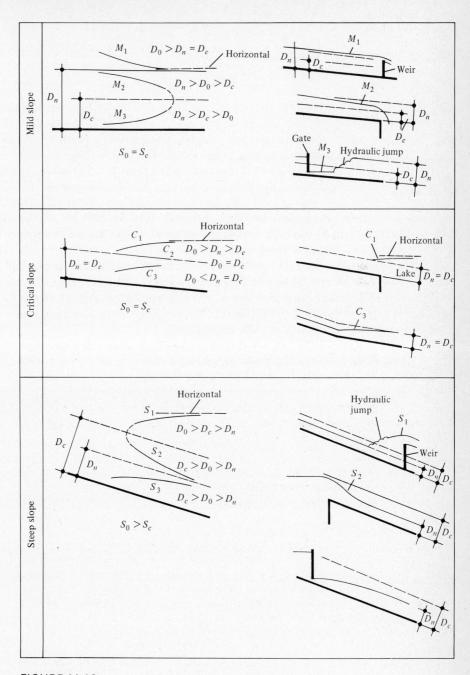

FIGURE 11.13

They are as follows:

Horizontal slope when s_0 is zero
Adverse slope when s_0 is negative

In theory then, for each of the five slope descriptions above there are three regions in which flow can be observed. It follows then that a total of 15 theoretical water-surface profiles are possible. These profiles, together with illustrations of practical applications, are shown in Fig. 11.13.

While this figure is for the most part self-explanatory, the following observations and explanations are presented here for further clarification.

Mild Slope ($s_0 < s_c$). The M_1 curve is generally very long and asymptotic to the horizontal and the line representing D_n. The M_2 and M_3 curves end in a sudden drop through the line representing D_c in the case of the former and a hydraulic jump in the case of the latter.

Critical Slope ($s_0 = s_c$). Since $D_n = D_c$ in this case, there is no region 2, and only two water-surface profiles exist, C_1 and C_3. The C_2 curve coincides with the water surface that corresponds to uniform flow at critical depth.

Steep Slope ($s_0 > s_c$). All curves are relatively short. S_1 is asymptotic to the horizontal, whereas S_2 and S_3 approach D_n.

Horizontal and Adverse-Slope Channels. In this instance, D_n is infinitely large and uniform flow cannot take place. Hence there are no H_1 or A_1 profiles.

EXAMPLE 11.11

A grass-lined trapezoidal channel has a bottom slope of 3.0 m/km, a bottom width of 2.0 m, and side slopes of 2 horizontal to 1 vertical. At point A in this channel the depth of flow is observed to be 1.5 m and the velocity is 0.90 m/s. Determine the general shape of the water-surface profile downstream from A.

SOLUTION

The solution of this problem requires the determination of the region in which the observed depth belongs and the type of channel slope. It will therefore be necessary to determine the rate of flow Q, the critical depth D_c, critical slope s_c, and the normal depth of flow D_n.

Grass-lined channels have a roughness coefficient n of 0.027, as found from Table 10.1.

The rate of flow can be determined as follows:
$$A = 1.5^2(1.33 + 2.0)$$
$$= 7.50 \text{ m}^2$$

and

$$Q = 7.50 \times 0.90$$
$$= 6.74 \text{ m}^3/\text{s}$$

D_c and D_n can now be determined as demonstrated earlier and as shown in tabular form below.

Trial D_n	$\dfrac{b}{D}$	K	$\dfrac{s^{1/2}}{n}$	Q
1.00	2.00	2.90	2.03	5.88
1.25	1.60	2.54		9.35
1.07	1.87	2.78		6.77

Hence, say, $D_n = 2.05$ m.
From Table 11.1 and Eq. (11.25),

Trial D_c	$\dfrac{b}{D}$	K_c	$Q^{2/5}$	D_c
1.0	2.0	0.395	2.15	0.85
0.9	2.22	0.386		0.86
0.8	2.50	0.374		0.80

Hence $D_c = 0.8$ m.
From the Manning equation, for $D = D_c = 0.8$ m,

$$s_c = \left(\frac{6.74 \times 0.027}{0.8^{8/3} \times 3.08}\right)^2$$

$$= 0.011$$

The following relationships have now been establishd.

1. If $s_o < s_c$, the channel has a mild slope.
2. If $D_o > D_n > D_c$, the flow is in Region 2.

Therefore, from Fig. 11.13, the water-surface profile is an M_1 profile.

EXAMPLE 11.12

A 600-mm-diameter asbestos cement sewer pipe has its invert at a slope of $s = 0.008$. The depth flow at the point A in this sewer is 215 mm when Therefore, from Fig. 11.13, the water-surface profile is an M profile.

SOLUTION

From Table 10.1 choose $n = 0.015$. The preliminary calculations determining D_n, D_c, and s_c are performed as illustrated earlier and summarized below.

$$Q_f = \left(\frac{0.008}{0.032}\right)^2$$

$$= 0.500 \text{ m}^3/\text{s}$$

$$= 500 \text{ L/s}$$

$$\frac{Q_p}{Q_f} = \frac{300}{500}$$

$$= 0.6$$

Hence from Table 10.5 for $K_p = 0.6$, $D/d = 0.67$, and

$$D_n = 600 \times 0.67$$

$$= 402 \text{ mm}$$

Trial D_c	$\dfrac{D}{d}$	K_c	$Q^{2/5}$	D_c
250	0.42	0.533	0.618	329
300	0.50	0.555		343
400	0.67	0.594		367
400	0.75	0.608		452
360	0.60	0.579		360

Hence, say, $D_c = 360$ mm.

To determine s_c, find Q_f for this pipe when the 300 L/s flow corresponds to D_c. From Table 10.5, for

$$\frac{D_c}{d} = \frac{450}{600}$$

$$= 0.75$$

$$K_p = 0.912$$

and

$$Q_f = \frac{300}{0.912}$$

$$= 329 \text{ L/s}$$

Therefore

$$s_c = 0.032 \times 0.329^2$$

$$= 0.0035$$

Consequently, the following observations can be made.

1. If $s_o > s_c$ the sewer has a *steep* slope.
2. If $D_n < D_o < D_c$, the flow is in Region 2.

Hence the water surface is an S_2 profile.

11.3 RAPIDLY VARYING FLOW

Transition Through Critical Depth

As mentioned earlier, when the flow changes from subcritical to supercritical, or vice versa, the water should theoretically become vertical. Although this does not take place, very pronounced variations in the water-surface profiles do occur. In the transition through the critical depth two possible conditions can exist:

1. The flow changes from the higher stage to the lower stage, from subcritical to supercritical. In this instance the change takes place with a generally smooth water surface and without undue turbulence and energy loss. Figure 11.13 shows examples of this type of flow.
2. The flow changes from the lower stage to the higher stage, supercritical to subcritical. This transition through critical depth is always accompanied by relatively high turbulence and energy losses. The phenomena is known as a hydraulic jump.

The Hydraulic Jump

The conditions under which a hydraulic jump will take place are shown in Fig. 11.14. The water approaches the jump at a high velocity and in supercritical conditions. When certain conditions in the channel exist that cause a reduction in velocity to subcritical, the water surface will jump to the subcritical depth. This jump is accompanied by high turbulence and rollers at the water surface. The degree of the turbulence depends on the relationship between the upstream and downstream depths and the Froude number immediately upstream of the jump. The high turbulence in the hydraulic jump, together with the considerable energy loss accompanying it, are two factors that make this flow condition an important tool for the civil engineer.

The hydraulic jump has many practical and useful applications. Among them are the following:

Reduction of the energy and velocity downstream of a dam or chute in order to minimize and control erosion of the channel bed.

Raising of the downstream water level in irrigation channels.

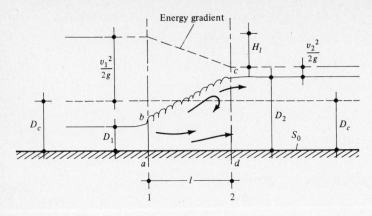

FIGURE 11.14

Acting as a mixing device for the addition and mixing of chemicals in industrial and water and sewage treatment plants. In natural channels the hydraulic jump is also used to provide aeration to the water for pollution control purposes.

Even though the hydraulic jump appears to be a wild phenomenon, with surface rollers and in some cases foaming water surfaces, there is nevertheless a regular and mathematical relationship between the upstream and downstream depths, D_1 and D_2, respectively.

This relationship can be derived by applying Newton's second law of motion to the prism of water in the jump, the prism *abcd* in Fig. 11.14. This prism is shown as a free-body diagram in Fig. 11.15. In order to simplify the discussions the channel is assumed to be rectangular and horizontal. The rate of flow per unit width of the channel is q.

According to Newton's second law of motion, the rate of change in momentum in the direction of flow from point 1 to point 2 must equal the algebraic sum of all the forces acting along the direction of flow on the prism of water between points 1 and 2.

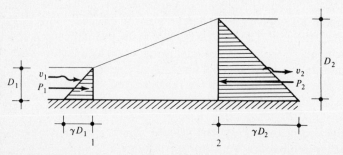

FIGURE 11.15

The forces acting on the prism of water in the jump in this instance are the hydrostatic forces at points 1 and 2, as shown in Fig. 11.15. Other forces acting on the prism along the direction of flow, are the friction between the water and the channel bottom and the air above the surface. Since the hydraulic jump takes place over relatively very short distances, these forces are very small and are generally considered to be insignificant. Hence Newton's second law as applied to the prism of water in Fig. 11.15 can be expressed as follows:

$$\frac{\gamma D_1{}^2}{2} - \frac{\gamma D_2{}^2}{2} = q(v_2 - v_1) \tag{11.38}$$

Substituting

$$\rho = \frac{\gamma}{g}$$

and

$$v = \frac{q}{D}$$

in Eq. (11.38) yields, after simplification,

$$\frac{q}{g} = D_1 D_2 \left(\frac{D_1 + D_2}{2}\right) \tag{11.39}$$

If the Froude number for the upstream flow F_{N1} is substituted in Eq. (11.39), the following expression for the relationship between D_1 and D_2 results.

$$\frac{D_2}{D_1} = \tfrac{1}{2}\{\sqrt{1 + 8F_{N1}^2} - 1\} \tag{11.40}$$

This expression is very convenient since it presents a dimensionless expression for the D_1/D_2 ratio.

If Eq. (11.40) is rearranged and if F_{N1}^2 is substituted by its value v^2/gD_1, the following relationships between the conjugate depths result.

$$D_1 = -\frac{D_2}{2} + \left(\frac{2v_2{}^2}{g} D_2 + \frac{D_2{}^2}{4}\right)^{1/2} \tag{11.41}$$

and

$$D_2 = -\frac{D_1}{2} + \left(\frac{2v_1{}^2}{g} D_1 + \frac{D_1{}^2}{4}\right)^{1/2} \tag{11.42}$$

Equations (11.39) to (11.42), it must be remembered, have been developed for rectangular channels and are therefore applicable only to such channels or to others with similar properties. Very wide trapezoidal channels and most natural channels can often be considered to be rectangular without introducing appreciable error.

For other channels, however, where the generalizations above do not apply, reasoning similar to that used for developing Eq. (11.39) leads to the following equation for the relationship between the upstream and downstream depth of any cross section.

$$\frac{Q^2}{g} = \frac{A_2\bar{D}_2 - A_1\bar{D}_1}{\dfrac{1}{A_1} - \dfrac{1}{A_2}} \tag{11.43}$$

where $\bar{D}_1$ and $\bar{D}_2$ are the heights to the center of gravity of the areas A_1 and A_2, respectively.

Equations (11.39) through (11.43) have been developed on the assumption that the channel bottom is horizontal. Most channels will, of course, have a bottom slope other than zero. If the channel slope is not zero, the channel bottom will form an angle θ with the horizontal. The only terms in Eq. (11.38) that are affected by this fact are those expressing the forces due to the hydrostatic pressures. Hence these forces would need to be multiplied by a value equal to $\cos\theta$. For most of the channel slopes, this factor will be nearly 1.0 and negligible. Only for steeply sloping aprons on dams and chutes must allowance be made for the bottom slope.

EXAMPLE 11.13

Water flows in a rectangular channel that is 1.50 m wide. At a certain point in the channel the slope of the channel bottom changes, causing a hydraulic jump to take place. Immediately prior to the jump, the depth of flow is 1.15 m and the velocity is 10.0 m/s. Determine the depth of water downstream of the jump.

SOLUTION

The Froude number upstream of the jump is, from (Eq. 11.8),

$$F_N = \frac{10.0}{(9.81 \times 1.15)^{1/2}}$$

$$= 2.98$$

Hence, from Eq. (11.40),

$$\frac{D_2}{D_1} = 0.5\{(1 + 8 \times 2.98)^{1/2} - 1\}$$

$$= 3.74$$

and

$$D_2 = 1.15 \times 3.74$$

$$= 4.31 \text{ m}$$

The rate of flow in this channel is

$$Q = 1.15 \times 10.0 \times 1.50$$

$$= 17.25 \text{ m}^3/\text{s}$$

and

$$q = \frac{17.25}{1.50}$$

$$= 11.5 \text{ m}^3/\text{s per meter of channel width}$$

Hence

$$D_c = \left(\frac{11.5^2}{9.81}\right)^{1/3}$$

$$= 2.38 \text{ m}$$

The latter confirms that the occurrence of a hydraulic jump was to be expected, since the flow passes from $D_1 = 1.15$ to $D_2 = 4.31$ through $D_c = 2.38$ m.

The velocity in the channel after the jump is

$$v_2 = \frac{11.5}{4.31}$$

$$= 2.67 \text{ m/s}$$

This value is considerably less than v_1, which is 10.0 m/s.

The specific energy head has also been reduced markedly, since

$$H_1 = \frac{10.0^2}{2 \times 9.81} + 1.15$$

$$= 6.25 \text{ m}$$

and

$$H_2 = \frac{2.67^2}{2 \times 9.81} + 4.31$$

$$= 4.67 \text{ m}$$

Upper and Lower Conjugate Depth

The depth immediately prior to the jump, D_1, and the depth just after the jump, D_2, are related to each other as defined by Eqs. (11.39) to (11.43). They are called the lower conjugate depth, D_1, and the upper conjugate depth, D_2.

For rectangular channels Eqs. (11.39) to (11.42) are easily and readily solved; if either of the conjugate depths is known, the other can be determined quickly. For channels with cross sections other than rectangular, the solution of Eq. (11.43) involves several related unknowns and requires trial-and-error methods. A somewhat less cumbersome graphical method for finding the conjugate depths is presented below.

If Eq. (11.38) is rearranged so that all the terms for point 1 are on the left and those for point 2 are on the right side of the equal sign, then

$$\frac{\gamma D_1}{2} + \rho q v_1 = \frac{\gamma D_2}{2} + \rho q v_2 \tag{11.44}$$

or

$$F_1 + M_1 = F_2 + M_2 \tag{11.45}$$

where F and M represent the forces due to the hydrostatic pressure and momentum, respectively.

In a generalized form, the sum $(F + M)$ can be expressed as follows:

$$F + M = \rho Q v + \gamma A \bar{D} \tag{11.46}$$

Since $v = Q/A$, Eq. (11.46) can be rearranged to read

$$F + M = \frac{Q^2}{gA} + A \bar{D} \tag{11.47}$$

where Q = flow in the channel

A = cross-sectional area of the flow

$\bar{D}$ = depth to the center of gravity of A.

If this equation is plotted for a constant value of Q versus values of the depth of flow, the graph of Fig. 11.16 results. As was the case for the specific

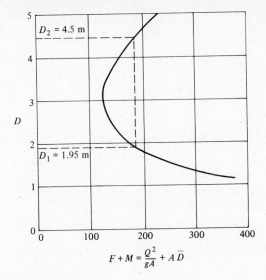

$$F + M = \frac{Q^2}{gA} + A\bar{D}$$

FIGURE 11.6

energy graph of Fig. 11.5, there are two values of D for each value of $F + M$. These are the upper and lower conjugate depths. To facilitate the construction of the $F + M$ curve, Tables 11.3 and 11.4 are included, to assist in finding values of $\bar{D}$, for trapezoidal and circular channels.

TABLE 11.3   Values of K_{cog} for Trapezoidal Channel

b/D	z								
	0.25	0.50	0.75	1.00	1.50	2.00	3.00	4.00	5.00
0.01	0.660	0.663	0.664	0.665	0.666	0.666	0.666	0.666	0.666
0.10	0.619	0.639	0.647	0.652	0.656	0.659	0.661	0.663	0.663
0.20	0.593	0.619	0.632	0.639	0.647	0.652	0.656	0.659	0.660
0.30	0.576	0.604	0.619	0.628	0.639	0.645	0.652	0.655	0.657
0.40	0.564	0.593	0.609	0.619	0.632	0.639	0.647	0.652	0.654
0.50	0.556	0.583	0.600	0.611	0.625	0.633	0.643	0.648	0.652
1.00	0.533	0.556	0.571	0.583	0.600	0.611	0.625	0.633	0.639
1.50	0.524	0.542	0.556	0.567	0.583	0.595	0.611	0.621	0.624
2.00	0.519	0.533	0.545	0.556	0.571	0.583	0.600	0.611	0.619
3.00	0.513	0.524	0.533	0.542	0.556	0.567	0.583	0.595	0.604
4.00	0.510	0.519	0.526	0.533	0.545	0.556	0.571	0.583	0.593
5.00	0.509	0.515	0.522	0.528	0.538	0.548	0.562	0.574	0.583
6.00	0.507	0.513	0.519	0.524	0.533	0.542	0.556	0.567	0.576
8.00	0.505	0.510	0.514	0.519	0.526	0.533	0.545	0.556	0.584
10.00	0.504	0.508	0.512	0.515	0.522	0.528	0.538	0.548	0.556
15.00	0.503	0.505	0.508	0.510	0.515	0.520	0.528	0.535	0.542
20.00	0.502	0.504	0.506	0.508	0.512	0.515	0.522	0.528	0.533

$\bar{D} = D K_{cog}$

TABLE 11.4 Values of K_{cog} for Circular Channels

D/d	0.00	0.01	0.02	0.03	0.04	0.05	0.06	0.07	0.08	0.09
0.00	0.000	0.001	0.012	0.018	0.024	0.030	0.036	0.042	0.048	0.054
0.10	0.060	0.066	0.071	0.077	0.083	0.089	0.095	0.101	0.107	0.113
0.20	0.118	0.124	0.130	0.136	0.142	0.147	0.153	0.159	0.165	0.171
0.30	0.176	0.182	0.188	0.193	0.199	0.205	0.210	0.216	0.222	0.227
0.40	0.233	0.238	0.244	0.249	0.255	0.261	0.266	0.271	0.277	0.282
0.50	0.288	0.293	0.299	0.304	0.309	0.315	0.320	0.325	0.330	0.336
0.60	0.341	0.346	0.351	0.356	0.361	0.366	0.371	0.376	0.381	0.386
0.70	0.391	0.396	0.400	0.405	0.410	0.414	0.419	0.423	0.428	0.432
0.80	0.437	0.441	0.445	0.449	0.453	0.457	0.461	0.465	0.469	0.472
0.90	0.476	0.479	0.482	0.485	0.488	0.491	0.494	0.496	0.498	0.499

$\bar{D} = DK_{cog}$

EXAMPLE 11.14

A trapezoidal channel has a bottom width of 2.0 m and side slopes of 2 horizontal to 1 vertical. An obstruction downstream in the channel causes a hydraulic jump to form. The upper conjugate depth of the jump is 4.5 m, and the mean velocity at the downstream end of the jump is 2.8 m/s. Determine the velocity just prior to the jump.

SOLUTION

For this channel,

$$\frac{b}{D_2} = 0.44$$

$$z = 2.0$$

and

$$A_2 = (0.44 + 2.0) \times 4.5^2$$

$$= 49.5 \text{ m}^2$$

and hence

$$Q = 49.5 \times 2.8$$

$$= 138.6 \text{ m}^3/\text{s}$$

Consequently, Eq. (11.47) will transform to

$$F + M = \frac{138.6^2}{9.81 \times A} + A\bar{D}$$

$$= \frac{1958}{A} + A\bar{D}$$

This relationship is plotted for various values of the channel depth as shown in Fig. 11.16. The depth corresponding to an upper conjugate depth of 4.50 m is 1.95, and can be found as illustrated in Fig. 11.16 and $D_1 = 1.95$ m.

An alternative method of solving this problem would have been to try different values of D_1, until a value that satisfies Eq. 11.43 is found.

If the value of $D_1 = 1.95$ is substituted in this equation, the following results

$$\frac{138.6^2}{9.81} = \frac{49.5 \times 2.86 - 11.7 \times 1.19}{1/11.7 - 1/49.5}$$

$$1958 \approx 1956$$

Location of the Hydraulic Jump

An abrupt change of depth from the lower to upper conjugate always accompanies the hydraulic jump. When either of these depths is known, the other can be found, as demonstrated in the previous section. A jump will occur, however, only when the depth of flow in the channel is exactly equal to the upper or lower conjugate depth.

A hydraulic jump can take place under two possible conditions of flow before and/or after the jump:

1. When one of the flows is uniform, as shown in Fig. 11.17a.
2. When both the upstream and downstream flows are nonuniform, as shown in Fig. 11.18.

In the case of the former, the location of the jump can be determined by finding the point where the nonuniform depth of flow equals the conjugate depth. In the event that both the upper and lower conjugate depths conform to nonuniform conditions, trial-and-error or graphical solutions are indicated.

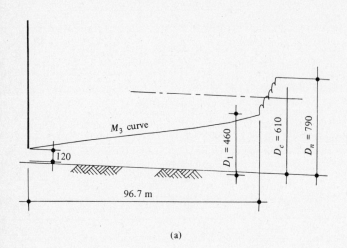

(a)

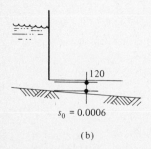

(b)

FIGURE 11.17

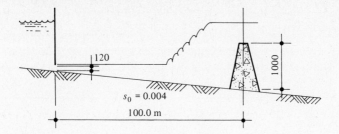

FIGURE 11.18

Examples of the solution of both types of problems are included in the next section.

Length of the Hydraulic Jump

Even though a great deal of research has been done regarding the length of the hydraulic jump, there is not a close agreement between the various sources, as to the exact length of the distance between D_1 and D_2. This is probably due to the rough nature of the flow conditions immediately prior to and especially after the jump, preventing accurate definition and measurement of the actual start and end of the jump.

For practical purposes, when the jump can be considered to be totally developed, lengths obtained from the graph in Fig. 11.19 can be used.

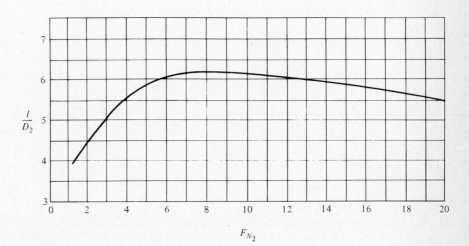

FIGURE 11.19

EXAMPLE 11.15

Figure 11.17b shows a rectangular concrete channel, 4.0 m wide, with a bottom slope of 0.0006. When the flow is 6.0 m³/s, will a hydraulic jump occur? If so, where?

SOLUTION

In order to solve the problem related to this channel, a number of preliminary calculations are required to establish basic flow data. These calculations are summarized below.

$$n=0.013 \qquad s_c=0.0011$$

$$q=1.5 \text{ m}^3/\text{s} \qquad D_n=0.79 \text{ m}$$

$$D_c=0.61 \text{ m} \qquad v_n=1.90 \text{ m/s}$$

On the basis of the above the following conclusions can be drawn.

1. The water leaves the gate at a depth less than D_c.
2. At some point downstream of the gate the flow must become uniform at a depth $D_n=0.79$ m, which is greater than D_c.
3. A hydraulic jump will occur.
4. The slope of the channel is less than s_c; hence the channel has a *mild* slope.
5. The flow approaching the jump is in Region 3.
6. The water-surface profile upstream of the jump will be an M_3 profile.

The channel slope in this instance is not horizontal; however, its angle with the horizontal is only 0.034°. Since the cosine of this angle has a value of approximately 1.0, no adjustments to the conjugate depth equations need to be made. Hence, from Eq. (11.41),

$$D_1 = -\frac{0.79}{2} + \left(\frac{2 \times 1.90^2 \times 0.79}{9.81} + \frac{0.79^2}{4}\right)^{1/2}$$

$$= 0.46 \text{ m}$$

Hence the jump will occur at the point where the upstream water-surface profile reaches a depth of 0.46. Determination of this point can be done by means of the water-surface profile calculations demonstrated earlier in this chapter. They are summarized in the table below.

D_D	D_U	D_{ave}	$\dfrac{b}{D_{ave}}$	K	l
0.46	0.40	0.43	9.3	7.19	11.5
0.40	0.35	0.375	10.67	9.52	14.2
0.35	0.30	0.325	12.31	11.14	15.1
0.30	0.25	0.275	14.55	13.35	15.6
0.25	0.20	0.225	17.78	16.56	15.7
0.20	0.15	0.175	22.81	21.61	15.4
0.15	0.12	0.135	19.63	28.38	8.7
Total l:					96.7

Therefore, the jump will commence at a point about 97 m down-stream of the gate. The corresponding water-surface profiles and the jump are shown in Fig. 11.17a.

EXAMPLE 11.16

Figure 11.18 shows a channel similar to that of Fig. 11.17. In this instance the channel-bottom slope is 0.004. In order to shorten the distance to the jump, a small concrete apron dam has been installed as shown. With the channel width 4.0 m and a rate of flow of 6.0 m³/s, determine the location of the hydraulic jump.

SOLUTION

As before, basic information regarding the flow in the channel is calculated. The results of these are as follows:

$$n = 0.013$$

$$q = 1.5 \text{ m}^3/\text{s}$$

$$D_c = 0.61$$

$$s_c = 0.0011$$

The water level downstream at the apron dam is controlled by the head on this dam. This head can be determined from Eq. (8.28). With $C = 3.5$,

$$H = \left(\frac{1.5}{3.5}\right)^{2/3}$$

$$= 0.57 \text{ m}$$

Hence the water depth just before the dam, D_{dam}, will be

$$D_{dam} = 1.57 \text{ m}$$

On the basis of the foregoing, the following conclusions can be made.

1. The water leaves the gate at a depth less than D_c.
2. At some point downstream the flow must reach a depth of 1.57 m, which is larger than D_c.
3. Hydraulic jump will occur.
4. The channel slope is a *steep* slope.
5. D_n is smaller than D_c, since the channel is *steep*.
6. The water surface upstream of the jump is in Region 3 and will have an S_3 profile.
7. The water surface downstream of the jump is in Region 1 and will have an S_1 profile.
8. Both the downstream and upstream flows adjacent to the jump are nonuniform.

The water-surface profiles for each of the flows, before and after the hydraulic jump, can now be calculated. The results are given in the tables below.

For the upstream profile,

D_D	D_U	D_{ave}	$\dfrac{b}{D_{ave}}$	K	l
0.12	0.22	0.18	22.22	20.98	43.8
0.22	0.32	0.28	14.28	13.18	47.0
0.32	0.42	0.38	10.52	9.37	55.0

For the downstream profile,

D_D	D_U	D_{ave}	$\dfrac{b}{D_{ave}}$	K	l
1.57	1.47	1.52	2.63	1.81	24.6
1.47	1.37	1.42	2.82	1.98	24.5
1.37	1.27	1.32	3.03	2.16	24.3
1.27	1.17	1.22	3.28	2.39	24.0
1.17	1.07	1.12	3.57	2.66	23.7

Both of these profiles have been plotted and labeled on Fig. 11.18a.

In order to determine the points where the upper conjugate and lower conjugate depths coincide, which would be where the hydraulic jump will take place, the following procedure can be used.

Using Eq. (11.41) or (11.42), find values of upper and lower conjugate depths related to this channel. These values have been tabulated below.

Lower Conjugate Depth D_1	Upper Conjugate Depth D_2
0.20	1.42
0.30	1.10
0.40	0.89

Shown on Fig. 11.18a is a plot of these values, the upper conjugate depths indicated by the line ab.

Since now, the line ab and the upstream water-surface profile represent a graph of corresponding values of D_2 and D_1, respectively, the point where the line ab intersects the downstream water surface profile will be the location of the hydraulic jump. The jump and its dimensions, derived from Fig. 11.19, are shown in Fig. 11.18a.

Civil Engineering Applications of Fluid Flow: Flow in Natural Channels

12.1 NATURAL CHANNELS

Channels formed by the natural erosion of the soil, as opposed to those created by machine or hand excavation, are natural channels. They include rivers, streams, and creeks and present flow patterns and forms that are even more complex than the regular channels discussed in previous chapters. This added complexity stems from the following special characteristics and problems related to natural-channel flow.

1. The cross sections are always of an irregular shape.
2. In any one cross section the roughness coefficient can vary considerably from one point to another.
3. Longitudinally, along the stream, no two cross sections are the same geometrically, and roughness coefficients vary constantly.
4. The slope of the channel bottom varies, sometimes considerably, along the entire length of the stream.
5. The hydraulic properties of the stream, particularly the roughness coefficient, vary in time, due to seasonal changes in the ground-cover and soil conditions.
6. Flows in natural channels are often nonuniform and unsteady.
7. When in flood, the stream flow takes place, for a large part, outside the channel proper.

Several of the points mentioned above are illustrated in Figs. 12.1, 12.2, and 12.3.

The study of the flow in natural channels is an important part of the civil engineering discipline. Through it the engineer is able to determine the water-surface elevations of the channel, under various conditions of flow for flood damage, navigation, and water supply studies. Further, this study provides information needed to determine the effects on the stream due to bridges, culverts, and other channel constrictions, as well as needed channel improvements, erosion control, and natural-habitat improvements.

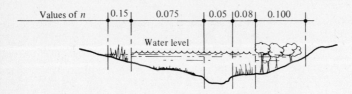

FIGURE 12.1

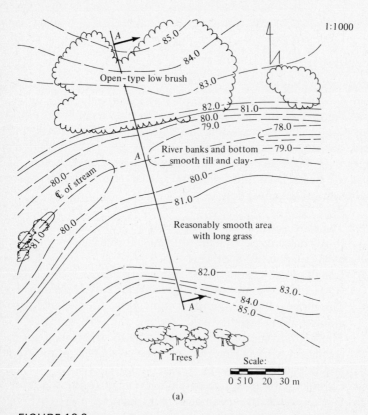

(a)

FIGURE 12.2a

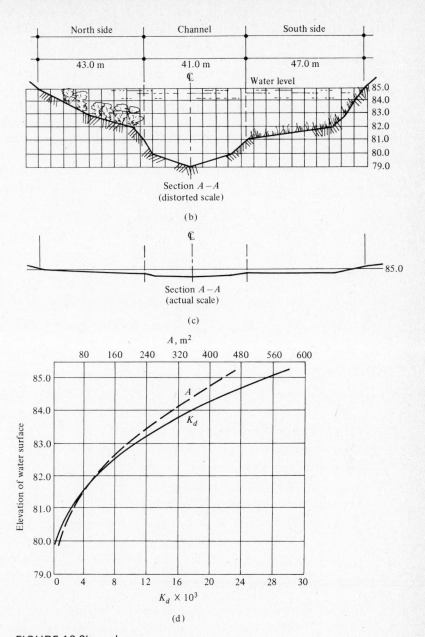

FIGURE 12.2b, c, d

The study of unsteady flows in natural channels is very complex, often associated with the field of hydrology and beyond the scope of this text. In this chapter natural channels are discussed only as related to steady uniform and nonuniform flows.

12.2   THE FLOW FORMULAS AND NATURAL CHANNELS

Manning's Formula

The most commonly used flow formula for the study of flow in open channels is the Manning formula. It is therefore the formula in use for natural channels as well. The extensive list of roughness coefficients developed for use with this formula, as shown in Table 12.1, is an indication of its acceptance in the industry.

For open-channel flow the formula is generally used in the form of Eq. (9.6), or

$$Q = \frac{AR^{2/3}s^{1/2}}{n}$$

A simplified form of this formula is

$$Q = K_n s^{1/2} \tag{12.1}$$

in which

$$K_n = \frac{AR^{2/3}}{n} \tag{12.2}$$

where Q = rate of flow in the channel
   A = cross-sectional area of the flow
   R = hydraulic radius
   s = slope of the energy gradient
   n = the Manning roughness coefficient
   K_n = a form factor related to the physical properties of the cross section

The Natural-Channel Cross Section

A typical natural stream is shown in Fig. 12.2. Shown in Fig. 12.2a is the plan of a reach of the stream, with its various ground covers on the basis of which n-coefficients can be estimated. In Figs. 12.2b and c are shown the cross section of the river at the point A. In Fig. 12.2b the cross section has been drawn with an exaggerated vertical scale for easier visualization and measurements. The vertical scale in this instance is one-fifth the horizontal scale. The actual cross section, drawn to a natural scale, with the horizontal and vertical scales equal, is shown in Fig. 12.2c.

TABLE 12.1   Recommended Values of *n* in Manning's Equation for Use with Natural Channels

Type of Channel and Description	Good	Fair	Poor
1. Minor streams (top width at flood stage <100 ft)			
a. Streams on plain			
1. Clean, straight, full stage, no rifts or or deep pools	0.025	0.030*	0.033
2. Same as above, but more stones and weeds	0.030	0.035	0.040
3. Clean, winding, some pools and shoals	0.033	0.040	0.045
4. Same as above, but some weeds and stones	0.035	0.045	0.050
5. Same as above, lower stages, more ineffective slopes and sections	0.040	0.048	0.055
6. Same as 4, but more stones	0.045	0.050	0.060
7. Sluggish reaches, weedy, deep pools	0.050	0.070	0.080
8. Very weedy reaches, deep pools, or floodways with heavy stand of timber and underbrush	0.075	0.100	0.150
b. Mountain streams, no vegetation in channel, banks usually steep, trees and brush along banks submerged at high stages			
1. Bottom: gravels, cobbles, and few boulders	0.030	0.040	0.050
2. Bottom: cobbles with large boulder boulders	0.040	0.050	0.070
2. Flood plains			
a. Pasture, no brush			
1. Short grass	0.025	0.030	0.035
2. High grass	0.030	0.035	0.050
b. Cultivated areas			
1. No crop	0.020	0.030	0.040
2. Mature row crops	0.025	0.035	0.045
3. Mature field crops	0.030	0.040	0.050
c. Brush			
1. Scattered brush, heavy weeds	0.035	0.050	0.070
2. Light brush and trees, in winter	0.035	0.050	0.060
3. Light brush and trees, in summer	0.040	0.060	0.080
4. Medium to dense brush, in winter	0.045	0.070	0.110
5. Medium to dense brush, in summer	0.070	0.100	0.160

Source: Reproduced from Ven Te Chow, *Open-channel Hydraulics* (New York: McGraw-Hill, 1959).

TABLE 12.1   (*Continued*)

Type of Channel and Description	Good	Fair	Poor
d. Trees			
1. Dense willows, summer, straight	0.110	0.150	0.200
2. Cleared land with tree stumps, no sprouts	0.030	0.040	0.050
3. Same as above, but with heavy growth of sprouts	0.050	0.060	0.080
4. Heavy stand of timber, a few down trees, little undergrowth, flood stage below branches	0.080	0.100	0.120
5. Same as above, but with flood stage reaching branches	0.100	0.120	0.160
3. Major streams (top width at flood stage >100 ft). The *n* value is less than that for minor streams of similar description, because banks offer less effective resistance.			
a. Regular section with no boulders or brush	0.025		0.060
b. Irregular and rough section	0.035		0.100

The latter sketch points out an important consideration. It indicates that even in the region of the river channel itself, excluding the flooded banks, the wetted perimeter and the horizontal distance are essentially the same. Certainly, within the usually available accuracy of cross-sectional drawings and surveys, this assumption will not produce any appreciable error.

If the wetted perimeter and the horizontal channel width are essentially the same, or

$$P \approx b \qquad (12.3)$$

it follows that

$$R = \frac{A}{b} \qquad (12.4)$$

However, A/b in a natural channel is also the average depth D_{ave} of the channel. Hence in many natural channel applications of the Manning equation the following substitution for the hydraulic radius R can be made.

$$R = D_{ave} \qquad (12.5)$$

Composite Cross Sections and the Form Factor

From a study of Figs. 12.1, 12.2, and 12.3, it is obvious that accurate calculations of the flow in a natural channel are not readily accomplished by a simple application of the Manning formula.

One approach to dealing with the variation in properties of cross sections and to allow for tne changing *n* coefficient would be to divide the cross section

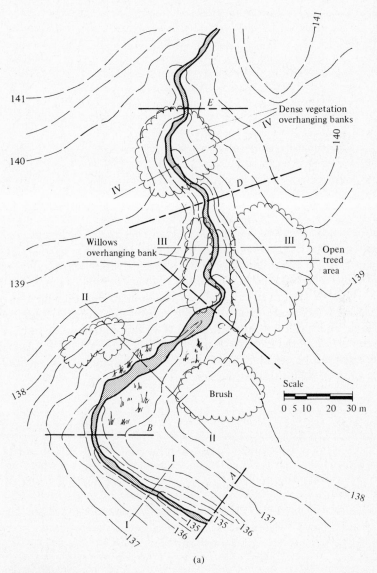

(a)

FIGURE 12.3a

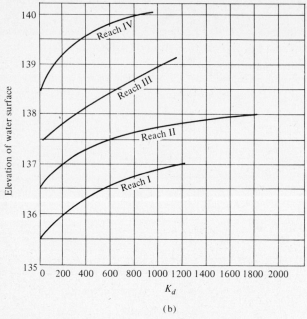

FIGURE 12.3b

into separate portions, each with its own roughness coefficient. For the cross section in Fig. 12.1, for instance, calculations of Q could be carried out separately for each of the five portions, with the different n values. The total Q for the channel could then be taken to be the sum of the Q's of the portions.

When a river is in flood, it can be assumed that the slope of the energy gradient is relatively uniform for the entire cross section. Therefore Q could also be obtained by means of Eq. (12.1). In this equation, the only variable for each portion of the cross section is K_n. Hence, values of K_n for each portion of the river can be calculated and summed for the entire cross section to obtain the value of K_n for the entire river.

EXAMPLE 12.1

If the slope of the energy gradient in the river of Fig. 12.2 is 0.0022, and n coefficients for the north bank, the central channel, and the south bank are 0.150, 0.030, and 0.075, respectively, determine the rate of flow in this stream, with the water level as shown.

SOLUTION

From the geometry of the cross section, values of A for each portion can be determined, by means of a planimeter, counting squares, or other methods. Keeping in mind the earlier discussion regarding the

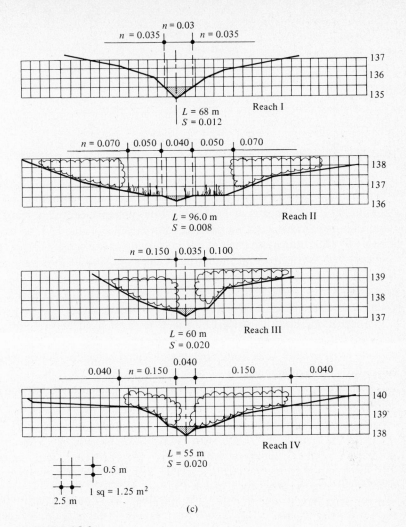

FIGURE 12.3c

value of the wetted perimeter, values of K_n for each portion can now be readily calculated, as tabulated below.

Channel Portion	n	A, m²	P, m	R, m	$R^{2/3}$	$K_n \times 10^3$
North bank	0.150	80	43	1.86	1.51	0.8
Central	0.030	220	41	5.37	3.06	22.5
South bank	0.075	135	47	2.87	2.02	3.6
Total K_n for stream, water level at 85.0 m=						26.9

Hence the estimated flow in this channel when the water surface is at elevation 85.0 is

$$Q = 26900 \times 0.0022^{1/2}$$
$$= 1260 \text{ m}^3/\text{s}$$

The n Coefficient

The selection of the proper n coefficient is a relatively difficult task and often results in widely varying flow estimates. The most common method of determining n is by means of Table 10.1. For convenience, the portion of this table relating to natural channels is here reproduced as Table 12.1. An examination of this table indicates values of n often varying over a range of 30 to 140 percent between maximum and minimum. Values of Q could thus vary proportionally. Much experience and care is necessary in order to ensure the proper selection of this coefficient. The descriptions in Table 12.1 are reasonably detailed, and when used conservatively, can yield acceptable results.

Another method, developed by the U.S. Geological Service, consists of selecting values of n from a series of photographs of river cross sections. Each of these cross sections has been labeled with an n coefficient, determined from actual flow measurements.

Recently a method developed by W. L. Cowan has been developed, based on the following formula.

$$n = (n_0 + n_1 + n_2 + n_3 + n_4)m_5 \qquad (12.6)$$

where n_0 is the basic n coefficient for a straight uniform channel, in the materials making up the channel walls and bottom, obtained from Table 12.1; n_1 to n_4 are correction factors to allow for variations in channel conditions; and m_5 is a correction factor to account for meander in the stream. Values of n_1 to n_4 and m_5 are shown in Table 12.2.

Equation (12.6) was developed on the basis of studies on approximately 50 small streams with a hydraulic radius in the vicinity of 5.0 m. Consequently it can be used confidently only for other streams with the same characteristics.

Natural Channels

When studying natural-channel flows, it is often convenient and necessary to divide the river into longitudinal portions, or reaches. Such reaches should be chosen to reflect a reasonably constant set of hydraulic characteristics throughout the length of the reach.

TABLE 12.2   Values of n_0, n_1, n_2, n_3, n_4, and m_5 for Determining the Roughness Coefficient in Manning's Formula for Natural Channels

		Channel Specifics	Values
n_0	Material of channel	Earth	0.020
		Rock cut	0.025
		Fine gravel	0.024
		Coarse gravel	0.028
n_1	Extent of irregularity	Smooth	0.000
		Minor	0.005
		Moderate	0.010
		Severe	0.020
n_2	Channel cross= section variations	Gradual	0.000
		Changing slightly	0.005
		Changing often	0.013
n_3	Channel obstructions	None	0.000
		Minor	0.013
		Extensive	0.025
		Severe	0.050
n_4	Vegetation	Low	0.008
		Medium	0.018
		High	0.040
		Very high	0.080
m_5	Meander	Minor	1.000
		Appreciable	1.150
		Severe	1.300

Depending on the type of study involved, the following are some of the major considerations that should dictate the reach selection:

1. The shape of the cross section should be reasonably constant throughout the reach length.
2. The n coefficients of the several portions of cross section should be constant, or nearly so.
3. The slope of the channel bottom in the reach should remain fairly uniform and must not contain steep and flat sections.

For illustration purposes, points A, B, C, D, and E have been indicated as possible reach boundaries in Fig. 12.3a. These reaches are hydraulically different from each other because reach AB has a straight channel, with a relatively steep bottom slope; reach BC is flat and open, with a marshy area; reach CD has a fairly regular cross section, with uniform open-type vegetation; and reach DE has dense growth overhanging the stream banks.

12.3   STEADY, UNIFORM FLOW IN NATURAL CHANNELS

Natural-Channel Flow

When considering the variations in the actual channel hydraulic characteristics inherent in even the most carefully chosen reach of a stream, it is evident that true uniform flow cannot exist. The changes in cross section, bottom slope, and n coefficient lead to continual changes in velocity and subsequent variations in the energy and hydraulic gradients.

However, in many instances, particularly during times when the flow is contained within the normal channel banks, or when the river is in full flood, uniform flow conditions are approached.

At such times, application of the Manning flow formula will yield estimates of the flows of an acceptable nature and accuracy.

Aids in Solving the Manning Formula for Natural Channels

From the work done in Example 12.1, it is evident that considerable calculations and work are involved in determining a river's capacity of flow. Particularly, if these estimates of flow are required at various water levels, the work becomes tedious and extremely time-consuming.

A convenient way to enable easy analysis of a particular reach of a stream is to develop for this reach a graph representing values of K_n at all water-level elevations expected in the stream. In this way it is necessary only to determine values for K_n at a number of convenient and representative water elevations and to plot them versus these levels.

Typical examples of these graphs, or K_n curve s, are shown in Figs. 12.2d and 12.3b. Application of Eq. (12.1) for Q can now be easily accomplished for any chosen depth of flow. The use and preparation of K_n curves is demonstrated in the following example.

EXAMPLE 12.2

For the stream in Fig. 12.2a develop a K_n curve and determine the uniform flow capacity at various water-surface elevations. Also, determine the water-surface elevation if the uniform flow is 525 m³/s and 250 m³/s.

SOLUTION

From the geometry of the cross section in Fig. 12.2b and in a manner as shown in Example 12.1, values of K_n are calculated at four elevations of the water level—namely, at 80.0 m, 81.0 m, 83.0 m, and 85.0 m. The K_n value for the last of these has already been determined in the previous example. The calculations for the other K values of the cross section are shown in tabular form below.

Channel Portion	n	A	P	R	$K_n \times 10^3$
Water level at 80.0 m					
Central	0.030	17	33	0.52	0.4
K_n for 80.0 m					0.4
Water level at 81.0 m					
Central	0.030	54	41	1.32	2.2
K_n for 81.0 m					2.2
Water level at 83.0 m					
North bank	0.150	18	23	0.78	0.1
Central	0.030	134	41	3.27	9.8
South bank	0.075	56	37	1.51	1.0
K_n for 83.0 m					10.9

The corresponding K_n curve is shown in Fig. 12.2d. Also shown on this graph are values of the cross-sectional area A. The latter is a handy addition to aid in determining average velocities, if required.

If the flow is uniform and $s = 0.0022$, as in Example 12.1, values for Q at various water-surface elevations can be readily determined, as follows: *Water surface at elevation 84.0.* From Fig. 12.2d,

$$K_n = 17500$$

$$A = 305 \text{ m}^2$$

From Eq. (12.1),

$$Q = 17500 \times 0.0022^{1/2}$$

$$= 820 \text{ m}^3/\text{s}$$

$$V_{ave} = \frac{820}{305}$$

$$= 2.7 \text{ m/s}$$

Water surface at elevation 82.5.

$$K_n = 8000$$

$$A = 150 \text{ m}^2$$

$$Q = 8000 \times 0.0022^{1/2}$$

$$= 375 \text{ m}^3/\text{s}$$

$$v = 2.50 \text{ m/s}$$

A procedure opposite to the above can be used to determine the water level from a given Q, as follows:
For Q = 525. From Eq. (12.1),

$$K_n = \frac{525}{0.0022^{1/2}}$$

$$= 11200$$

From Fig. 12.2*d*, the water-surface elevation is 83.1 m.
For Q = 250. Again, from Eq. (12.1),

$$K_n = \frac{250}{0.0022^{1/2}}$$

$$= 5300$$

Hence the water-surface elevation is 82.3 m.

12.4   STEADY, NONUNIFORM FLOW IN NATURAL CHANNELS

Water-Surface Profiles

As mentioned earlier, almost all natural channel flow is nonuniform. Changes in channel slope, cross section, or n coefficient and obstructions such as bridges and culverts, checkdams, and the like almost always create retarded or accelerated flow. Often it is necessary to determine the profile of the water surface in such cases. For instance, the installation of a bridge or culvert could create a certain amount of backup in the stream under certain conditions of flow. Such a back up could result in flooding of lands upstream of the proposed structure. The knowledge of the actual amount of flooding to be expected will be of great help in choosing a practical and economical waterway opening for the bridge or culvert.

Similarly, a stream discharging into a reservoir with fluctuating water levels could create flood damages upstream. During time of high reservoir water levels and peak flow conditions in the stream this damage could prove to be extensive. Therefore, knowledge of the backwater surface profile is essential not only in the design of the reservoir but also in establishing its operation policies.

Methods of Determining Water-Surface Profiles

The determination of the actual water-surface profile in natural channels is based upon the same principle as those discussed for regular channels. The basic relationships established there, apply as well to natural channels and are expressed in the form of Eqs. (11.34) and (11.35); that is,

$$l = \frac{\dfrac{v_2}{2g} + D_2 - \dfrac{v_1}{2g} - D_1}{s_0 - s}$$

$$l = \frac{H_2 - H_1}{s_0 - s}$$

where H, v, and D are specific energy head, velocity, and depth of flow, respectively, at the upstream and downstream points of the reach; s_0 is the slope of the channel bottom; s is the slope of the energy or hydraulic gradient; and l is the length of the reach. These formulas, combined with the previously discussed form factors, K_n, provide a convenient method of determining backwater-surface profiles.

In the case of regular channels, where the channel cross section is constant, the calculations for backwater profiles started with an assumed upstream or downstream water level for the reach, and with this depth the reach length l was calculated. In a natural channel the reaches are predetermined as discussed earlier. Then by means of successive approximation and the use of Eqs. (11.34) and (11.35), the upstream and downstream water elevations of each reach can be determined.

The method is best described by the following step-by-step procedure.

Step 1. Divide the stream in convenient reaches.

Step 2. Construct K_n curves for each reach.

Step 3. Starting with a known water surface elevation—say, the downstream water elevation of Reach 1—estimate the water-surface elevation at the center of this reach.

Step 4. With the estimated water surface and its corresponding K_n value, determine the slope of the hydraulic gradient by means of Eq. (12.1).

Step 5. With the value of s so obtained, compute the upstream and downstream water-surface elevations and compare with the known elevation.

Step 6. If agreement between the known and computed elevation is found, proceed to Reach 2. If the known and computed elevations vary by more than an acceptable margin of error, adjust the estimated water-surface elevation at the center of the reach and repeat Steps 4 and 5.

Step 7. Continue this operation through all the reaches and plot the water-surface profile.

EXAMPLE 12.3

For the river in Fig. 12.3a, determine the water-surface profile and extent of flooding, assuming a culvert is to be constructed at A. When the design flow of the stream equals 100 m³/s, the water level at the upstream side of the culvert will be 136.0 m. The river is in a silty-soil area, and unless otherwise indicated, the soil cover is medium long grass.

SOLUTION

The work indicated in Step 1 above was discussed earlier and the boundaries of the proposed reaches are as shown on Fig. 12.3a.

The cross sections of each reach, located at a point considered to be most representative of that reach, are also located on this figure. The actual cross sections are shown in Fig. 12.3c. As discussed earlier, the K_n curves for each cross section were computed and plotted in Fig. 12.3b.

In this instance, the known water-surface elevation is the elevation at the culvert, or the downstream elevation of Reach I. Starting with this reach, the work of Steps 3, 4, 5, and 6 can now be completed. The corresponding calculations are shown in tabular form below.

Reach	$\dfrac{l}{2}$, m	Assumed Elevation at Center, m	K_n	s	Fall, m	Elevation		
						Upstream, m	Downstream, m	
I	34.0	136.5	500	0.04	1.36		135.1	NG
		136.75	750	0.018	0.60		136.15	NG
		136.65	690	0.021	0.70		135.95	OK
	Reach I profile					137.4	136.0	
II	48.0	138.0	1800	0.0031	0.15		137.85	NG
		137.75	900	0.012	0.59		137.16	NG
		137.80	1100	0.008	0.40		137.4	OK
	Reach II profile					138.2	137.2	
III	30.0	138.75	810	0.015	0.54		138.3	NG
		138.70	780	0.016	0.49		138.21	OK
	Reach III profile					139.2	138.2	
IV	27.5	139.75	500	0.04	1.1		138.65	NG
		140.00	800	0.015	0.43		139.6	NG
		139.85	650	0.024	0.65		139.2	OK
	Reach IV profile					140.50	139.2	

The corresponding water-surface elevations are shown as the flood line on Fig. 12.3c.

In order to clarify the calculations in the work table above, the calculations for Reach I are set out in detail below.

Initially assumed elevation at center of the reach: 136.5 m
From Fig. 12.3b,

$$K_n = 500$$

From Eq. (12.1),

$$s = \left(\frac{100.0}{500}\right)^2$$
$$= 0.04$$

Difference in elevation of water surface between center and end of the reach:

$$\text{Fall} = 0.04 \times 34.0 = 1.36 \text{ m}$$

Downstream water surface elevation: $136.5 - 1.36 = 135.14$
The known downstream elevation is $136.0 \neq 136.14$
Therefore, assume a new water-surface elevation at the center and try again.

It should be noted that the allowable difference between the trial elevation and the known elevation in this case was set at 0.1 m or 100 mm.

Graphical Method of Determining Water-Surface Profiles

Many applications exist where the starting water-surface elevation is not fixed but fluctuates for various design conditions. These fluctuating water levels occur in cases where the river discharge is controlled by a reservoir or lake water level. Here it is necessary to know the effects on upstream water surfaces at various combinations of outlet elevations and river flows, as dictated by design considerations or seasonal conditions.

Other instances where fluctuating water levels downstream play a role is in the design of culvert or bridge waterway openings. The larger waterway openings generally create a greater cost of construction. Economic comparison therefore of bridge or culvert size with upstream floods will result in the requirement of investigating several downstream design water levels.

In these cases it would be necessary to repeat the trial-and-error solution from Example 12.3 for each of the design water levels, which could prove to be tedious and time-consuming. If a sufficient number of design elevations are to be considered, the work can be facilitated and shortened by considering the following.

1. In any reach, for a certain Q there is only one slope of the hydraulic gradient corresponding to a specific water-surface elevation. Hence for any downstream water level in a reach there is only one corresponding upstream water level.
2. The upstream water level of one reach is the downstream water level of an adjacent reach, and vice versa.

For instance, for Reach I in the previous example, the corresponding upstream water level when the downstream elevation is 136.0 m, is 137.4 m. The latter can be read directly off the work table of the example. In a similar manner, several other corresponding pairs of elevations can be found for this reach. When these values are plotted with the downstream elevation as the abscissa

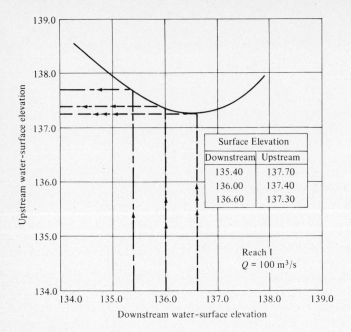

Surface Elevation	
Downstream	Upstream
135.40	137.70
136.00	137.40
136.60	137.30

Reach I
$Q = 100$ m³/s

Downstream water-surface elevation

FIGURE 12.4

or x axis and the corresponding upstream elevation as the ordinate or y-axis, the graph of Fig. 12.4 results. From this graph it is now possible to readily determine any pair of corresponding watersurface elevations, as shown.

EXAMPLE 12.4

Figure 12.5a is a set of K_n curves for five reaches of a stream. The design Q is 375.0 m³/s. Determine the water-surface profile for this stream when the downstream water level of Reach I is (a) 158.0 m and (b) 160.0 m.

SOLUTION

Figure 12.5b is a plot of water levels for each reach of the stream. However, it will be noted that the downstream elevations of the even-numbered reaches are plotted as ordinates, while the downstream elevations of the uneven numbered reaches are plotted on the x axis. In this way, remembering that the upstream end of one reach is the downstream of the next, the water=surface profile can be determined quickly for any starting water level, as shown in the figure.

For case (a), determining the downstream water surface of Reach I at 158.0 m involves the following process.

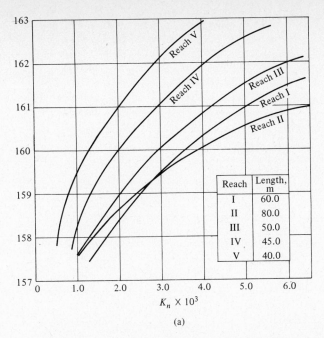

(a)

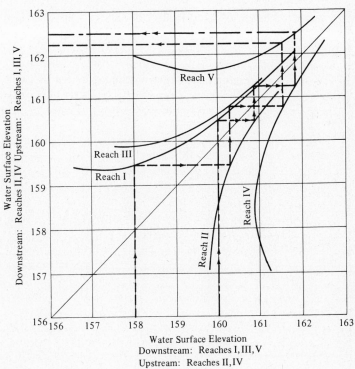

(b)

FIGURE 12.5

461

Reach I downstream=158.0; hence from the graph, upstream 159.5.
Reach II downstream=upstream of Reach I; downstream 159.5,
hence from graph, upstream 160.3.
Reach III downstream=upstream of Reach II=160.3; downstream
160.3, hence from graph upstream 160.8.
and so forth.

Therefore the water-surface profile of this stream is, for case (a),

Reach	Downstream Elevation	Upstream Elevation
I	158.0	159.5
II	159.5	160.3
III	160.3	160.8
IV	160.8	161.6
V	161.6	162.3

For case (b),

Reach	Downstream Elevation	Upstream Elevation
I	160.0	160.5
II	160.5	160.9
III	160.9	161.2
IV	161.2	161.9
V	161.9	162.5

For some further clarification of the work involved, the table below
sets out the steps involved in calculating pairs of corresponding downstream and upstream water surface elevations. These calculations are here
shown only for Reach I.

Reach	$l/2$	Elevation Q Center	K_n	s	Fall	Elevation Upstream	Downstream
I	30.0	158.5	2080	0.033	1.0	159.5	157.5
		159.0	2500	0.023	0.7	159.7	158.3
		159.0	2500	0.023	0.7	159.7	158.3
		160.0	3550	0.011	0.3	160.3	159.7
		161.0	5070	0.005	0.2	161.2	160.8

13

Flow of Water Through Soils

13.1 SOIL WATER

Types of Soil Water

During the excavation of a hole in granular soils, three distinctly different soil moisture conditions are generally encountered. Although these various moisture conditions will be more readily observed in granular soils, they also exist in other types of soils, such as silts and clays.

Unless recent precipitation has moistened the top layers of soil, these layers will be dry and essentially without soil moisture. The granular and silty soils will be noncohesive and runny. As the excavation continues deeper, the soil will become more and more laden with moisture, resulting in increased cohesion. At this early stage the water will not freely run out of the soil. However, a handful of the soil can be squeezed to release some of the water in it.

Further excavation will lead to the point where the water in the soil will readily run out of it and the bottom of the excavation will be inundated. At this point the groundwater table has been reached. Between the beginning of the excavation and the end, three forms of soil water have been encountered. They are described in some detail below.

463

Attached Water

Very early in the excavation, in the top layers of soil, a certain amount of water is present in the soil, even though the soil appears to be dry. This moisture in the soil is generally referred to as *attached water*. It is in fact moisture adhering to the soil particles as a damp film. Attached water consists of water left behind from the percolation or infiltration of rainwater into the soil. It can also be the result of movement or recession of other types of soil water.

The amount of attached water in any soil, will depend on the time since the soil was wetted, how long the dry period has been, and also the size of the soil particles. Obviously, the smaller the soil particles, the larger surface area there is available for soil moisture to adhere to.

Attached water can be removed from the soil through drying or the application of pressure. It is a soil water that does not move within the soil of its own volition but remains stationary until external forces cause it to leave the soil.

Capillary Water

As already mentioned in Chapter 2, liquids, though generally considered to be unable to resist internal tensile forces, nevertheless exhibit a property referred to as surface tension. This property is readily observed at the formation of a meniscus in a tube containing water.

It is this surface tension that is responsible for the presence of the second type of water or soil moisture encountered in the excavation. It is because of surface tension between the surfaces of the soil particles and the water that *capillary action* will take place, and water will be drawn upward into the soil. Consider the much enlarged soil particles as shown in Fig. 13.1a. Soil water will form a meniscus, as shown at A and redrawn in Fig. 13.1b. The surface tension σ will create a force, F, pulling upwards on the small volume of water contained by the small space between the soil particles. The small volume of water in the space is thus subjected to two forces, the force G due to the force of gravity and the force F created by the surface tension.

As the interstices between soil particles become smaller, the force G will also decrease. The force F, on the other hand, will increase because the surface

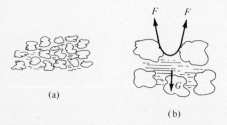

(a)

(b)

FIGURE 13.1

area of the soil particle increases. When F becomes larger than G the water will be drawn upwards into the soil.

This phenomenon is referred to as capillary rise or capillary action in the soil. Once equilibrium between F and G has been obtained, *capillary water* will be stationary. Water held in the soil by capillary action therefore will not flow unless the surface tension forces are interfered with. This can be accomplished by means of subjecting the soil to pressure or the addition of certain chemicals.

Generally, capillary water is encountered a certain distance above the groundwater or other forms of free soil water. The thickness of the capillary layer is dependent upon the soil type and particularly the relationship between grain size and soil interstices.

Free Soil Water

At the end of the excavation, water is encountered in the soil that runs relatively freely into the excavation and keeps the bottom inundated. Obviously, this is water that is not held into the soil by any force and is allowed to run freely between the soil particles. Generally this water is referred to as *groundwater*. Free groundwater in the soil and subsoil, in many forms, is illustrated in Fig. 13.2.

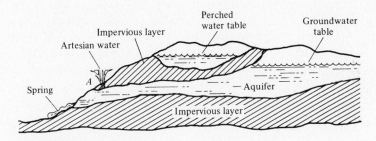

FIGURE 13.2

13.2 GROUNDWATER AND AQUIFERS

Groundwater Table

The surface of the groundwater immediately below the ground surface is called the *groundwater table*. It is also the surface of the groundwater that is subject to atmospheric pressure.

The elevation of the groundwater table fluctuates with the seasons, whether dry or wet conditions exist, the atmospheric pressure, and the rate of withdrawal or replenishment of the groundwater.

Perched Water Table and Water Lenses

As shown in Fig. 13.2, when an area of groundwater is situated in a pan or depression, underlain by an impervious soil layer, the water table in such an area could well be some distance above the normal water table of the surrounding area. Such an elevated water surface is called a *perched water table*.

When waters of different densities are encountered in the soil, such as near the ocean where salt water infiltrates and supports layers of fresh water, *water lenses* are encountered, as shown in Fig. 13.3.

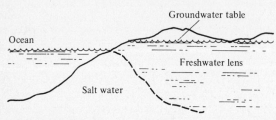

FIGURE 13.3

Artesian Groundwater

Under certain conditions, as shown in Fig. 13.2, groundwater is collected and confined between two impervious soil layers. When this is the case, the soil water can be under considerable pressure, or what is referred to as an artesian condition. A well drilled through the top impervious layer, as shown at A in Fig. 13.2 is an artesian well. Due to the high hydrostatic head in the groundwater at A, the water will rise in the well and in some instances flow freely on the surface.

Aquifers

Any water-bearing soil stratum is called an *aquifer*. Unconfined aquifers are those with their water surfaces at the water table and open to atmospheric pressure. The water surface of an unconfined aquifer therefore fluctuates, as mentioned earlier. Confined aquifers are those in which the water is held between two impervious soil layers.

13.3   FLOW OF WATER THROUGH SOILS

Groundwater Movement

Generally the surface of the groundwater table is not horizontal; instead, it slopes towards a river, lake, or ocean. This sloping groundwater table results in different energy heads at various points along the water surface. Naturally, water will flow from points with higher energy head to points with lower energy head. Therefore water in underground aquifers will always be in motion and flowing from one point to another. The only exception to this phenomenon would be the case of *perched water table* situations, where perfectly horizontal water surfaces can exist.

The rate of flow of water in aquifers depends not only on the amount of energy head difference between two points but also on the ease with which the water can move through the soil.

Soil Porosity

The *porosity* of a soil sample is expressed as the percentage that the volume of the pores takes up of the total volume of the sample. Porosity, therefore, is an indication of the amount of space there is available in a soil for the storage and or flow of water. It appears self-evident, for example, that a coarse gravel having a porosity of 30 percent, will provide more space and consequently less resistance to the flow of water than a fine sand having a porosity of 15 percent.

Although the above statement is generally true, in the case of some granular materials, the statement that more pores provides for better flow is not applicable to all soils. Certain clays, for instance, can have porosities as high as 55 percent, and yet it is a well-known fact that the flow of water through clays is very slow and in some cases essentially nonexistent. Therefore, even though the porosity is in itself an indication of the space available for the storage or flow of water, it is not necessarily a proper indicator of the flow capabilities of a soil.

Permeability

The ease with which a water flows through a soil, is expressed by the *permeability coefficient* of that soil. Many authorities in soil moisture discuss this property of a soil and use various names and units for its expression. Some of the more common forms of expressing permeability are hydraulic conductivity, the field coefficient of permeability, and the coefficient of transmitability.

    1.   *Hydraulic conductivity* (k).   The hydraulic conductivity is expressed as follows:

$$k = \frac{\beta \gamma}{\mu} \tag{13.1}$$

where β is a proportionality coefficient. γ and μ are the specific weight and absolute viscosity of the water, respectively. If the specific weight is expressed in kN/m^3 and the viscosity in $Pa \cdot s$, and β is dimensionless, k will have units of m/s, or velocity.

2. *The field coefficient of permeability (P).* The field coefficient of permeability is the actual quantity of water flowing through a unit area of the aquifer, when the slope of the hydraulic gradient is 1.0. Since this is expressed in terms of volume per time per unit area, the field coefficient of permeability will also have units of distance per time, or velocity.

3. *The coefficient of transmittability.* The coefficient of transmittability is the actual amount of water flowing in an aquifer through an area, consisting of a vertical slice of the aquifer, one unit in width and equal in height to the total height of water in the aquifer and the slope of the hydraulic gradient equal to 1. Units of the coefficient of transmittability are also those of velocity.

The coefficient of permeability, when referred to in this text, will have units of m/s and be an indication of the velocity of flow through a soil.

For certain applications its value can be approximated by the following expression, developed by Hazen from experiments on clean sands.

$$k = (D_{10})^2 \tag{13.2}$$

where k is expressed in cm/s and D_{10} is the effective size of the soil in millimeters. Because it was developed from experimental research for a specific type of soil, this equation cannot be considered to be applicable to all soils.

The permeability of a soil can be determined by means of laboratory tests, using permeability meters. The results of such test, when applications in the field of underground water flow are considered, must be used with some discretion and caution. The behavior of water in a soil in a laboratory cannot be expected to be the same as when the soil is in its actual location in the aquifer.

The permeability of a soil depends on several factors, including the viscosity and temperature of the water, the pressure to which the soil is subjected, the cross-sectional area of the interstices between the soil particles, and the porosity of the soil.

The latter two properties of a soil account for the fact that·soils with very high porosities but very fine particle sizes, such as most clays, are not conducive to easy flow of water. This is because the very small interstices between the particles are filled with attached water, which, as explained earlier, does not flow.

The permeability of various soils has been reported by many authorities, and considerable variation among the findings exist. Table 13.1 is a partial summary of the more common values of the permeability of certain soils.

Darcy's Law Regarding the Flow of Water in Soils

In the latter half of the nineteenth century, Henry Darcy, the French hydrologist mentioned in Chapter 7, formulated a law relating to the flow of water through soils, which is still in use today and bears his name. The law

TABLE 13.1   Permeability Coefficients of Certain
Soil Types as Suggested by Some Authorities

Soil Type	Permeability Coefficient, m/day
Clay	0.0001–0.2
Clay-silt	0.0001–0.30
Silt	0.05 –0.80
Very fine sand	0.0001–20.0
Fine sand	1.0–20.0
Medium sand	5.0–100.0
Coarse sand	20.0–450.0
Fine gravel	500–1000
Gravel	150–9000
Gravel and sand	8–700

states: "The velocity of the flow of water through a soil is directly related to the hydraulic gradient."

In equation form this can be expressed as follows:

$$v = ks \qquad (13.3)$$

where v is the velocity of flow, s the slope of the hydraulic gradient, and k is a proportionality factor—the coefficient of permeability, in fact. Since s, the slope of the hydraulic gradient, is dimensionless, it follows that as previously stated, k has the same dimensions as v.

Considering the flow of water through a section of soil with a cross-sectional area A, and remembering that

$$Q = Av$$

it follows that

$$Q = Aks \qquad (13.4)$$

where Q is the capacity of flow through the soil. Obviously, the units of Q will depend on the units of k and A.

The velocity in Eq. (13.3) is generally referred to as the filter velocity or the Darcy velocity in the soil. It is not the actual velocity of the water as it flows through the interstices of the soil. This latter velocity will actually be considerably larger than the filter velocity.

EXAMPLE 13.1

A sand layer has a groundwater table with a surface slope of 1:3000. The permeability of the sand is 20 m/day and the porosity is 30 percent. Determine the flow of water through a 1.0 m² section of this soil. Determine also the filter or Darcy velocity and the actual velocity of flow.

SOLUTION

From Eq. (13.4),

$$Q = 1.0 \text{ m}^2 \times 20 \text{ m/day} \times \frac{1}{3000}$$

$$= 0.0067 \text{ m}^3/\text{day}$$

$$= 6.7 \text{ L/day}$$

and

$$\text{Darcy } v = \frac{0.0067 \text{ m}^3/\text{day}}{1.0 \text{ m}^2}$$

$$= 0.0067 \text{ m/day}$$

In order to determine the actual velocity, it must be remembered that since the porosity of the soil is 30 percent, only 30 percent of the total cross-sectional area is made up of interstices, allowing water to flow. Therefore the actual cross-sectional area of flow is

$$A_{act} = 1.0 \text{ m}^2 \times 0.30$$

$$= 0.30 \text{ m}^2$$

and

$$v_{act} = \frac{0.0067 \text{ m}^3/\text{day}}{0.30 \text{ m}^2}$$

$$= 0.022 \text{ m/day}$$

13.4 WELLS

Ideal Wells and the Dupuit Formula

Also in the latter part of the nineteenth century, the French philosopher Jules Dupuit developed a formula for the flow of underground water to a well that is approximately correct. The formula is based on theoretical assumptions and conditions that in real applications are seldom, if ever, duplicated exactly. Nevertheless, judicious use of the Dupuit formula and the theory involved forms the basis of most of today's practical work related to the flow of water in aquifers to a well, or wells.

An understanding of the development of the Dupuit formula and its limitations is very important in the understanding of the principles related to groundwater flow and its use as a resource.

In order to develop the Dupuit formula, consider the theoretical circular island of Fig. 13.4a, with the dimensions as shown. Assume the island to be perfectly circular, with a radius r_e, and the soil making up the island to be perfectly homogeneous, with a constant permeability coefficient k. Prior to pumping water from the well, located in the exact center of the island, the groundwater level in the island is assumed to be horizontal, as shown by the line AB in the figure. When water is pumped from the well, the water level in it will drop and a hydraulic inequilibrium will be created, causing water to flow toward the well. The flow to the well, because of the theoretical conditions imposed earlier, will run in equal amounts from all directions to the well. Consequently the velocity of flow from all points radially toward the well will be equal from all directions, as indicated in Fig. 13.4b.

During this process, the water table will be lowered to a greater degree

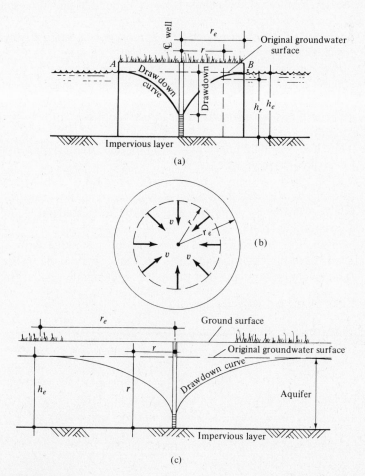

FIGURE 13.4

near the well than at the outskirts of the island. It will take on the shape shown in Fig. 13.4a. The surface of the free water in this instance is called the Dupuit or drawdown curve. Above this surface is still an area with considerable water in the soil. However, this water is either attached water or capillary water, or both, and because of its nature it will not flow.

The water surface will therefore be at the level of the water in the well at the center of the island and at the water surface at the circumference of the island. If pumping is continued at a certain rate, the water level in the well will continue to drop and the drawdown curve will continue to deform to suit the new water level in the well. Eventually a point may be reached where the amount of water pumped out of the well is equal to the amount of water entering the well from the aquifer. When this point is reached, equilibrium has been achieved in the aquifer. Further pumping at this rate will not produce any more changes in the Dupuit or drawdown curve.

In the development of his formula, Dupuit made the following assumptions.

1. The soil of the island is perfectly homogeneous, and the permeability is constant throughout.
2. The island is circular, and the well is located in its center.
3. There is no flow in the aquifer prior to pumping, and the groundwater table in the island is horizontal.
4. The well extends in depth to the impervious layer at the bottom of the aquifer.
5. The well is being pumped at such a rate that equilibrium conditions as described above exist.

If all of the above conditions are met, the amount of flow toward the well at the points a distance r from the well can be calculated using the Darcy formula, given in Eq. (13.4).

At a distance r from the well, the cross-sectional area in the aquifer through which flow toward the well takes place is equal to a right circular cylinder with radius r and height h_r, the depth of water in the aquifer at that point. Hence

$$A = \pi r \times h_r \tag{13.5}$$

Therefore, since $Q = vA$, the flow to the well will be

$$Q = \pi r \times h_r \times k \times s \tag{13.6}$$

The value of s to be used in Eq. (13.6) will be the slope of the water surface, or the drawdown curve. It is not constant, but varies as r changes. Taking this fact into consideration and applying rules of differentiation to Eq. (13.6), the Dupuit formula results. This formula reads

$$Q = \frac{\pi k (h_r{}^2 - h_e{}^2)}{\ln (r/r_e)} \tag{13.7}$$

or, using common logarithms,

$$Q = \frac{1.36k(h_r^2 - h_e^2)}{\log(r/r_e)} \qquad (13.8)$$

In both formulas, Q is the capacity of flow into the well. Since it is assumed that the well is being pumped at equilibrium, Q also represents the pumping rate. The units of Q are dependent on the units of k, h, and r. For instance, if h and r are expressed in meters and the permeability coefficient k in meters per day Q will be in cubic meters per day.

The Dupuit formula can also be applied to an aquifer instead of to a circular island. If the island is sufficiently large in diameter, it is easily visualized that the water surface far enough away from the well will not change. In other words, the drawdown curve will not reach the outskirts of the island, and r_e will be as shown in Fig. 13.4c.

It must be remembered that a number of theoretical preconditions were assumed in the development of the formula. In reality, these conditions will rarely be met because aquifers do not consist of homogeneous soils, nor do they have perfectly horizontal groundwater surfaces. Often aquifers are encroached upon by impervious soil strata, or intersected by rivers or lakes. Further, in many instances wells do not penetrate the entire aquifer depth. All of the above, or indeed any one of them, can significantly alter the flow to the well from the value anticipated by the formula.

The primary function of the Dupuit formula is as a preliminary tool in estimating well or aquifer capacities. For such applications, experience has shown that acceptable results can be obtained from its use.

EXAMPLE 13.2

In Fig. 13.5 is shown the layout of a well, A, and two wells used as observation wells, B and C, during a pumping test. When equilibrium was reached in the aquifer—in other words, when the water level in the observation wells remained constant while the test well was being pumped at a constant rate—the rate of pumping was 35 L/s and the drawdown in the observation wells B and C, was 3.5 m and 1.0 m, respectively. Well A is 500 mm in diameter and extends downward to the bottom of the aquifer, which is 35 m deep. Determine (a) the value of the permeability coefficient of the aquifer; (b) the radius of influence r_e; and (c) the drawdown in well A.

SOLUTION

For use in the Dupuit formula, the following information is given:

$$Q = 35 \text{ L/s} = 3024 \text{ m}^3/\text{d}$$

$$h_r = 35.0 \text{ m} - 1.0 \text{ m} = 34.0 \text{ m at well } C$$

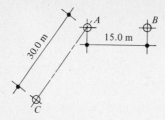

FIGURE 13.5

$$h_r = 35.0 \text{ m} - 3.5 \text{ m} = 31.5 \text{ m at well } B$$

$$h_e = 35.0 \text{ m}$$

$$r = 50.0 \text{ m for well } C$$

$$r = 20.0 \text{ m for well } B$$

Therefore the flow of water through the right circular cylinder through well B is, according to the Dupuit formula, Eq. (13.7),

$$Q = \frac{\pi k (31.5^2 - 35.0^2)}{\ln(30/r_e)}$$

Since equilibrium pumping conditions exist, it follows that the amount of water flowing through the cylinder at B must equal the amount of water being removed from the test well, or 3024 m³/day. Therefore

$$3024 \text{ m}^3/\text{day} = \frac{\pi k (31.5^2 - 35.0^2)}{\ln(30/r_e)}$$

Similar reasoning will result in an expression for the flow at C, as follows:

$$3024 \text{ m}^3/\text{day} = \frac{k (34.0^2 - 35.0^2)}{\ln(15.0/r_e)}$$

In the above two equations the left sides are equal to each other, and hence the right sides must also be equal, or

$$\frac{\pi k (31.5^2 - 35.0^2)}{\ln(15/r_e)} = \frac{\pi k (34.0^2 - 35.0^2)}{\ln(30/r_e)}$$

and

$$(\ln 30.0 - \ln r_e)(-233.0) = (\ln 15.0 - \ln r_e)(-69.0)$$

$$233.0 \ln r_e - 69.0 \ln r_e = 605.6$$

$$\ln r_e = 3.69$$

$$r_e = 40.2 \text{ m}$$

Substitution of this value of r_e in either of the equations for wells B or C will lead to finding the value of k, because

$$3024 \text{ m}^3/\text{day} = \frac{\pi k (31.5^2 - 35.0^2)}{\ln(15/40.2)}$$

or

$$3024 \text{ m}^3/\text{day} = \frac{\pi k (34.0^2 - 35.0^2)}{\ln(30/40.2)}$$

from which

$$k = 4.07 \text{ m/day}$$

In order to find the drawdown at the well, the flow through the right circular cylinder with a diameter equal to the well diameter can be considered. Applying reasoning similar as that for wells B and C above and inserting the values found above for k and r_e, the Dupuit formula yields

$$3024 \text{ m}^3/\text{day} = \frac{\pi \times 4.07 (h_w^2 - 35.0^2)}{\ln(0.25/40.2)}$$

Solving this equation for h_w will provide

$$h_w = 5.1 \text{ m}$$

and

$$\text{Drawdown} = 35.0 \text{ m} - 5.1 \text{ m}$$
$$= 29.9 \text{ m}$$

Radius of Influence

The value of r_e, the radius of influence in the Dupuit formula, varies depending on the rate of pumping. Once, as is assumed in the development of the formula, equilibrium is reached in the pumping rate, the replenishment of flow into the area near the well must be equal to this rate. Theoretically, this will result in a steady state in the drawdown curve as well. Consequently a constant ratio between the pumping rate and the radius of influence can be assumed. This ratio C is therefore

$$C = \frac{r_e}{Q} \tag{13.9}$$

where Q is the rate of flow in m^3/day.

In the absence of other information, an estimated value of r_e can be found as suggested by Sichardt, as follows,

$$r_e = 3000d\sqrt{k} \qquad (13.10)$$

where d is the drawdown in the well in meters and k is the permeability coefficient in meters per second.

EXAMPLE 13.3

Determine the estimated maximum capacity of the well in Example 13.2.

SOLUTION

An investigation of Fig. 13.4a indicates that as the pumping rate in the well increases, the water level in the well will drop. Eventually, if the pumping rate is allowed to become sufficiently large, the water level in the well will drop to the bottom. At this point the pump will run out of water. Consequently, maximum pumping will take place immediately before the drawdown in the well equals the depth of the well. In terms of the Dupuit formula, it can therefore be said that the maximum pumping rate is reached when $h_w = 0$. In the case of the well in this example,

$$Q_{max} = \frac{\pi \times 4.07 \times (0 - 35^2)}{\ln(0.25/r_e)}$$

From Eq. (13.9) and the information obtained in the previous example,

$$C = \frac{40.2}{3014}$$

$$= 0.013$$

Consequently the radius of influence r_e when Q is maximum is

$$r_e = 0.013 Q_{max}$$

Substituting this value of r_e in the Dupuit formula above will yield

$$Q_{max} = \frac{\pi \times 4.07 \times (0 - 35^2)}{\ln(0.25/0.013 Q_{max})}$$

or

$$2.95 Q_{max} - (Q_{max} \times \ln Q_{max}) = -15663$$

This equation can be solved for Q_{max} by trial and error, as illustrated in the following table.

Trial Q	$2.95Q_{max}$	$(Q_{max} \times \ln Q_{max})$	Left Side of Formula	Result
4000	11800	33176	−21376	Too high Q
3500	10325	28562	−18237	Too high Q
3300	9753	26736	−16410	Too high Q
3100	9145	24921	−15776	Too high Q
3050	8998	24470	−14472	Too low Q

Obviously, a correct value of Q_{max} will fall between 3050 m³/day and 3100 m³/day. Any further refinement of the value for Q_{max} is not necessary. It would, in fact, be misleading, given the approximations inherent in the Dupuit formula, to indicate that values more accurate can be estimated.

Consequently, it may be stated that the estimated maximum capacity of the well is 3100 m³/day, or 36.0 L/s.

Wells Not Penetrating the Entire Aquifer Depth

The Dupuit formula has been developed on the basis that the flow into the well is in a horizontal direction as shown in Fig. 13.4a. If the well does not reach the bottom limit of the aquifer, the flow into the well will more or less take on a pattern as shown in Fig. 13.6. This type of flow pattern would naturally result in an increase in flow into the well. Elaborate formulas have been developed to take this condition into account. Their application requires knowledge of the degree of penetration by the well. This information is not often available and is sometimes expensive to obtain.

Keeping in mind the nature and limitations of the Dupuit formula, it is seldom economical or warranted to determine the degree of penetration. Experience indicates that if the bottom of the aquifer is a considerable distance

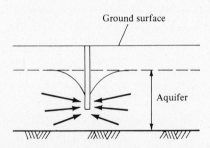

FIGURE 13.6

below the bottom of the well, the estimated value of Q from the Dupuit formula can be safely increased by 20 percent.

Rate of Flow from Several Wells

When several wells operate simultaneously in proximity in the same aquifer, the cone of depression or drawdown curves of these wells will interfere with each other, as shown Fig. 13.7. Under such conditions the hydraulic gradient of the flow to individual wells will be asymmetrical about the well centerline. It is readily observed then that the flow into such a well will be less than the estimated Q from the Dupuit formula.

Figure 13.8 shows general plan of a number of wells located near a point A. For a situation such as this the Dupuit formula can be written as follows:

$$Q = \frac{\pi k d_a(2d_w - d_a)}{\ln r_e - (1/n)\ln(x_1 x_2 x_3 \cdots x_n)}$$

(13.11)

where Q = capacity of water pumped from the well

d_a = drawdown at the point A

d_w = drawdown in the well

n = number of wells being pumped

k = the permeability coefficient

r_e = radius of influence

and $x_1, x_2, \ldots, x_n$ are the distances from the wells to the point A.

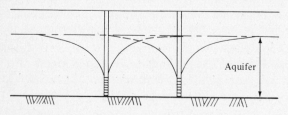

FIGURE 13.7

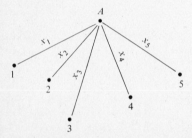

FIGURE 13.8

Applications and Limitations of the Dupuit Formula

The Dupuit formula, as demonstrated before, was developed on the basis of purely mathematical considerations and ideal conditions. The accuracy of the results obtained from the formula will therefore depend entirely on how closely the actual field conditions and aquifer approximate the ideal conditions assumed. Nevertheless, if sufficient test pumping is performed—combined with careful observations of the drawdown in the areas surrounding the well— estimates of the well's capacity can be obtained that will be sufficiently accurate for most preliminary-study purposes. It must, of course, be remembered that such testing must also relate to the assumptions made by the formula.

It is, for instance, rare that test pumping is carried out until equilibrium of flow into and out of the aquifer and well are reached. In many cases true equilibrium is reached only after prolonged periods of pumping.

Other methods have been developed to predict the flows from wells in such conditions. These methods all rely on extensive field testing and often require considerable expenditures of funds.

The main advantage of using the Dupuit formula is that it is relatively simple to apply, with a limited amount of field observation and testing required. The results obtained, though not totally accurate, are certainly useful in many instances and in most cases provide sufficient information as to the desirability and need for further expenditures of test funds.

Two specific applications for which the formula is useful are the determination of the aquifer permeability coefficient and the estimating of water-pumping capacities for dewatering operations. Examples of each of these applications are given below.

EXAMPLE 13.4

Determine the permeability coefficient of an aquifer based on the following test pumping information: test well diameter 250 mm, aquifer depth $h_e = 20.0$ m, pumping rate at a drawdown of 1.0 m is 5.0 L/s, and the pumping rate at a drawdown of 4.0 m is 15.0 L/s.

SOLUTION

Writing the Dupuit formula, in the form of Eq. (13.8), for each of pumping rates will yield

$$432 \text{ m}^3/\text{day} = \frac{1.36k \times (19.0^2 - 20.0^2)}{\log(0.125/r_e)}$$

and

$$1296 \text{ m}^3/\text{day} = \frac{1.36k \times (16.0^2 - 20.0^2)}{\log(0.125/r_e)}$$

It must be remembered that the values of r_e in the above equations are not equal to each other because the pumping rates vary. Values of r_e can be established from Eq. (13.9).

When pumping at a rate of 5 L/s,

$$r_e = 432C$$

and when pumping at the rate of 12 L/s,

$$r_e = 1296\,C$$

Since the pumping rate of 15 L/s is 3.0 times the pumping rate of 5 L/s, if the Dupuit equation for the 5-L/s condition is multiplied by this factor of 2.4, the right sides of both equations will become equal and the following relationship will result.

$$\frac{3.0 \times 1.36k \times (19.0^2 - 20.0^2)}{\log(0.125/432C)} = \frac{1.36k \times (16.0^2 - 20.0^2)}{\log(0.125/1296C)}$$

which reduces to

$$\log C = -\frac{39.5}{27.0}$$

and

$$C = 0.16$$

Consequently, for $Q = 5.0$ L/s,

$$r_e = 15.0 \text{ m}$$

and, for $Q = 12$ L/s,

$$r_e = 44.0 \text{ m}$$

Substituting each of these values for r_e in the appropriate form of the Dupuit formula will result in a formula with only one unknown, k.

$$432 = \frac{1.36k \times (19.0^2 - 20.0^2)}{\log(0.125/15.0)}$$

or

$$1296 = \frac{1.36k \times (16.0^2 - 20.0^2)}{\log(0.125/44.0)}$$

Solving either of these equations will yield 16.9 m/day.

EXAMPLE 13.5

Figure 13.9 shows a proposed arrangement of well points, intended to be used to dewater a sewer trench. Determine the amount of water that will need to be pumped from the well points in order to keep the trench dry in the vicinity of the point A. Assume k to be 3.5 m/day.

SOLUTION

A careful examination of the figure leads to the conclusion that it is reasonable to assume that only the six well points nearest A need to be considered. This assumption is also in keeping with common practice, where only the well points nearest the excavation are operated. Therefore only well points a, b, c, d, e, and f are considered to be operational.

In order to use Eq. (13.11), the equation related to the flow from several wells, values for d_a and d_w are required. The former, in this example, can be directly found from the study of the cross section of Fig. 13.9. The water at point A must be lowered a distance of 3.50 m, and $d_a = 3.5$ m.

The value of d_w depends on the type of well points or other data related to the project. For instance, if the well points are driven only a distance of, say, 8.0 m, a value of $d_w = 7.5$ m would appear reasonable. If the well points are at different elevations, other values of d_w should be used.

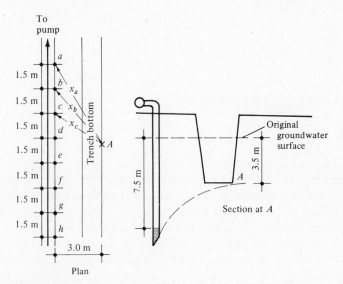

FIGURE 13.9

Information for use in conjunction with Eq. (13.10) is then

$$k = 3.5 \text{ m/d}$$

$$d_a = 3.5 \text{ m}$$

$$d_w = 7.5 \text{ m}$$

$$x_a = x_f = (3.0^2 + 1.75^2)^{1/2} = 3.47 \text{ m}$$

$$x_b = x_e = (3.0^2 + 1.75^2)^{1/2} = 3.47 \text{ m}$$

$$x_c = x_d = (3.0^2 + 0.75^2)^{1/2} = 3.09 \text{ m}$$

$$r_e = 3000 \times 7.5 \times (0.0000405)^{1/2} \qquad \text{from Eq. (13.10)}$$

$$= 143.0 \text{ m}$$

Therefore, from Eq. (13.11),

$$Q = \frac{\pi \times 3.5 \times 3.5 (2 \times 7.5 - 3.5)}{\ln 143.0 - (1/6) \ln (4.8 \times 3.47 \times 3.09 \times 3.09 \times 3.47 \times 4.8)}$$

$$= 121.0 \text{ m}^3/\text{d}$$

$$Q = 1.5 \text{ L/s}$$

Theoretically, such a rate of pumping would make the drawdown curve pass just through the point A. Actually, remembering that if the well points extend only partially into the aquifer, the real pumping capacity can be expected to increase by some 20 percent, for the given values of drawdown. Consequently, a pumping rate of 1.8 to 2.0 L/s should be suggested.

13.5   FLOW NETS AND FLOW IN SOILS

Flow Nets

The general principles and development of flow nets were discussed in an earlier chapter. Flow nets are especially useful in the study of flow of water in soils. They can be used in estimating the water loss through dikes, earth embankments, and dams used for retaining water in reservoirs or artificial lakes. In the case of sheet piling and other water-retaining structures, the use of flow nets is often instrumental in determining the velocity at which the water seeps under or around them. Often this velocity is critical on the downstream side of the structure, since it can lead to unstable soil conditions, piping problems, and other difficulties, which could result in complete failure of the structure.

In the case of gravity dams and other similar structures, the pressure exerted by the water beneath them, the *uplift pressure*, is an important design consideration. Flow nets are a very useful tool in determining the actual uplift-pressure diagram acting on them. They can also be used to study the effects of cutoff walls and impervious upstream and downstream blankets on the shape of the flow net for this type of structure.

General Application of Flow Nets to Soil Water

Water flowing through a soil under or around a structure behaves in most instances in a manner similar to free-flowing water in a pipe or other conveyance. Only the velocities of flow are generally so low that the Reynolds number is almost always below 2000. Hence these flows are usually considered to be laminar.

Flow through a soil can therefore be represented as shown in Fig. 13.10, with smooth flow lines and equipotential lines, as discussed earlier. Regular flownets, such as those in Fig. 13.10, will exist only however if the following two major conditions are satisfied.

1. The soil must be homogeneous in its hydraulic characteristics; that is, the permeability coefficient must be constant.
2. The soil must be anisotropic; that is, the permeability when the flow is in the horizontal direction must be equal to the permeability when the flow is in the vertical direction.

If either of the two conditions above are not met, the flow nets can become so complex that their analysis is beyond the scope of this text.

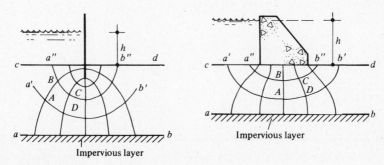

FIGURE 13.10

Determination of the Rate of Flow Through Soils

A flow net consists of equipotential lines and flow lines drawn in such a way as to form squares, as shown by $ABCD$ in Fig. 13.10. In this square, shown enlarged in Fig. 13.11, BC and AD are flow lines and AB and CD are equipotential lines. Assume that the square of Fig. 13.11 represents a prism of soil,

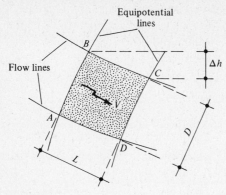

FIGURE 13.11

bounded by the pairs of flow lines and equipotential lines and having a width of one unit. Then the distance Δh is the drop in energy head through the prism, and the slope of the energy gradient in the prism will be

$$s = \frac{\Delta h}{\Delta L} \qquad (13.12)$$

Hence, the velocity of flow, according to Darcy's formula, or Eq. (13.3), will be

$$v = k \times \frac{\Delta h}{\Delta L} \qquad (13.13)$$

where k is the permeability coefficient of the soil.

The rate of flow Δq through the prism is then

$$\Delta q = A \times k \times \frac{\Delta h}{\Delta L} \qquad (13.14)$$

and, since the prism has a unit width and $A = \Delta D \times 1.0$, this reduces to

$$\Delta q = \Delta D \times k \times \frac{\Delta h}{\Delta L} \qquad (13.15)$$

However, the prism has been drawn so that $ABCD$ is a square, and therefore $\Delta D = \Delta L$. Hence

$$\Delta q = k \times \Delta h \qquad (13.16)$$

For each pair of equipotential lines, the value of Δh will vary. However, if h is the pressure drop across the entire section of the soil studied, the average value of Δh could be expressed as

$$\Delta h = \frac{h}{N_p} \tag{13.17}$$

where N_p is the number of spaces between equipotential lines.

Also, the total flow through such an earth section will be the sum of the flows through all the flow tubes, or sections bounded by flow lines. Hence the following expression for flow through a soil, derived from a flow net, results.

$$q = k \times h \times \frac{N_f}{N_p} \tag{13.18}$$

where N_f is the number of spaces between flow lines. Since this formula was developed for the prism in Fig. 13.11, with unit width, it must be remembered that the flow q from Eq. (13.18), represents the flow through one unit width only. For total flow through a structure, q must be multiplied by the length of the structure.

EXAMPLE 13.6

A sheet-piling wall is 20.0 m long and constructed to retain water as shown in Fig. 13.12. The permeability coefficient of the pervious layer into which the piling penetrates is $k = 0.0024$ m/s. The pervious layer is 2.5 m deep and the head of water at the upstream side of the piling is 1.5 m. Determine the amount of seepage under the wall if it penetrates the pervious layer to a depth of 0.75 m.

SOLUTION

The flow net for this wall has been sketched in Fig. 13.12. From the sketch it can be determined that the number of spaces between equipotential lines, N_p, is 18 and the number of spaces between flow lines, N_f, is 10.

Hence, from Eq. (13.16), the seepage per meter of wall is

$$q = 0.0024 \times 1.5 \times \frac{10}{18}$$

$$= 0.0020 \text{ m}^3/\text{s per linear meter of wall}$$

$$= 0.002 \times 3600 \times 24$$

$$= 172.8 \text{ m}^3/\text{day per linear meter of wall}$$

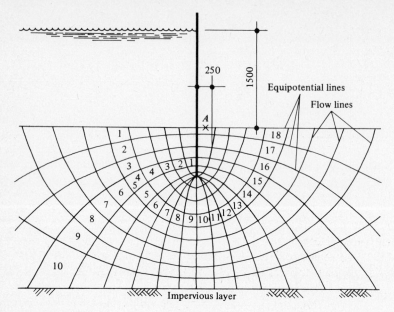

250

1500

Equipotential lines

Flow lines

A

Impervious layer

FIGURE 13.12

or

$$Q = 172.8 \times 20$$

$$= 3460 \text{ m}^3/\text{day}$$

for the entire wall.

Stability of Soils Under Seepage Conditions

In hydraulic structures such as those illustrated in Figs. 13.10 and 13.12, water enters the soil at one side of the structure and exits on the opposite side. The flowing water exerts a force on the soil particles through which it flows. This force is generally referred to as the drag force of a flowing fluid and is exerted in the direction of flow. Hence all the soil particles in these structures are subjected to two forces: the drag force mentioned above and the force of gravity acting on the particle. These forces are shown in Fig. 13.13.

At the point of entry A, the force of gravity G and the drag D by the water are both in the same direction and combine in the resultant force F. At this point the resulting force F tends to push the particle into the soil and, in a sense, provide some compaction of the soil. On the other hand, at the point C, where the water exits the soil, G and D act in opposite directions. If D becomes larger than G, the soil particle will be carried away by the water flow. Once this process

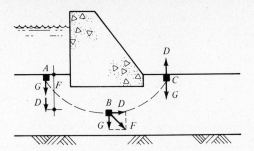

FIGURE 13.13

has started, more and more particles will be swept away and a condition known as "piping" could occur. As the particles in the downstream portion of the structure are removed, the velocity of flow will increase, and so will D. Therefore the movement of soil particles will become progressively more pronounced, and sudden failure of the structure can result.

The net force to which the soil particles are subjected is dependent upon the following factors.

1. The velocity of the flow of seeping water
2. The size of the soil particles
3. The density of the soil
4. The density and viscosity of the water

The velocity of flow of the seeping water can, at any point in the soil, be determined from the flow net, as demonstrated in the earlier discussion of flow nets in general and as shown in the following example.

EXAMPLE 13.7

In the case of Example 13.6, determine the velocity with which the water exits the soil at point A in Fig. 13.12.

SOLUTION

The flow under the sheet piling was found to be 172.8 m³/day per linear meter of piling. This flow takes place in 10 flow tubes, bounded by flow lines. Hence the average flow per flow tube is

$$q = \frac{172.8}{10}$$

$$= 17.3 \text{ m}^3/\text{day}$$

At the point A the width of the flow tube can be measured from the flow net. This width is 0.25 m, as indicated in Fig. 13.12. Hence, the area through which the flow q exits in the vicinity of A is

$$A = 0.25 \times 1.0$$

$$= 0.25 \text{ m}^2$$

Consequently, the velocity of the flow is

$$v = \frac{17.3}{0.25}$$

$$= 69.2 \text{ m/day}$$

or

$$v = 0.0008 \text{ m/s}$$

Settling Velocity of Soil Particles

A commonly used method of determining the stability of the soil under seepage conditions, relies on Stoke's Law of the settling of spherical particles in liquids. This law can be expressed in the following form:

$$v_s = \frac{gd^2}{18\mu}(\gamma_s - \gamma_f) \tag{13.19}$$

where v_s = settling velocity of a spherical particle in a liquid
  d = diameter of the soil particle
  μ = dynamic viscosity of the fluid
  γ_s = mass density of the soil particles
  γ_f = mass density of the fluid

For most applications in civil engineering, and particularly those related to seepage of water through soils, the temperature of the water will be in the vicinity of 6°C to 10°C. At this temperature the density of water varies only marginally and can be taken as 1000 kg/m³. Using this constant in Eq. (13.18) and substituting the kinematic viscosity for the dynamic viscosity yields the following.

$$v_s = \frac{gd^2}{18v}(s - 1) \tag{13.20}$$

where v is the kinematic viscosity of the water and s is the relative density of the soil particles.

Obviously, if the velocity with which a particle will settle in a fluid is greater than the upward velocity at which the fluid moves, the net motion of the particle will be downward. On the other hand, if the reverse is true, the particle will be carried upward.

Therefore, comparison of the exit velocity of the seepage water in a water-retaining structure with the settling velocity of the soil particles involved will provide an indication as to the stability of the soil on the downstream side of the structure.

EXAMPLE 13.8

If the soil particles downstream of the sheet-piling wall of Example 13.7 are sand, with a mass density of 2650 kg/m³ and an average minimum grain size of 0.2 mm, will the soil be stable?

SOLUTION

The relative density of the sand particles is

$$s = \frac{2650}{1000}$$

$$= 2.65$$

If the water temperature is assumed to be between 6°C and 10°C, the kinematic viscosity of the water will be, from Table 2.1, 0.140×10^{-5} m²/s; hence the estimated settling velocity of the sand particles is, from Eq. (13.20),

$$v_s = \frac{9.81 \times 0.0002^2}{18 \times 0.140 \times 10^{-5}} \times (2.65 - 1.0)$$

$$= 0.026 \text{ m/s}$$

Hence the settling velocity far exceeds the upward velocity of the seepage water of 0.0008 m/s and no piping or washout is to be expected.

Pressure Diagrams from Flow Nets

The use of flow nets to determine pressure diagrams for flowing water was demonstrated in an earlier chapter. In the case of pressure diagrams related to seepage water, the same principles and basic method indicated there apply as well.

Pressure diagrams in structures such as dams and other water-retaining walls, where uplift forces are important, provide useful design information. They are also very good tools in analyzing the effectiveness of uplift-reducing devices, such as cutoff walls and upstream and downstream impervious blankets as discussed in an earlier chapter.

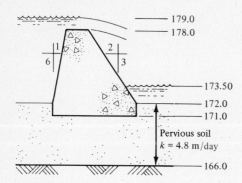

FIGURE 13.14

EXAMPLE 13.9

Determine the uplift pressure diagram for the dam of Fig. 13.14. Also, construct an uplift pressure diagram for this dam when cutoff walls have been installed at the heel and the toe of the dam, extending 2.5 and 1.0 m below the base of the dam, respectively.

SOLUTION

The flow nets for this problem are shown in Fig. 13.15a and b. The figures have been constructed following the procedures outlined in this chapter and previously.

In Fig 13.15a, when the dam has no cutoff walls, there are nine divisions between the equipotential lines. In accordance with this, the total net head on the dam is divided in 9 equal portions, each representing a pressure head of

$$\Delta h = \frac{h}{9}$$

These divisions are shown as horizontal lines across the dam in Fig 13.15a. The heel of the dam, the point m, is located approximately in the middle of the equipotential division 2. Therefore the pressure at this point on the dam is represented by the distance mm'. At the points a, b, and so on the pressure heads will also be aa', bb', etc. On this basis the pressure diagram can be constructed as shown in Fig. 13.15a.

A similar procedure can be applied when the cutoff walls are in place. The pressure diagram in this instance is shown in Fig. 13.15b.

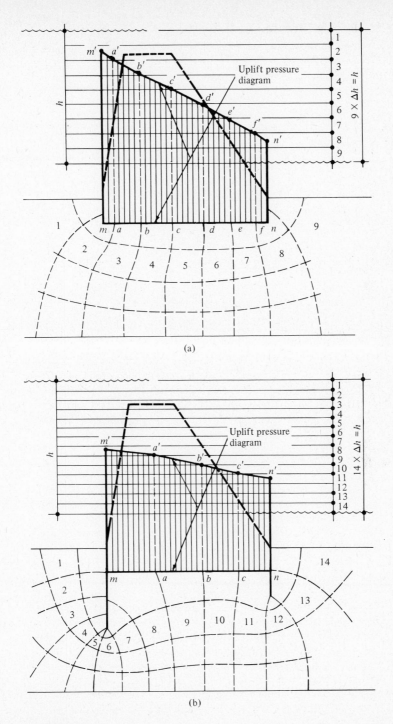

FIGURE 13.15

Measurement of Flowing Water

14.1 FLOW MEASUREMENT

Almost all devices used for the measurement of the flow of fluids rely on some of the principles discussed in this text. Whether the measurement deals with pressure flow or open-channel flow, the principles of energy loss and the measurement of pressure differential play an important part in the theory involved in water-flow meters.

Even so, some measurement techniques rely simply on mechanical devices, such as the turbine and current meters. Still other methods use the observation of dyes or chemical concentration as a means for determining the rate of flow.

14.2 PRESSURE-FLOW MEASUREMENT

The Pitot Tube

One of the simplest devices for measuring the rate of flow is the *pitot tube*, invented by Henry Pitot about 1730. The pitot tube measures the velocity head in a flowing fluid. It is mostly used in pressure-flow applications, even though it is equally adaptable to many types of open-channel flow.

Figure 14.1 shows the principles involved in this measurement device as related to open-channel flow. The velocity of the flowing water will force some water into the tube to a distance h above the water surface as shown. At a certain point the water in the tube will become stationary and h will remain constant

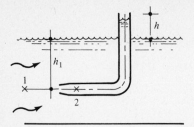

FIGURE 14.1

until the velocity changes. If Bernoulli's theorem is expressed for the point 1, which is in the stream of water and some distance upstream of the mouth of the tube, and the point 2, which is inside the tube, where the water has become stationary, the following expression results.

$$\frac{v_1{}^2}{2g} + \frac{p_1}{\gamma} + z_1 = \frac{v_2{}^2}{2g} + \frac{p_2}{\gamma} + z_2 + h_l$$

In this expression,

$v_1{}^2/2g$ is the velocity head of the flowing water in the channel; If the point 1 is far enough upstream that no turbulence is felt due to presence of the tube, this velocity will be the true velocity at point 1.

p_1/γ is the pressure head at the point 1, or h_1.

z_1 and z_2 are equal to each other here and thus will be canceled out.

$v_2{}^2$ is equal to zero, since at the point 2 the water is stationary.

p_2/γ is the pressure head at the point 2, or $h_1 + h$.

h_l will be very small, and hence negligible, especially if the upstream edge of the tube is constructed to ensure a minimum amount of turbulence.

Consequently, the Bernoulli expression will reduce to

$$\frac{v_1{}^2}{2g} + h_1 = h_1 + h \tag{14.1}$$

or

$$h = \frac{v_1{}^2}{2g} \tag{14.2}$$

Hence h is a measure of the velocity head of the stream. In general, Eq. (14.2) can be rearranged to read

$$v = \sqrt{2gh} \tag{14.3}$$

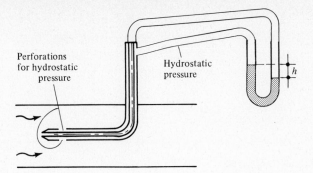

Perforations
for hydrostatic
pressure

Hydrostatic
pressure

h

FIGURE 14.2

In Eqs. (14.2) and (14.3), h is the difference in head between points 1 and 2. When this principle is applied to pressure flow, the same relationship will result. In pressure-flow applications it will thus be necessary to measure the actual pressure head at point 1. A piezometer or manometer in the pipe near the pitot tube could be used to accomplish this. In actual practice, the piezometer and the pitot tube are often combined by surrounding the pitot tube with a perforated tube, as shown in Fig. 14.2. The orifices in the outer tube serve as ports for the measurement of pressure head. The outer tube and the pitot tube are then connected to opposite ends of a differential manometer. The velocity relationship of Eq. (14.3) will now read

$$v = \sqrt{2gh(s_g - 1)} \qquad (14.4)$$

where s_g is the relative density of the gage fluid.

EXAMPLE 14.1

If the gage liquid is mercury ($s = 13.6$) and h is 45 mm in Fig. 14.2, determine the velocity and the rate of flow of water in the pipe, which is 250 mm in diameter.

SOLUTION

From Eq. (14.4),

$$v = \sqrt{2 \times 9.81 \times 0.045(13.6 - 1)}$$

$$= 3.33 \text{ m/s}$$

and the rate of flow Q is as follows:

$$Q = 3.33 \times \frac{0.25^2 \times \pi}{4}$$

$$= 0.164 \ \text{m}^3/\text{s}$$

$$= 164 \ \text{L/s}$$

Differential-Head Meters

When the cross-sectional area in a pipe is reduced and the flow is steady, the velocity will increase. The increased velocity produces an increase in velocity head and, in accordance with the Bernoulli principle, this must result in a decrease in pressure head. The above principle underlies several of the more common flow meters, namely the *orifice plate*, the *flow nozzle*, and the *venturi meters*.

THE ORIFICE PLATE

The theoretical considerations used in the orifice plate and other differential-flow meters are illustrated in Fig. 14.3.

Liquid flowing from point 1 to point 3 in Fig. 14.3 will accelerate through the restricted opening of the orifice plate. Once the liquid has passed the plate, it will reestablish the original velocity; hence at point 3 the velocity will be the same as at point 1.

Writing the Bernoulli expression between points 1 and 2 and keeping in

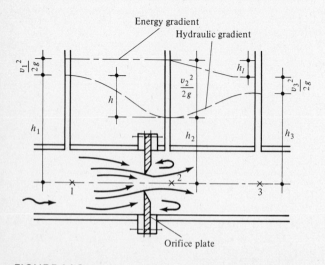

FIGURE 14.3

mind that z_1 and z_2 are equal, yields

$$\frac{v_1{}^2}{2g} + h_1 = \frac{v_2{}^2}{2g} + h_2 \tag{14.5}$$

Point 2 is located at the vena contracta of the orifice flow, and consequently the rate of flow at this point can, in accordance with principles discussed in an earlier chapter, be expressed as follows:

$$Q_2 = Cav_2 \tag{14.6}$$

where C is a coefficient allowing for the difference in area between the flow stream at the vena contracta and the orifice area, and a is the cross-sectional area of the orifice plate opening.

With steady-flow conditions Q_2 must also be the rate of flow of Q of the pipe, or

$$Q_2 = Q_1 = Q_3 \tag{14.6}$$

Hence

$$Av = Cav_2 \tag{14.7}$$

where A is the cross-sectional area of the pipe and v is the average uniform velocity of flow in the pipe.

If the ratio of the cross-sectional areas of the pipe and the orifice is called k_0, or

$$k_0 = \frac{A}{a} \tag{14.8}$$

and substituted in Eq. (14.7), then

$$v_1 = \frac{Cv_2}{k_0} \tag{14.9}$$

Replacing v_1 in Eq. (14.5) with its value from Eq. (14.9), and rearranging and simplifying, produces

$$v_2 = \left(\frac{2gh}{1 - (C/k_0)^2}\right)^{1/2} \tag{14.10}$$

Therefore the rate of flow Q in the pipe will be, from Eq. (14.6),

$$Q = a\left(\frac{2gh}{1 - (C/k_0)^2}\right)^{1/2} \tag{14.11}$$

or

$$Q = Ma\sqrt{2gh} \tag{14.12}$$

where

$$M = \frac{C}{(1 - (C/k_0)^2)^{1/2}} \tag{14.13}$$

Values of M are related to the ratio A/a, the orifice coefficient C, and the location of the points where the pressure heads are measured to determine h. M values range from 0.72, for ratios of A/a of 2.0, to 0.60, when $A/a = 20.0$. Table and graphs relating values of M to the upstream Reynolds number have been produced from experimental investigations, by many authorities, some of which are listed in the bibliography.

A study of the energy and hydraulic gradients of Fig. 14.3 indicates that there is a certain amount of energy loss h_l between points 1 and 3. This head loss is caused by the turbulence created by the orifice plate and is not recoverable. Orifice-plate-type meters, therefore, require additional energy expenditure. The additional energy requirements increase gradually as the velocity of flow increases.

The Flow Nozzle

The flow nozzle, shown in Fig. 14.4, is an apparatus very similar to the orifice plate. However, the orifice in this instance is a short tube with well-rounded upstream entrance conditions. In this way, turbulence is reduced somewhat and energy head losses are less than those for the orifice plate.

The principles of operation are exactly as those mentioned above, and the rate of flow is calculated by means of Eq. (14.12). Comments regarding values of the coefficient M made earlier apply here as well. For approximations, the value of M can be taken to be 0.97.

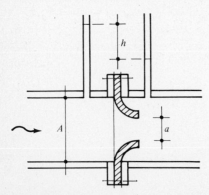

FIGURE 14.4

THE VENTURI METER

The sudden enlargement encountered downstream in flow-nozzle meters is responsible for considerable turbulence, resulting in energy losses and some sacrifice in accuracy. These problems have been minimized by the venturi tube or meter. As shown in Fig. 14.5, the venturi meter uses gradual contraction and enlargement leading to and away from the meter's throat.

Operation principles for this meter are as discussed above, and the rate of flow is calculated by means of Eq. (14.12). The values of M for venturi meters are all nearly equal to 1.0 and depend on the ratio of the throat diameter to the pipe diameter and the angle of the inlet and outlet cones.

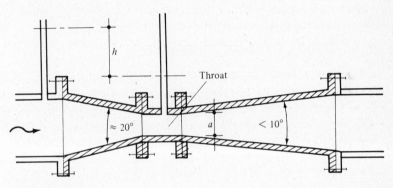

FIGURE 14.5

Other Methods of Pipe-Flow Measurement

Several other methods of pipe-flow measurement exist. Some of these use sophisticated electronic devices to sense and record the rate of flow. Since the principles of operation of these methods are generally described fully in the manufacturers' literature, they will not be covered here.

Two other types of meters using hydraulics principles and properties of flow are the variable area, or *rotameter*, and the *turbine meter*.

The rotameter, shown in Fig. 14.6, relies on the drag exerted by the fluid, flowing upward, on a float. This takes place in a conical-shaped transparent glass

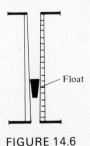

FIGURE 14.6

tube. As the velocity increases, the float is carried upward, until the enlarged area of the cone reduces the velocity to the point that it allows the cone to become stationary. Any further change in velocity will again cause the float to move until equilibrium between the drag force and the gravitational force on the float is reestablished. The glass tube is calibrated and graduated in convenient units of flow.

The turbine meter has vanes, or propellers, which are made to rotate by the flowing fluid. The number of revolutions per unit of time increases as the velocity increases. The number of revolutions is usually counted electronically and compared with a calibration chart to determine the velocity and/or the rate of flow.

14.3   OPEN-CHANNEL-FLOW MEASUREMENT

Open-channel-flow measurement can be accomplished by various methods. Turbine-type meters or *current meters* having impellers or cups, as shown in Fig. 14.7, are often employed in river-flow measurement or river gaging. In some cases estimates of river flow can be made by monitoring concentrations of a known chemical deposited in the flow upstream.

For regular channels and, in some instances, smaller natural streams the most common method of flow measurement is by means of weirs or measuring flumes.

Weirs

Most *measuring weirs* consist of thin plates, with various types of cross sections. The general formula for the flow over weirs was developed earlier and is in the form of Eq. (8.27), or

$$Q = ClH^{3/2}$$

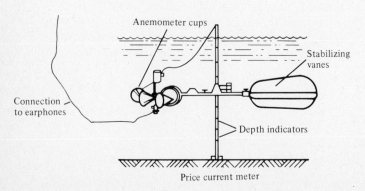

FIGURE 14.7

where Q = rate of flow over the weir

$\quad C$ = a weir coefficient

$\quad l$ = length of the weir crest

$\quad H$ = energy head on the weir

The C coefficient depends on, among other factors, the weir type, the energy head, and the weir height.

Plate weirs are either full width or contracted weirs depending on whether or not the weir crest extends across the entire width of the channel.

FULL-WIDTH WEIRS

A full-width thin-plate weir is illustrated in Fig. 14.8. Such a weir is often referred to as the Rehbock weir. Two important considerations must be taken into account when measuring the rate flow with this type of weir.

First, since the weir extends across the entire cross section of the channel, the area below the overflow nappe is isolated from the atmosphere. The flowing water from the lower nappe will entrain some of the air below it and create a partial vacuum in this space. As a result, the flow will be more or less pulled over the weir by this partial vacuum, producing conditions of flow not anticipated or included in the development of Eq. (8.27). In some instances the flow will cling to the downstream face of the weir.

The second consideration concerns the measurement of H, for use in Eq. (8.27). It applies to other types of weirs as well. As indicated in Fig. 14.8b, a longitudinal section through a Rehbock weir, H includes the height of water above the weir crest H', as well as the velocity head. In practice this is often neglected, since the velocity head is usually very small relative to H, so that for most applications

$$H \approx H' \tag{14.14}$$

The measurement of H' must take place at a point sufficiently far upstream of the weir that the water surface has not yet been affected by the weir flow Generally this distance is taken to be 3 to 4 times the weir height, d.

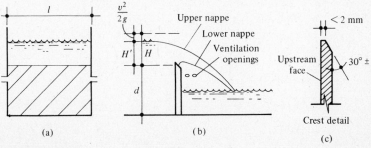

FIGURE 14.8

Under these conditions, the value of C, or the Rehbock coefficient for full-width weirs, is

$$C = 1.78 + \frac{0.24H}{d}$$

(14.15)

EXAMPLE 14.2

A rectangular channel 1.5 m wide has installed in it a Rehbock weir that is 0.75 m high. The weir crest is at an elevation of 178.465 m, and the water level about 3.0 m upstream of the weir is observed to be 178.847 m. Determine the rate of flow in this channel. Estimate also the error made by neglecting the velocity head in the value of H.

SOLUTION

From the given information,

$$H' = 178.874 - 178.465$$

$$= 0.409 \text{ m}$$

Neglecting the approach velocity head and using Eq. (14.15).,

$$C = 1.78 + \frac{0.24 \times 0.409}{0.75}$$

$$= 1.91$$

Hence, from (Eq. 8.27),

$$Q = 1.91 \times 1.5 \times 0.409^{3/2}$$

$$= 0.75 \text{ m}^3/\text{s}$$

$$= 750 \text{ L/s}$$

The upstream velocity with this flow is

$$v = \frac{0.75}{1.5 \times (0.75 + 0.409)}$$

$$= 0.43 \text{ m/s}$$

and

$$\frac{v^2}{2g} = 0.009 \text{ m}$$

Therefore H should have been

$$H = 0.409 + 0.009$$

$$= 0.418 \text{ m}$$

Then

$$C = 1.78 + \frac{0.24 \times 0.418}{0.75}$$

$$= 1.91$$

and the corrected flow will be

$$Q = 1.91 \times 1.5 \times 0.418^{3/2}$$

$$= 0.774 \text{ m}^3/\text{s}$$

$$= 774 \text{ L/s}$$

or approximately 3 percent more than the value calculated previously.

FULLY CONTRACTED WEIRS

When the edges of the weir are such that they are far enough away from the sides of the channel to ensure that the nappe can fully contract and will not be interfered with, the weir is said to be fully contracted. Four specific examples of such weirs are shown in Fig. 14.9.

The rate of flow for these weirs is calculated by means of Eq. (8.27) or special forms thereof. The C coefficients will, of course, vary for each type of weir. The following are the special considerations regarding each of the four weirs of Fig. 14.9.

For the rectangular weir of Fig. 14.9a,

$$C = 0.616\left(1.0 - 0.1\frac{H}{l}\right) \tag{14.16}$$

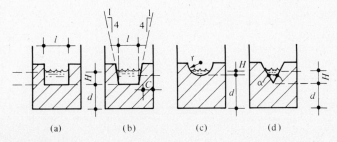

(a)　　　(b)　　　(c)　　　(d)

FIGURE 14.9

For the Cipoletti weir of Fig. 14.9*b*,

$$C = 1.86$$

when $l \geq 3H$, $d > 3H$, and $c > 2H$.

For the circular weir of Fig. 14.9*c*, Eq. (8.27) transforms to

$$Q = 2.8H^2\sqrt{r} \tag{14.17}$$

when $H < r$.

For the triangular weir or V-notch weir of Fig. 14.9*d*, Eq. (8.27) reads

$$Q = 1.41 \tan\frac{\alpha}{2} \times H^{5/2} \tag{14.18}$$

The V-notch or triangular weir is especially adaptable to measuring relatively small flows.

Measuring Flumes

Measuring flumes are channel restrictions created by narrowing the channel width or by raising the channel bottom, or both. They operate in accordance with the principles related to critical depth flow discussed in an earlier chapter. It was pointed out there that when a channel is constricted beyond a certain width or depth, the flow upstream must back up in order to maintain steady-flow conditions. In this way the measuring flumes are the venturi tubes of open-channel flow. Fig. 14.10 shows the flow conditions under which the average flume operates.

The rate of flow in a flume is expressed as follows:

$$Q = 1.71CID^{3/2} \tag{14.19}$$

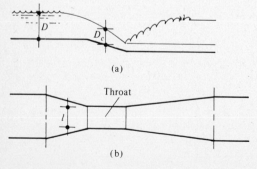

(a)

Throat

(b)

FIGURE 14.10

where l = throat width

    D = depth of flow upstream of the weir

    C = the flume coefficient

The values of C are very nearly equal to 1.0. Hence, for many practical applications, Eq. (14.19) can be written as

$$Q = 1.71lD^{3/2} \qquad (14.20)$$

In most cases the flow will be such that it passes through the critical depth in the throat. After passing through the throat the flow will return to subcritical and a hydraulic jump will form in the downstream flared section, as shown in Fig. 14.10a. If a hydraulic jump is not present, Eq. (14.20) cannot be used for determining the flow in the flume.

River-Stage/Discharge Curves

As indicated in Chapter 12, when dealing with the flow in natural channels, the use of the Manning formula in the form of Eq. (12.1) is helpful in estimating the rate of flow in a river. This formula reads

$$Q = K_n s^{1/2}$$

where K_n is a form factor dependent on the hydraulic characteristics of the stream cross section and s is the slope of the energy gradient. This relationship and the K_n curves that can be developed for a river reach can be used to estimate the flow in the river from measurements of the river-stage or water-surface elevation.

Using the above form of the Manning equation and appropriate K_n curves will provide curves of stage versus discharge for a stream as demonstrated in the following example.

EXAMPLE 14.3

Assuming the flow in the reach to be uniform, determine the stage/discharge curve for the reach of the stream in Example 12.2. Assume the channel bottom slope to be 0.0022 m/m.

SOLUTION

From the graph of the K_n values for various elevations of the water surface, of Fig. 12.2d and using the Manning formula, the following values for corresponding Q's can be obtained.

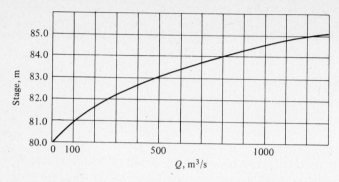

FIGURE 14.11

Stage, m	K_n	$s^{1/2}$	Q, m³/s
80.0	400	0.047	19
81.0	2200		103
83.0	10900		510
85.0	26900		1260

The corresponding curve is shown in Fig. 14.11. This type of graph will provide a rate of flow at any river stage—or vice versa, the river stage for any rate of flow.

It must be emphasized, however, that this graph was developed on the assumption that the flow in the reach is uniform, a rare occurrence in natural channels. Ideally, stage discharge curves should be calibrated for various rates of flow and slopes of the water surface by means of actual flow measurements. Such a project is not readily accomplished, since it would take many years to complete.

River-Flow Measurements

One of the most commonly used methods of measuring river flows, for purposes such as those mentioned above and many others, is the *area-velocity method*. In this method, the river cross section is divided into a number of convenient sections, and velocity measurements in each section are taken by means of a current meter. The product of the velocity so measured and the area of the section is the rate of flow in the section. The total river flow is then the sum of the flows in each section.

Velocity measurements are taken in each section at points 0.2 and 0.8 of the average depth of the section. From numerous experiments, it has been found that the average of the velocities at these depths is most representative of the average velocity of the section. When the depth of the stream is insufficient to warrant this procedure, the velocity measurement is performed at 0.6 of the average depth of the section.

Index